PICTORIAL KEY TO GENERA OF PLANT-PARASITIC NEMATODES

PICTORIAL KEY TO GENERA OF PLANT-PARASITIC NEMATODES

W. F. MAI, with

H. H. LYON, photographer

FOURTH EDITION, REVISED

COMSTOCK PUBLISHING ASSOCIATES a division of
CORNELL UNIVERSITY PRESS | Ithaca & London

First Revision, 1962
Second Revision, 1964
Supplement to Second Revision, 1966
Third Revision, 1968

First published 1975 by Cornell University Press.
Published in the United Kingdom by Cornell University Press Ltd.,
Ely House, 37 Dover Street, London W1X 4HQ.
Second printing, 1982.

International Standard Book Number 0-8014-0920-9
Library of Congress Catalog Card Number 74-14082
Printed in the United States of America by Vail-Ballou Press, Inc.

The paper in this book is acid-free, and meets the guidelines for permanence and durability of the Committee on Production Guidelines for Book Longevity of the Council on Library Resources.

CONTENTS

PREFACE

This pictorial key is intended primarily for the use of students and others who wish to identify nematodes. The descriptive key, the accompanying photographs, the generic descriptions, and a paragraph or more listing the characteristics of genera will aid such users in learning to differentiate plant-parasitic nematodes from other types of nematodes that also occur in soil and plant tissue and to identify the plant-parasitic nematodes as to genus. The descriptive key and the pictures serve as a guide in the analysis of the morphology and anatomy of a specimen and in its tentative assignment to a genus. The descriptions of the various genera and the characteristics cited aid in the verification of the initial determination. *The description, of course, should always be consulted before assigning a nematode to a particular genus.* To identify species it is necessary to consult one or more of the taxonomic publications cited in the Selected References, for each genus; each citation of a publication containing a species key is marked with an asterisk.

Included in the key are all genera having one or more species proved to be parasitic to plants. Several genera, such as *Telotylenchus* and *Pratylenchoides*, found around plant roots but not proved to be plant parasitic, are included because of their very close relationship to plant-parasitic species of other genera. Also included are genera such as *Psilenchus*, which, although not proved to be parasitic to higher plants, often are found in large numbers in soil samples taken from around plant roots.

Although found around plant roots, certain recently described or rare genera, especially those difficult for a student to differentiate from a more widely distributed or economically important genus, are not included in the key. Each of these genera, however, is discussed briefly in the General Characteristics section of the genus to which it is considered most closely related and, although the generic description is omitted, the publication containing this description is cited. The key is arranged so that tentative identification of a specimen of such a genus will indicate that it belongs to the closely related genus, and information in the General Characteristics section of that genus will indicate the possibility that such a specimen belongs to a related recently described or rare genus. In addition, each subgenus is mentioned in the General Characteristics section of the genus under which it is classified. Further information may be obtained about any recently described or rare genera and subgenera by referring to the section of the Selected References that includes papers covering its most closely related genus. Descriptions of most genera are included in Golden (1971). This and other publications cited in this Preface are noted in the General References.

The literature cited for each genus in the key is listed alphabetically by genus and is included in the Selected References. The General References lists publications on the taxonomy and morphology of nematodes, particularly plant-parasitic nematodes, and techniques related to this subject.

The key is based on morphological characteristics of mature females. Males and juveniles may be used to verify an identification, but only females should be used to positively identify a nematode as to genus.

The techniques used to extract nematodes from soil and plant parts and to prepare them for observation may greatly alter these specimens and may make identification easy or more difficult. For example, in centrifugal flotation extraction, certain kinds of nematodes will shrink if immersed for extended periods in sugar solutions. Also the appearance of specimens may be influenced by the care they receive prior to, during, and after extraction; in general, exposure to high temperature, desiccation, and storage in water too long before being killed reduces the lifelike appearance of nematodes. Specimens recently killed by gentle heat (65C for 2 min) and mounted in water appear more lifelike, permitting better observation of characters than do killed and fixed specimens, particularly those mounted in glycerine. In temporary or permanent mounts, the omission of coverslip supports may result in the flattening of a specimen and thus influence measurements of taxonomic characters. Because of the considerable variation in morphology among individual specimens of the same genus or species and the difficulty in observing taxonomic characters of certain specimens, at least four to six specimens should be observed before making a positive identification. Techniques for the extraction of nematodes from soil and roots, killing, fixing, and making temporary and permanent mounts are given in a manual by J. F. Southey (1970).

The characters that I use in this book to separate genera are those most easily recognizable by students, and scanning electron-microscope pictures of plant-parasitic nematodes are included to aid students in visualizing certain difficult characters. Selected terms used in the descriptive key and in the generic descriptions are defined in the glossary. Numbers occurring in parentheses after definitions refer to plates picturing the characters defined.

Because the organization of this key is based primarily on characters judged to be most easily observed by students, related nematodes may not appear together. To emphasize natural relationships, two diagrammatic outlines are included of the Tylenchida genera in which each genus in the key is classified according to order, superfamily, family, and subfamily. The first of these classifications was published by Allen and Sher in 1967 and the other by Golden in 1971. Other classifications of the Tylenchida, differing from the two selected for this publication, have also been published. There is, in addition, a 1971 classification by V. R. Ferris of that part of the Dorylaimida with genera having plant-parasitic species.

For convenience, in addition to the pictorial key, a section giving the descriptive portion of the key—without pictures—is included. When taxonomic characters are thoroughly understood, so that pictures of nematodes are not needed, identification to genus can be made more rapidly with this key than with the pictorial key. An index makes it possible to locate the picture of a representative of a specific genus, the description and general characteristics of that genus, and a listing of publications concerning that genus.

Part of the descriptive key is based on "A Key to the Mature Females of the Tylenchoidea," an unpublished paper prepared by A. L. Taylor. Some of the terms in the Glossary of Nematological Terms are from a list prepared by H. J. Jensen. Permission to photograph line drawings from several publications and to use photographs that have appeared in other publications is gratefully acknowledged; specific acknowledgment is given in the legends to the plates.

Some of the citations of early taxonomic references were obtained from *Check List of the Nematode Superfamilies Dorylaimoidea, Rhabditoidea, Tylenchoidea, and Aphelenchoidea* by A. D. Baker (1962).

The substantial help of T. H. Kruk is recognized for his work in accumulating and organizing taxonomic information for previous editions of this key. The very valuable assistance of Janice Foster Stuttle and Nancy Austin in preparing reference lists, checking literature citations, and helping in innumerable other ways is greatly appreciated. The role of Rodrigo Tarté in reviewing the key and making suggestions for its improvement is gratefully acknowledged.

WILLIAM F. MAI

Ithaca, New York

TWO NATURAL CLASSIFICATIONS OF TYLENCHIDA AND A PART OF DORYLAIMIDA

Tylenchida

The following classification of the order Tylenchida is from:
Golden, A. Morgan. 1971. Classification of the genera and higher categories of the order Tylenchida (Nematoda). *In:* B. M. Zuckerman, W. F. Mai, and R. A. Rohde, eds. Plant parasitic nematodes, Vol. 1. Academic Press, New York and London, pp. 191–232.

Classification of the Tylenchida Thorne 1949

A. Suborder Tylenchida (Orley 1880) Geraert 1966

1. Superfamily Tylenchoidea (Orley 1880) Chitwood and Chitwood 1937

Family Tylenchidae Orley 1880

Subfamily Tylenchinae (Orley 1880) Marcinowski 1909

Genera: *Tylenchus* Bastian 1865; *Cephalenchus* (Goodey 1962 n. rank; *Aglenchus* (Andrassy 1954) Meyl 1961; *Malenchus* Andrassy 1968; *Miculenchus* Andrassy 1959; *Chitinotylenchus* (Micoletzky 1922) Filipjev 1936

Subfamily Psilenchinae Paramonov 1967

Genera: *Psilenchus* de Man 1921; *Neopsilenchus* Thorne and Malek 1968; *Basiria* Siddiqi 1959; *Basiroides* Thorne and Malek 1968; *Clavilenchus* (Jairajpuiri 1966) Thorne and Malek 1968

Subfamily Ditylenchinae n. subf.

Genera: *Ditylenchus* Filipjev 1936; *Pseudhalenchus* Tarjan 1958

Subfamily Anguininae Paramonov 1962

Genera: *Anguina* Scopoli 1777; *Paranguina* Kirjanova 1955; *Subanguina* Paramonov 1967

Subfamily Dactylotylenchinae Wu 1969

Genus: *Dactylotylenchus* Wu 1968

Subfamily Tylodorinae Allen and Sher 1967

Genus: *Tylodorus* Meagher 1963

Subfamily Sychnotylenchinae (Paramonov 1967) n. rank

Genera: *Sychnotylenchus* Ruhm 1956; *Neoditylenchus* Meyl 1961

Family Tylenchorhynchidae (Eliava 1964) n. rank

Subfamily Tylenchorhynchinae Eliava 1964

Genera: *Tylenchorhynchus* Cobb 1913; *Tetylenchus* Filipjev 1936; *Geocenamus* Thorne and Malek 1968; *Nagelus* Thorne and Malek 1968

Subfamily Trophurinae Paramonov 1967

Genera: *Trophurus* Loof 1956; *Macrotrophurus* Loof 1958

Family Dolichodoridae (Chitwood and Chitwood 1950) Skarbilovich 1959 (in part)

Subfamily Dolichodorinae Chitwood and Chitwood 1950

Genera: *Dolichodorus* Cobb 1914; *Brachydorus* de Guiran and Germani 1968

Family Belonolaimidae (Whitehead 1959) n. rank
Subfamily Belonolaiminae Whitehead 1959
Genera: *Belonolaimus* Steiner 1949; *Morulaimus* Sauer 1966; *Carphodorus* Colbran 1965
Subfamily Telotylenchinae Siddiqi 1960
Genera: *Telotylenchus* Siddiqi 1960; *Trichotylenchus* Whitehead 1959
Family Pratylenchidae (Thorne 1949) Siddiqi 1963
Subfamily Pratylenchinae Thorne 1949
Genera: *Pratylenchus* Filipjev 1936; *Radopholoides* de Guiran 1967; *Hoplotylus* s'Jacob 1959
Subfamily Radopholinae Allen and Sher 1967
Genera: *Radopholus* Thorne 1949; *Hirschmanniella* Luc and Goodey 1963; *Zygotylenchus* Siddiqi 1963; *Pratylenchoides* Winslow 1958
Family Hoplolaimidae (Filipjev 1934) Wieser 1953
Subfamily Hoplolaiminae Filipjev 1934
Genera: *Hoplolaimus* Daday 1905; *Aorolaimus* Sher 1963; *Scutellonema* Andrassy 1958; *Peltamigratus* Sher 1963
Subfamily Rotylenchinae n. subf.
Genera: *Rotylenchus* Filipjev 1936; *Helicotylenchus* Steiner 1945
Subfamily Aphasmatylenchinae Sher 1965
Genus: *Aphasmatylenchus* Sher 1965
2. Superfamily Heteroderoidea (Filipjev 1934) n. rank
Family Heteroderidae (Filipjev 1934) Skarbilovich 1947
Subfamily Heteroderinae Filipjev 1934
Genus: *Heterodera* Schmidt 1871
Subfamily Meloidogyninae Skarbilovich 1959
Genera: *Meloidogyne* Goeldi 1887; *Hypsoperine* Sledge and Golden 1964
Family Nacobbidae (Chitwood and Chitwood 1950) n. rank
Subfamily Nacobbinae Chitwood and Chitwood 1950 (in part)
Genus: *Nacobbus* Thorne and Allen 1944
Subfamily Rotylenchulinae Husain and Khan 1967
Genus: *Rotylenchus* Linford and Oliveira 1940
3. Superfamily Criconematoidea (Taylor 1936) Geraert 1966
Family Criconematidae (Taylor 1936) Thorne 1949
Genera: *Criconema* Hofmanner and Menzel 1914; *Criconemoides* Taylor 1936; *Bakernema* Wu 1964; *Hemicriconemoides* Chitwood and Birchfield 1957
Subfamily Hemicycliophorinae Skarbilovich 1959
Genera: *Hemicycliophora* de Man 1921; *Caloosia* Siddiqi and Goodey 1963
Family Paratylenchidae (Thorne 1949) Raski 1962
Subfamily Paratylenchinae Thorne 1949
Genera: *Paratylenchus* Micoletzky 1922; *Gracilacus* Raski 1962; *Cacopaurus* Thorne 1943
Family Tylenchulidae (Skarbilovich 1947) Kiryanova 1955
Subfamily Tylenchulinae Skarbilovich 1947
Genera: *Tylenchulus* Cobb 1913; *Trophotylenchulus* Raski 1957
Subfamily Sphaeronematinae Raski and Sher 1952
Genera: *Sphaeronema* Raski and Sher 1952; *Trophonema* Raski 1957
4. Superfamily Atylenchoidea (Skarbilovich 1959) n. rank
Family Atylenchidae Skarbilovich 1959
Subfamily Atylenchinae Skarbilovich 1959
Genera: *Atylenchus* Cobb 1913; *Eutylenchus* Cobb 1913
5. Superfamily Neotylenchoidea (Thorne 1941) Jairajpuri and Siddiqi 1969
Family Neotylenchidae (Thorne 1941) Thorne 1949
Subfamily Neotylenchinae Thorne 1941
Genera: *Hexatylus* Goodey 1926; *Deladenus* Thorne 1941; *Gymnotylenchus* Siddiqi 1961
Family Paurodontidae (Thorne 1941) Massey 1967
Subfamily Paurodontinae Thorne 1941

Genera: *Paurodontus* Thorne 1941; *Paurodontoides* Jairajpuri and Siddiqi 1969; *Paurodontella* Husain and Khan 1968; *Stictylus* Thorne 1941
Subfamily Misticiinae Massey 1967
Genera: *Misticius* Massey 1967; *Anguillonema* Fuchs 1938
Family Nothotylenchidae (Thorne 1941) Jairajpuri and Siddiqi 1969
Subfamily Nothotylenchinae Thorne 1941
Genera: *Nothotylenchus* Thorne 1941; *Thada* Thorne 1941; *Nothanguina* Whitehead 1959; *Dorsalla* Jairajpuri 1966; *Sakia* S. H. Khan 1964.
Subfamily Boleodorinae Khan 1966
Genera: *Boleodorus* Thorne 1941; *Boleodoroides* Mathur, Khan, and Prasad 1966
Subfamily Halenchinae Jairajpuri and Siddiqi 1969
Genus: *Halenchus* Cobb 1913
Family Ecphyadophoridae Skarbilovich 1959
Subfamily Ecphyadophorinae Skarbilovich 1949
Genera: *Ecphyadophora* de Man 1921; *Ecphyadophoroides* Corbett 1964
B. Suborder Aphelenchina (Fuchs 1937) Geraert 1966
1. Superfamily Aphelenchoidea (Fuchs 1937) Thorne 1949
Family Aphelenchidae (Fuchs 1937) Steiner 1949
Subfamily Aphelenchinae (Fuchs 1937) Schuurmans Stekhoven and Teunissan 1938
Genus: *Aphelenchus* Bastian 1965
Family Aphelenchoididae (Skarbilovich 1947) Paramonov 1953
Subfamily Aphelenchoidinae Skarbilovich 1947
Genus: *Aphelenchoides* Fischer 1894 (Other genera are insect associates.)
Subfamily Rhadinaphelenchinae Husain and Khan 1967
Genus: *Rhadinaphelenchus* Goodey 1960
Subfamily Seinurinae Husain and Khan 1967 (No plant parasites.)
Family Paraphelenchidae (Goodey 1951) Goodey 1960
Subfamily Paraphelenchinae Goodey 1951 (No plant parasites.)
Family Anomyctidae Goodey 1960 (No plant parasites.)

The following classification of the Tylenchida Thorne 1949 is from:
Allen, M. W., and S. A. Sher. 1967. Taxonomic problems concerning the phytoparasitic nematodes. *In:* Annual Review of Phytopathology, Vol. 5. Annual Reviews, Palo Alto, California, pp. 247–264.

Classification of the Tylenchida Thorne 1949
Superfamily Tylenchoidea Chitwood and Chitwood 1937
Family Tylenchidae Oerley 1880
Subfamily Tylenchinae (Oerley 1880) Marcinowski 1909
Genera: *Anguina* Scopoli 1777; *Paranguina* Kirjanova 1955; *Ditylenchus* Filipjev 1934; *Neoditylenchus* Meyl 1961; *Pseudhalenchus* Tarjan 1958; *Psilenchus* de Man 1921; *Sychnotylenchus* Ruhm 1956; *Tylenchus* Bastian 1865; *Trophurus* Loof 1956
Subfamily Tylenchorhynchinae Eliava 1964
Genera: *Tylenchorhynchus* Cobb 1913; *Tetylenchus* Filipjev 1936; *Macrotrophurus* Loof 1958; *Chitinotylenchus* (Micoletzky 1922) Filipjev 1936
Subfamily Dolichodorinae Chitwood and Chitwood 1950
Genus: *Dolichodorus* Cobb 1914
Subfamily Telotylenchinae Siddiqi 1960
Genera: *Telotylenchus* Siddiqi 1960; *Carphodorus* Colbran 1966; *Trichotylenchus* Whitehead 1960
Subfamily Belonolaiminae Whitehead 1960
Genera: *Belonolaimus* Steiner 1949; *Morulaimus* Sauer 1966
Subfamily Pratylenchinae Thorne 1949
Genera: *Pratylenchus* Filipjev 1934; *Hoplotylus* s'Jacob 1959

Subfamily Nacobbinae Chitwood and Chitwood 1950
Genus: *Nacobbus* Thorne and Allen 1944
Subfamily Radopholinae n. subfam.
Genera: *Radopholus* Thorne 1949; *Pratylenchoides* Winslow 1958; *Hirschmanniella* Luc and Goodey 1962; *Zygotylenchus* Siddiqi 1963
Subfamily Hoplolaiminae Filipjev 1934
Genera: *Hoplolaimus* Daday 1905; *Helicotylenchus* Steiner 1945; *Rotylenchus* Filipjev 1934; *Scutellonema* Andrassy 1958; *Peltimigratus* Sher 1963; *Aorolaimus* Sher 1963
Subfamily Aphasmatylenchinae Sher 1965
Genus: *Aphasmatylenchus* Sher 1965
Subfamily Rotylenchoidinae Whitehead 1958
Genus: *Rotylenchoides* Whitehead 1958
Subfamily Rotylenchulinae n. subfam.
Genus: *Rotylenchulus* Linford and Oliveira 1940
Subfamily Tylodorinae n. subfam.
Genus: *Tylodorus* Meagher 1963

Family Heteroderidae Thorne 1949

Subfamily Heteroderinae Filipjev 1934
Genera: *Heterodera* Schmidt 1871; *Hypsoperine* Sledge and Golden 1964; *Meloidodera* Chitwood, Hannon, and Esser 1956; *Meloidogyne* Goeldi 1887; *Meloidoderita* Poghossian 1966

Family Tylenchulidae Kirjanova 1955

Subfamily Tylenchulinae Skarbilovich 1947
Genus: *Tylenchulus* Cobb 1913
Subfamily Sphaeronematinae Raski and Sher 1952
Genera: *Sphaeronema* Raski and Sher 1952; *Trophonema* Raski 1957

Family Criconematidae Thorne 1949

Subfamily Criconematinae Taylor 1936
Genera: *Criconema* Hofmanner and Menzel 1914; *Criconemoides*[1] Taylor 1936; *Bakernema* Wu 1964; *Hemicriconemoides* Chitwood and Birchfield 1957
Subfamily Hemicycliophorinae Skarbilovich 1959
Genera: *Hemicycliophora* de Man 1921; *Caloosia* Siddiqi and Goodey 1963
Subfamily Paratylenchinae Thorne 1939
Genera: *Paratylenchus* Micoletzky 1922; *Cacopaurus* Thorne 1943

Family Atylenchidae Skarbilovich 1959

Subfamily Atylenchinae Skarbilovich 1959
Genera: *Atylenchus* Cobb 1913; *Eutylenchus* Cobb 1913

Family Neotylenchidae (Thorne 1941) Thorne 1949

Subfamily Neotylenchinae Thorne 1941
Genera: *Deladenus* Thorne 1941; *Gymnotylenchus* Siddiqi 1961; *Hexatylus* Goodey 1926; *Neotylenchus* Steiner 1931; *Scytaleum* Andrassy 1961
Subfamily Nothotylenchinae Thorne 1941
Genera: *Anguillonema* Fuchs 1938; *Basiliophora* Husain and Khan 1965; *Dorsalia* Jairajpuri 1966; *Halenchus* Cobb 1933; *Nothanguina* Whitehead 1959; *Nothotylenchus* Thorne 1941; *Thada* Thorne 1941
Subfamily Paurodontinae Thorne 1941
Genera: *Paurodontus* Thorne 1941; *Stictylus* Thorne 1941
Subfamily Boleodorinae Khan 1964
Genera: *Boleodorus* Thorne 1941; *Boleodoroides* Mathur, Khan, and Prasad 1966
Subfamily Ecphyadophorinae Skarbilovich 1959
Genera: *Ecphyadophora* de Man 1921; *Ecphyadophoroides* Corbett 1964

[1] This genus contains forms that have been placed in the genera *Criconemella, Discocriconemella, Lobocriconema*, and *Nothocriconema* of DeGrisse and Loof 1965, *Macroposthonia* de Man 1880, and *Neocriconema* Diab and Jenkins 1965. The authors follow the view of Raski and Golden 1966, and Tarjan 1966, in rejecting these proposed genera until more evidence is presented to support the separation of species groups into discrete genera.

Family Sphaerulariidae (Lubbock 1861) Skarbilovich 1947
Subfamily Sphaerulariinae (Lubbock 1861) Pereira 1931
Genera: *Sphaerularia* Dufour 1837; *Sphaerulariopsis* Wachek 1955; *Tripius* Chitwood 1935
Subfamily Allantonematinae Pereira 1931
Genera: *Allantonema* Leukart 1884; *Aphelenchulus* Cobb 1920; *Bovienema* Nickle 1963; *Bradynema* zur Strassen 1892; *Chondronema* Christie and Chitwood 1931; *Contortylenchus* Ruhm 1956; *Dotylaphus* Andrassy 1953; *Heterotylenchus* Bovien 1937; *Howardula* Cobb 1921; *Metaparasitylenchus* (Wachek 1955) Nickle 1967; *Neoparasitylenchus* Nickle 1967; *Parasitylenchoides* Wachek 1955; *Parasitylenchus* Micoletzky 1922; *Proparasitylenchus* (Wachek 1955) Nickle 1967; *Protylenchus* Wachek 1955; *Scatonema* Bovien 1932; *Sulphuretylenchus* (Ruhm 1956) Nickle 1967; *Helionema* Brzeski 1962 (tentative)
Subfamily Fergusobiinae Goodey 1963
Genus: *Fergusobia* Currie 1937
Subfamily Iotonchinae T. Goodey 1953
Genus: *Iotonchium* Cobb 1920
Family Myenchidae Pereira 1931
Genera: *Myenchus* Schuberg and Schroeder 1904; *Myoryctes* Eberth 1863
Superfamily Aphelenchoidea Thorne 1949
Family Aphelenchidae Steiner 1949
Genus: *Aphelenchus* Bastian 1865
Family Aphelenchoididae Paramonov 1953
Genera: *Aphelenchoides* Fisher 1894; *Bursaphelenchus* Fuchs 1937; *Cryptaphelenchoides* J. Goodey 1960; *Cryptaphelenchus* (Fuchs 1937) Ruhm 1954; *Ektaphelenchus* (Fuchs 1937) Skrjabin, et al. 1954; *Entaphelenchus* Wachek 1955; *Laimaphelenchus* Fuchs 1937; *Megadorus* J. Goodey 1960; *Paraseinura* Timm 1960; *Parasitaphelenchus* Fuchs 1929; *Peraphelenchus* Wachek 1955; *Rhadinaphelenchus* J. Goodey 1960; *Ruehmaphelenchus* J. Goodey 1963; *Seinura* Fuchs 1931; *Tylaphelenchus* Ruhm 1956
Family Paraphelenchidae J. Goodey 1960
Genera: *Paraphelenchus* Micoletzy 1925; *Metaphelenchus* Steiner 1943
Family Anomyctidae J. Goodey 1960
Genus: *Anomyctus* Allen 1940

Dorylaimida

The following classification of part of the Dorylaimida Pearse 1942 (includes only genera with plant-parasitic species) is from:
Ferris, Virginia R. 1971. Taxonomy of the Dorylaimida. *In:* B. M. Zuckerman, W. F. Mai, and R. A. Rohde, eds. Plant-parasitic nematodes. Vol. I. Morphology, anatomy, taxonomy, and ecology. Academic Press, New York, pp. 163–189.

Superfamily	*Family*	*Genus*
Dorylaimoides	Longidoridae	*Longidorus*
Dorylaimoidea	Longidoridae	*Xiphinema*
Diptherophoridae	Trichodoridae	*Trichodorus*

KEY TO GENERA OF PLANT-PARASITIC NEMATODES (WITHOUT PICTURES)

(Based mainly on characteristics of adult females)

1. Stylet absent . . . .
   Stylet present 2
2. Two-part esophagus, no valvulated apparatus, anterior part slender, posterior part glandular and muscular; stylet usually without basal swellings . . . . *DORYLAIMIDA* 3
   Three-part esophagus usually with a valvulated metacorpus (median bulb) followed by a slender isthmus and glandular basal bulb; stylet usually with basal knobs 6
3. Stylet short, curved; body short and thick (0.45 to 1.5 mm long) . . . . *TRICHODORUS*
   Stylet long, straight, tapering to a long slender point with long extensions; body long and slender 4
   Stylet straight, usually not very long. (Includes a large number of genera most of whose feeding habits are not definitely known. Group contains no known pathogens.) . . . . *A LARGE NUMBER OF GENERA*
4. Stylet extensions with sclerotized basal flanges; guiding ring near base of stylet just anterior to junction of stylet and stylet extensions . . . . *XIPHINEMA*
   Stylet extensions without basal flanges; guiding ring near apex of stylet 5
5. Amphid openings minute, slitlike; amphids consisting of large pouches which almost encircle the head . . . . *LONGIDORUS*
   Amphid openings wide, sublabial extending at least halfway across the neck at that point; amphid pouches funnel- to stirrup-shaped . . . . *PARALONGIDORUS*
6. Dorsal esophageal gland outlet in metacorpus, anterior to valve or in that position when median bulb absent (usually difficult to see); metacorpus very large, often appears nearly as wide as the diameter of the body . . . . *APHELENCHOIDEA* 7
   Dorsal esophageal gland outlet in procorpus (usually can be seen more readily in recently prepared water mounts than in glycerine mounts); metacorpus moderate to reduced in size (less than 3/4 body width) . . . . *TYLENCHOIDEA* 9
7. a = less than 80, vulval flap absent; vagina normal 8
   a = around 100; vulva with wide overlapping flap; vagina curved . . . . *RHADINAPHELENCHUS*
8. Tail of female blunt; lateral field with 6 to 14 incisures; male with bursa and gubernaculum . . . . *APHELENCHUS*
   Tail of female usually conoid, often with 1 or more sharp points at the terminus (mucronate); lateral field with 2 to 4 incisures; male without bursa or gubernaculum . . . . *APHELENCHOIDES*
9. Head with setae. No plant parasites . . . . *ATYLENCHUS* *EUTYLENCHUS*
   Head without setae. Numerous plant parasites 10
10. Metacorpus absent or reduced; if reduced, no sclerotized valve . . . . . *NEOTYLENCHIDAE*
    Example of family . . . . *NOTHANGUINA*
    Example of family . . . . *NOTHOTYLENCHUS*
    Metacorpus with sclerotized valves present (usually can be seen more readily in recently prepared water mounts than in glycerine mounts) 11

11. Mature female greatly enlarged (pear-shaped, lemon-shaped, kidney-shaped or saccate); found in roots of plants, either embedded or attached by neck; some occur as cysts in the soil 12
    Mature females vermiform, may be slender to slightly swollen 23
12. Mature female bodies soft, elongate-saccate, or kidney-shaped with tail; except for *Sphaeronema* which is spherical in shape without a tail 13
    Mature females becoming cysts or remaining soft bodied; pyriform-saccate, spheroid or lemon-shaped, usually without a tail 18
13. Mature female has 2 ovaries . . . . . *ROTYLENCHULUS*
    Mature female has 1 ovary 14
14. Excretory pore located in normal position, near nerve ring 15
    Excretory pore located posterior to nerve ring 16
15. Mature female subspherical; cuticle marked with coarse reticulate pattern; may have a prominently protruding vulva, subterminal in position . . . . . *SPHAERONEMA*
    Mature female spiral, thickened; without protruding vulva . . . . . *TROPHONEMA*
16. Circumoral elevation present in females and larvae . . . . . *TROPHOTYLENCHULUS*
    Circumoral elevation absent 17
17. Excretory pore near vulva . . . . . *TYLENCHULUS*
    Excretory pore located near basal region of esophagus . . . . . *NACOBBUS*
18. Females with irregular body annules around perineum (perineal pattern); excretory pore at level with stylet or close behind it; lip region with 2 lateral lips wider than 4 sublateral lips. Second-stage larval stylet less than 20μ; weakly developed labial framework; usually marked galling of the host roots . . . . . *MELOIDOGYNE*
    Females with no irregular body annules around perineum; excretory pore posterior to median bulb; lip region with 2 lateral lips narrower than 4 sublateral lips. Second-stage larval stylet usually more than 20μ; well-developed labial framework. Usually no galling of the host roots 19
19. Vulva subequatorial, cuticle annulated . . . . . *MELOIDODERA*
    Vulva terminal or subterminal. Cuticle annulated or lace like 20
20. Cuticle annulated . . . . . *CRYPHODERA*
    Cuticle with lacelike pattern 21
21. Cyst stage. Vulva terminal, anus dorsal, not on vulva lip; or vulva sunken into terminal vulval cone with anus on upper inside of dorsal vulval lip 22
    No cyst stage. Vulva and anus terminal, on prominence . . . . . *ATALODERA*
22. Vulva terminal, usually on prominence. Anus dorsal, not on vulva lip. Second-stage larvae stylet less than 30μ . . . . . *HETERODERA*
    Vulva sunken into terminal vulva cone. Anus on upper inside of dorsal vulva lip. Second-stage larvae stylet more than 38μ . . . . . *SARISODERA*
23. Tail equal to or longer than 6 times the anal body diameter (tail filiform, with pointed or clavate terminus) 24
    Tail generally less than 6 times the anal body diameter. However, if longer, tail is cylindroid rather than filiform 28
24. Female with 2 ovaries 25
    Female with 1 ovary 26
25. Stylet without basal knobs, no cephalic sclerotization, tail filiform with usually clavate terminus . . . . . *PSILENCHUS*
    Stylet with basal knobs; heavy cephalic sclerotization; tail filiform with pointed terminus . . . . . *BRACHYDORUS*
26. Esophagus criconematoid; thick cuticle, coarsely annulated . . . . . *CALOOSIA*
    Esophagus tylenchoid, thin cuticle, not coarsely annulated 27
27. Long stylet (s = 2.5 or more) . . . . . *TYLODORUS*
    Short stylet (s = 2.5) . . . . . *TYLENCHUS*
28. One ovary (vulva usually located in posterior third of body) 29
    One ovary (vulva located near center of body); lip region conical, not annulated; female tail tip rounded, cuticle of tail swollen . . . . . *TROPHURUS*
    Two ovaries (vulva located near center of body) 44
29. Procorpus and metacorpus not swollen and combined into a large valvular bulb 30
    Procorpus and metacorpus swollen and combined into a large valvular bulb 37
30. Stylet delicate (15μ or less in length); tail acute or subacute 31
    Stylet strong (generally more than 15μ in length); tail tapering or bluntly rounded 33
31. Ovary with oocytes in 1 or 2 lines, not arranged around a rachis. Mature female, slender or stout 32

Ovary with multiple rows of oocytes arranged around a rachis. Mature female, mostly obese. Found in galls in leaves or flower parts . . . . *ANGUINA*

32. Ovary with 1 or more flexures, moderately stout forms. Found in root galls of Gramineae . . . . *SUBANGUINA*

    Ovary outstretched, slender forms. Found in bulbs, stems, leaves and tubers . . . . *DITYLENCHUS*

33. s = 1.5 or more; tail generally 1.5 times anal body diameter or shorter. . . . . *ROTYLENCHOIDES*

    s = less than 1.5; tail generally longer than 1.5 times anal body diameter 34

34. Esophagus overlaps intestine ventrally . . . . *PRATYLENCHUS*

    Esophagus overlaps intestine dorsally 35

35. Lip region low, generally rounded; stylet knobs flattened anteriorly; marked sexual dimorphism . . . . *RADOPHOLOIDES*

    Lip region, high conoid; stylet knobs sloping anteriorly or indented; males present or absent 36

36. Female body swollen; posterior part of stylet knobs sloping anteriorly; marked sexual dimorphism . . . . *ACONTYLUS*

    Female body slender; each stylet knob tapering anteriorly to a dentate tip; males unknown . . . . *HOPLOTYLUS*

37. Mature female without extra cuticle or sheath 38

    Mature female with extra cuticle or sheath 40

38. Cuticle with prominent retrorse annules 39

    Cuticle without prominent retrorse annules 41

39. Annules of female with spines, scales, plates, or stalked appendages on posterior margins . . . . *CRICONEMA*

    Annules of female with smooth or crenate posterior margins . . . . *CRICONEMOIDES*

40. Stylet knobs rounded, sloping posteriorly; cuticle usually with more than 200 annules . . . . *HEMICYCLIOPHORA*

    Stylet knobs anchor-shaped, with anterior projections; cuticle usually with less than 200 annules . . . . *HEMICRICONEMOIDES*

41. Annules of female without membranous structures on posterior margins 42

    Annules of female without membranous structures on posterior margins . . . *BAKERNEMA*

42. Cuticle of female ornamented with minute tubercules . . . . *CACOPAURUS*

    Cuticle of female not ornamented with minute tubercules 43

43. Female stylet 36μ or less . . . . *PARATYLENCHUS*

    Female stylet 45 to 120μ . . . . *GRACILACUS*

44. s = 2.5 or more 45

    s = generally less than 2.5 49

45. Esophageal glands not enclosed within a bulb, usually unequal in length, overlapping intestine 46

    Esophageal glands enclosed within a bulb, usually not overlapping intestine 47

46. Average body length usually 1.75 mm or more . . . . *BELONOLAIMUS*

    Average body length usually less than 1.75 mm 48

47. Lip region continuous . . . . *MACROTROPHURUS*

    Lip region set off by distinct constriction . . . . *DOLICHODORUS*

48. Lateral field with 4 incisures . . . . *MORULAIMUS*

    Lateral field with 2 incisures . . . . *CARPHODORUS*

49. Phasmids absent . . . . *APHASMATYLENCHUS*

    Phasmids present 50

50. Tail generally less than 1.5 times anal body diameter 61

    Tail 1.5 or more times as long as anal body diameter 51

51. Esophageal glands usually unequal in length, overlapping the intestine dorsally or latero-ventrally 52

    Esophageal glands usually enclosed within a bulb; if not enclosed, then of about equal length, and therefore considered as not overlapping the intestine 59

    [It is possible that certain conditions may cause the lengthening of either the dorsal or the subventral glands which may give the impression of overlapping. Therefore, several specimens should be carefully observed in relation to this character. Some confusion may arise with careful observation of the esophagus because the extent of the variation of this and other morphological characters has not been properly studied in many of the nematode genera described.]

52. No cephalic framework or moderately developed; female head not low or flattened . . . 53

Well-developed cephalic framework; female head low, rounded or flattened 56

53. Well-developed stylet; lateral field with 4 incisures 54
    Slender stylet with diverging basal knobs; lateral field with 3 incisures . . . . *TRICHOTYLENCHUS*
54. Female tail cylindroid with round terminus 55
    Female tail elongate conoid with blunt terminus . . . . *TELOTYLENCHUS*
55. Anterior portion of stylet asymmetrical; tail rather short with broadly rounded terminus . . . . *HISTOTYLENCHUS*
    Anterior portion of stylet asymmetrical; female tail with broadly rounded to bulbous terminus with strongly thickened cuticle . . . . *TELOTYLENCHOIDES*
56. Esophagus overlapping intestine dorsally 57
    Esophagus overlapping intestine ventrally 58
57. Short overlap. No marked sexual dimorphism . . . . *PRATYLENCHOIDES*
    Long overlap. Marked sexual dimorphism . . . . *RADOPHOLUS*
58. Tail tip mucronate . . . . *HIRSCHMANNIELLA*
    Tail tip not mucronate . . . . *ZYGOTYLENCHUS*
59. Female tail not acute 60
    Female tail acute or subacute . . . . *TETYLENCHUS*
60. Female tail conoid, with terminus usually bluntly rounded . . . . . . *TYLENCHORHYNCHUS*
    Female tail cylindroid, with terminus broadly rounded, and strongly thickened cuticle . . . . *PARATROPHURUS*
    [*Paratrophurus lobatus* Loof 1970 has overlapping glands, and for this reason it was placed in *Telotylenchoides*, by Siddiqi (1971). A more accurate decision about the correct placement of this species in either of the 2 genera can only be made when the extent of the variation of the esophageal glands is properly studied and its validity as a taxonomic character in this case is established.]
61. Phasmids enlarged 63
    Phasmids small, porelike 62
62. Esophagus overlapping intestine typically dorsally and laterally; lip region with or without annulation or striation; dorsal esophageal gland opening usually less than 1/4 of the stylet length behind stylet knobs . . . . *ROTYLENCHUS*
    Esophagus overlapping intestine typically ventrally; lip region without longitudinal striation; dorsal esophageal gland opening usually 1/4 or more of the stylet length behind stylet knobs . . . . *HELICOTYLENCHUS*
63. Both phasmids located posterior to vulva 64
    One phasmid located anterior to vulva and 1 posterior to vulva 65
64. Phasmids opposite or nearly opposite each other in region of anus; lip region with transverse striae . . . . *SCUTELLONEMA*
    Phasmids not opposite each other, anterior to anus; lip region without striae . . . . *PELTAMIGRATUS*
65. Spear knobs with distinct anterior projections; with 4 or fewer incisures areolated throughout length of lateral field . . . . *HOPOLAIMUS*
    Spear knobs rounded or without distinct anterior projections; with 4 incisures areolated at phasmids and anteriorly . . . . *AOROLAIMUS*

KEY TO GENERA OF PLANT-PARASITIC NEMATODES (WITH PICTURES)

Plate 1. Anterior regions of nematodes belonging to three orders. A. Tylenchida, with a stomatostylet and a 3-part esophagus with a valvulated metacorpus; B. Rhabditida, with a stoma instead of a stylet and a 2-part esophagus with a valvulated basal region; C. Dorylaimida, with an odontostylet and a 2-part esophagus without a valvulated region.

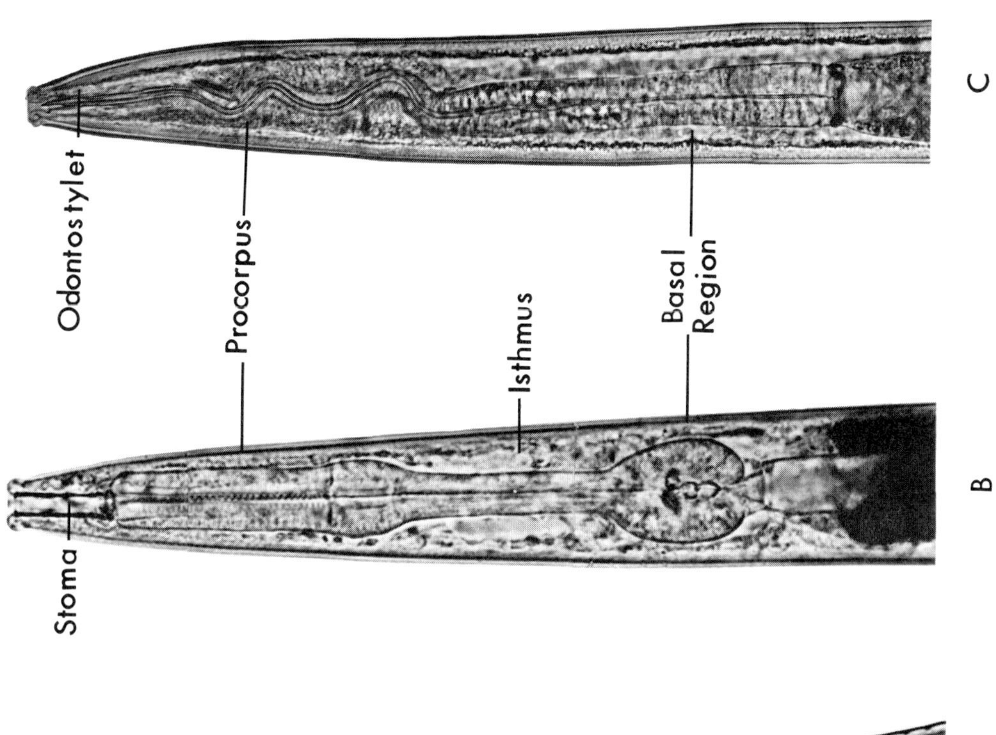
Odontostylet
Procorpus
Basal Region
C
Stoma
Isthmus
B

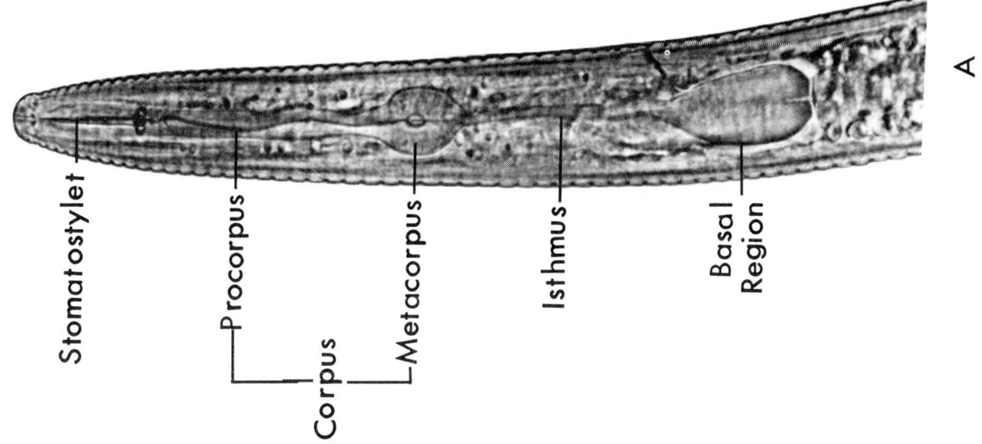
Stomatostylet
Procorpus
Corpus
Metacorpus
Isthmus
Basal Region
A

1. Stylet absent . . . . (Plate 2)

Most nematode species encountered in soil and plant tissue and which do not possess protrusible stylets belong to the superfamily Rhabditoidea. In addition, scattered genera belong to the orders Chromadorida and Enoplida.

Buccal capsules not armed with a protrusible stylet may possess 1 or more teeth. The buccal capsule is variable in size and shape. In some nematodes, especially rhabditoids, the buccal capsule is divisible into 3 sections: an anterior end enclosed by the lips, the vestibule or cheilostom; a middle and longest portion, the protostom; and a small terminal chamber, the telostom. The walls of stoma, called rhabdions, are divided into cheilorhabdions, protorhabdions, and telorhabdions.

Plate 2. Without protrusible stylets. A. *Panagrolaimus*. B. *Rhabditis*. C. *Mononchus*. a. stoma. b. glottoid apparatus.

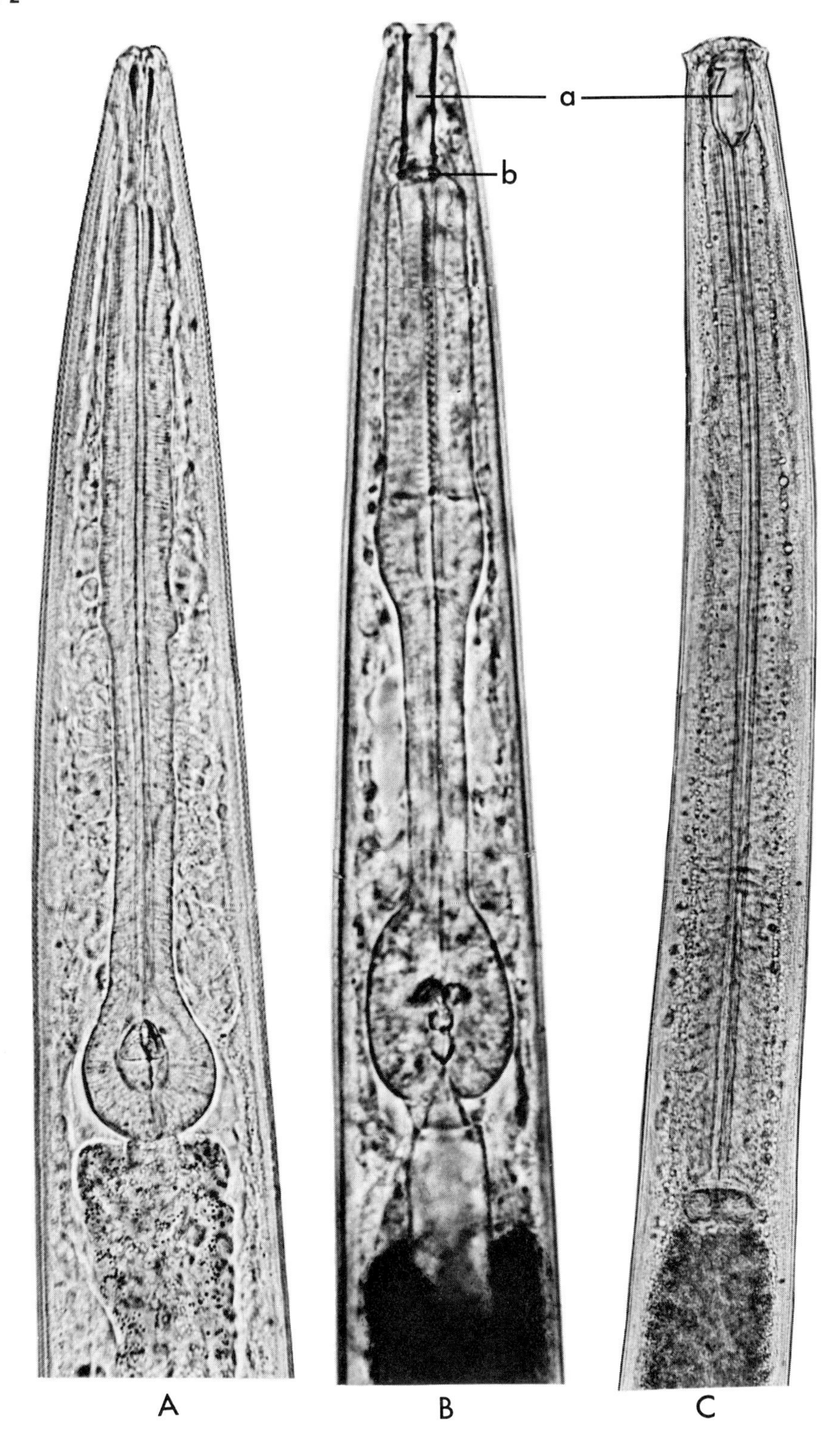
a
b
A
B
C

1. Stylet present . . . . (Plate 3) 2

Nematodes possessing protrusible stylets belong to either the order Tylenchida or the order Dorylaimida. The general types of stylets present in the Tylenchida are called stomatostylets while those in the Dorylaimida are called odontostylets. The stomatostylet is thought to have developed gradually through evolution by the coming together of the sclerotization of the buccal capsule. In the Dorylaimida the stylet represents a modified and enlarged tooth that possibly originates in the wall of the esophagus.

Plate 3. A–C. Possessing stomatostylets (*Hoplolaimus* sp., *Meloidodera* sp., and *Belonolaimus* sp., respectively). D. Possessing an odontostylet (*Dorylaimus* sp.).

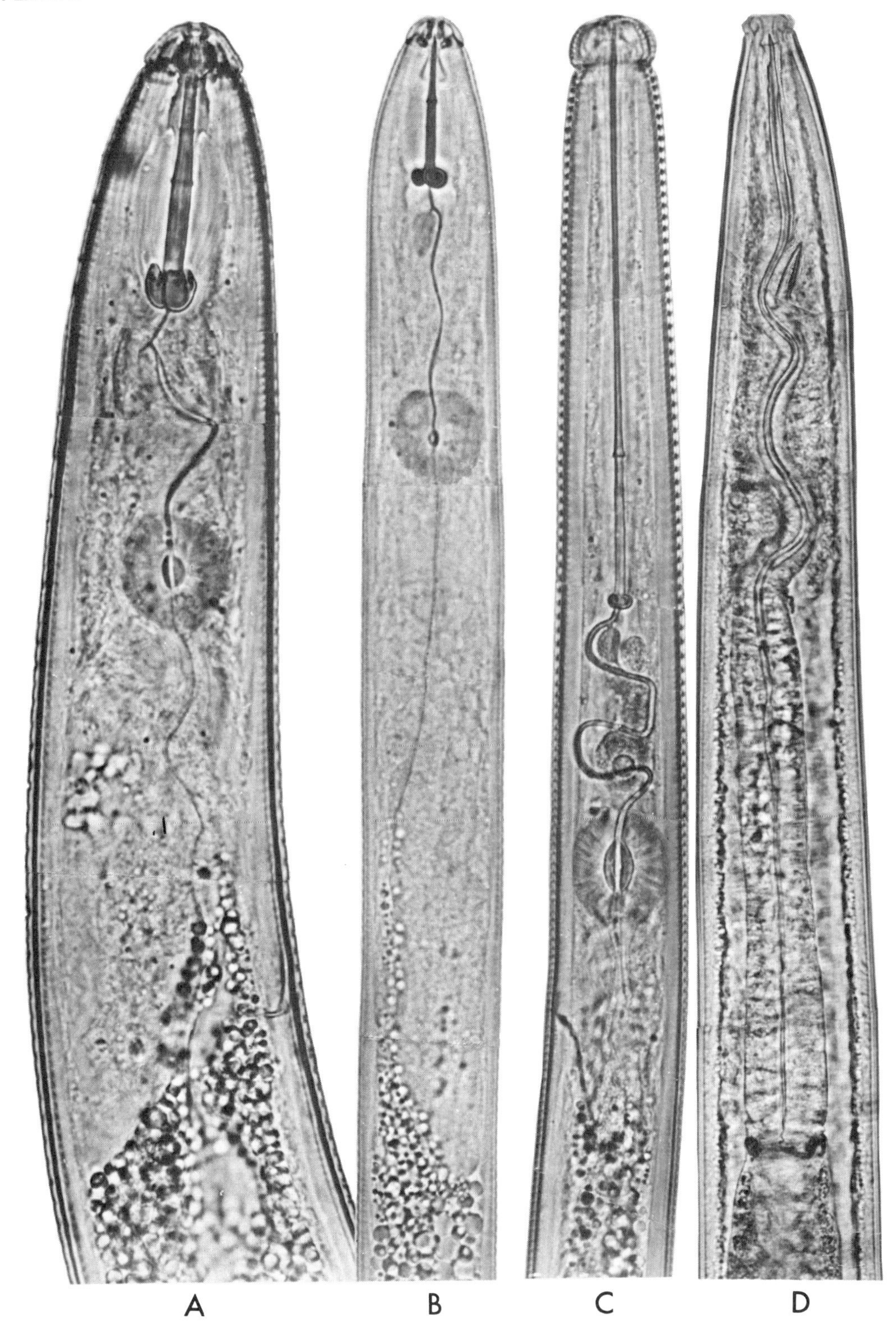
A
B
C
D

2. Two-part esophagus, no valvulated apparatus, anterior part slender, posterior part glandular and muscular; stylet usually without basal swellings . . . . (Plate 4A) DORYLAIMIDA 3
Three-part esophagus usually with a valvulated metacorpus (median bulb) followed by a slender isthmus and glandular basal bulb; stylet usually with basal knobs . . . . . (Plate 4B) TYLENCHIDA 6

Dorylaimida

Most individuals belonging to this order are free-living nematodes inhabiting soil and fresh water. Three genera include species which have been shown to be pathogenic to plant roots.

Esophagus consists of a slender anterior portion, sometimes with small muscular swellings, followed by an expanded portion that may be reduced to a simple valveless bulb. Stoma is provided with an axial stylet or mural tooth. Aperture of stylet is located dorsally. No setae or caudal spinneret. Excretory pore generally absent or rudimentary.

Tylenchida

With the exception of degenerate males of a few species all individuals possess a stoma armed with a protrusible stylet. The basal portion of the esophagus is either bulbar or lobe-like, without a sclerotized valvular apparatus. Cuticle is marked by striae which usually are interrupted on the lateral fields by incisures. Excretory pore is conspicuous and is usually located near the latitude of the nerve ring.

Plate 4. A. Anterior portion of *Dorylaimus* sp. B. Anterior portion of *Tylenchorhynchus* sp. a. guiding ring. b. stylet. c. developing stylet. d. lumen of esophagus. e. basal enlarged portions of esophagus. f. esophago-intestinal valve. g. stylet with basal knobs. h. metacorpus. i. valve of metacorpus. j. striation. k. annule. l. excretory pore. m. basal bulb of esophagus.

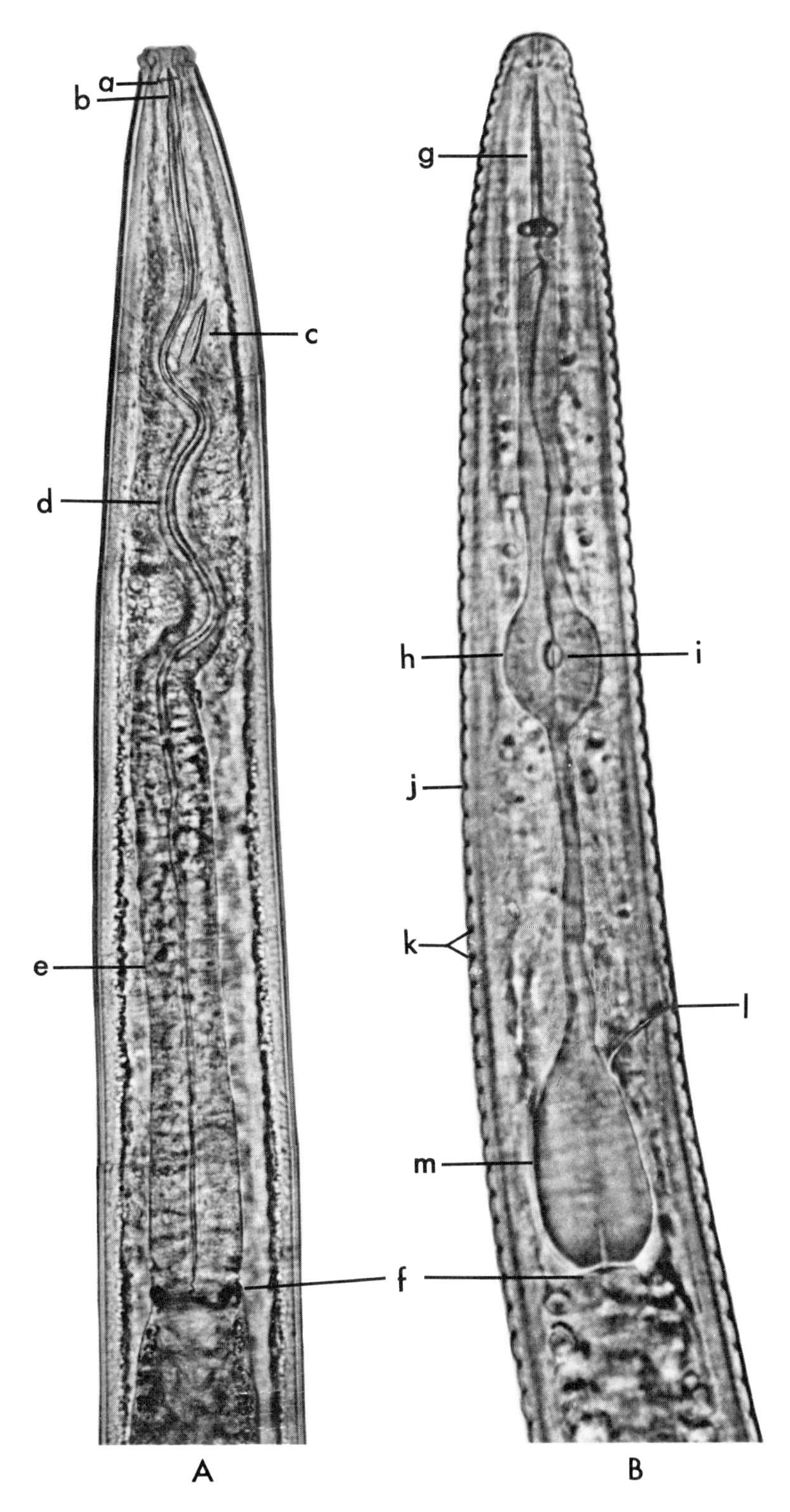
a
b
c
d
e
f
g
h
i
j
k
l
m
A
B

Plate 5. Heads and tails of nematodes. A. Face view of *Pratylenchus fallax*. B. Face view of *P. loosi*. C. Near profile view of head of *P. loosi*. D. Tail tip and lateral field with phasmid of *P. andinus*. E. Tail tip and lateral field, with phasmid of *P. penetrans*. F. Everted spicules of *Heterodera rostochiensis* male. (Scanning electron micrographs were taken by D. C. M. Corbett and Sybil A. Clark, Rothamsted Experimental Station.)

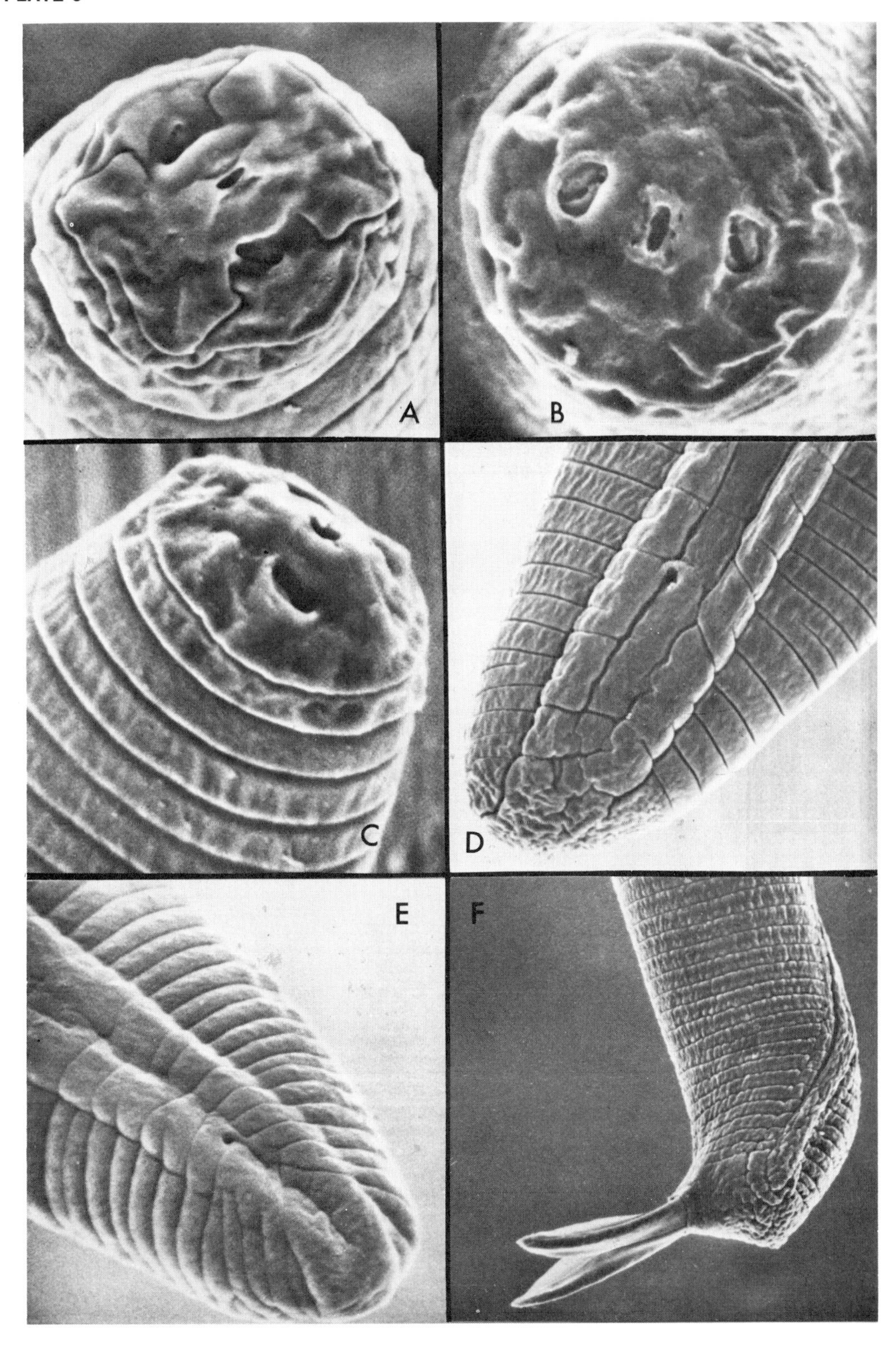
A
B
C
D
E
F

3. Stylet short, curved; body short and thick (0.45–1.5 mm long) . . . . (Plate 6) *TRICHODORUS*
   Stylet long, straight, tapering to a long slender point with long extensions; body long and slender . . . . (Example: Plate 7) 4
   Stylet straight, usually not very long. (Includes a large number of genera most of whose feeding habits are not definitely known. Group contains no known pathogens.) . . . . A LARGE NUMBER OF GENERA

Trichodorus Cobb 1913

Type species: *Trichodorus primitivus* (de Man 1880) Micoletzky 1922

Description: Trichodorinae. Plump nemas with blunt, rounded-to-acute tails, and thick cuticle. Spear, or onchiostyle, dorsally arcuate, slender, tripartite in its middle sector. Amphids elongate, pocketlike, with ellipsoid apertures. Sensillae arranged in fiberlike bundles almost adjacent to amphids. Esophagus with a pyriform basal bulb containing 3 large and 2 very small gland nuclei. Ovaries 2, except in *Trichodorus monohystera*, reflexed when 2 are present. Testis single, outstretched. Supplements ventromedian, an anal pair not present. Males of certain species with bursae, the only instance in which this organ is known among the Dorylaimoidea. Gubernaculum present. (Mostly from Thorne 1961 with an addition from Bird 1967.)

General Characteristics

Trichodorus spp. are widely distributed throughout the world, having been found in 14 countries on 5 continents. At present there are approximately 30 generally recognized species. The host range of *Trichodorus* spp. appears to be wide, but the host of a majority of the species have not been studied.

Nematodes belonging to *T. christiei* were first called "stubby root" nematodes, but approximately 9 additional species have since been shown to cause this symptom complex. External feeding at tips and sides of young succulent roots results in smaller root systems with fewer and shorter rootlets in affected than in unaffected roots. Six species are known to transmit the tobacco-rattle virus and 3 species pea early-browning virus. Two other genera of the Dorylaimida, *Xiphinema* and *Longidorus*, contain species which are pathogenic to plants and transmit plant viruses.

Under a dissecting microscope, aids in identifying *Trichodorus* spp. are the curved spear and overall cylindrical shape tapering at the anterior end. Male and female tails are short and bluntly rounded with a subterminal anus. Allen (1957) states: "The cuticle is often observed to wrinkle as the nematode moves giving the impression that the cuticle is rather loose. Fixed specimens frequently exhibit a swollen condition of the cuticle, this is apparently due to separation of the cuticle from the body during fixation."

The tail shape of *Trichodorus acutus*, described by Bird (1967), was convex conoid anteriorly, then conoid to an acute or slightly rounded terminus. Bird amends the generic description to include specimens with tail shapes.

Plate 6. Trichodorus christiei. A. Showing swollen conditions of cuticle, 435 X. B. Not exhibiting swollen condition of cuticle, 435 X. a. curved stylet.

a
A
B

4. Stylet extension with sclerotized basal flanges; guiding ring near base of stylet just anterior to junction of stylet and stylet extensions . . . . (Plate 7) *XIPHINEMA*
   Stylet extensions without basal flanges; guiding ring near apex of stylet . . . . . . (Example: Plate 8) 5

Xiphinema Cobb 1913

Type species: *Xiphinema americanum* Cobb 1913

Description: Longidorinae. Spear greatly attenuated with long extensions bearing basal flanges. Guiding ring located near base of spear. Esophagus beginning as a slender, coiled tube which is straight only when the spear is extruded. This slender portion suddenly expands to form the elongate basal bulb which usually is about 3 times as long as the neck width. Dorsal esophageal gland nucleus at extreme anterior end of bulb. Intestinal cells packed with coarse refractive granules. Prerectum present. Vulva transverse. Ovaries 1 or 2, reflexed. Spicula with lateral guiding pieces. Supplements consisting of an adanal pair and a ventromedian series. Testes 2, Dorylaimoid. (From Thorne 1939.)

General Characteristics

Xiphinema spp., commonly called dagger nematodes, are found around the roots of numerous plant species in many parts of the world. Only a few of the more than 30 species are known to be pathogenic to plants. Field observations indicate that results of future research will show certain species of *Xiphinema* are pathogens of major importance to numerous crops, especially woody plants. Three other genera of the Dorylaimida, *Longidorus, Paralongidorus,* and *Trichodorus,* contain species pathogenic to plants.

Cohn and Sher (1972) proposed the following 8 subgenera of the genus *Xiphinema*: *Radiphinema, Krugiphinema, Xiphinema, Elongiphinema, Halliphinema, Basiphinema, Rotundiphinema,* and *Diversiphinema*. A key to the subgenera of *Xiphinema* is presented, plus a list of 50 species in the genus. The subgenera are separated primarily on the basis of gonad structure, body size, and tail shapes.

Hewitt *et al.* (1958) reported that the soil-borne virus, fan leaf of grape, was transmitted by *X. index,* this being the first experimental evidence that a plant virus is transmitted by a nematode. Several plant viruses, including arabis mosaic, strawberry latent ringspot, and tobacco ringspot, are transmitted by species of this genus.

Under the dissecting microscope, one important diagnostic character for most species is the long, thin shape of the body. The long stylet, flanged at the base, is present on individuals of all described species. When at rest these nematodes assume the shape of a wide C.

Plate 7. Xiphinema americanum. A. Mature female, 255 X. B. Anterior portion of female, 350 X. C. Anterior portion of female, 600 X. a. guiding ring. b. flanges. c. developing stylet. d. anus. e. tail. f. stylet extension.

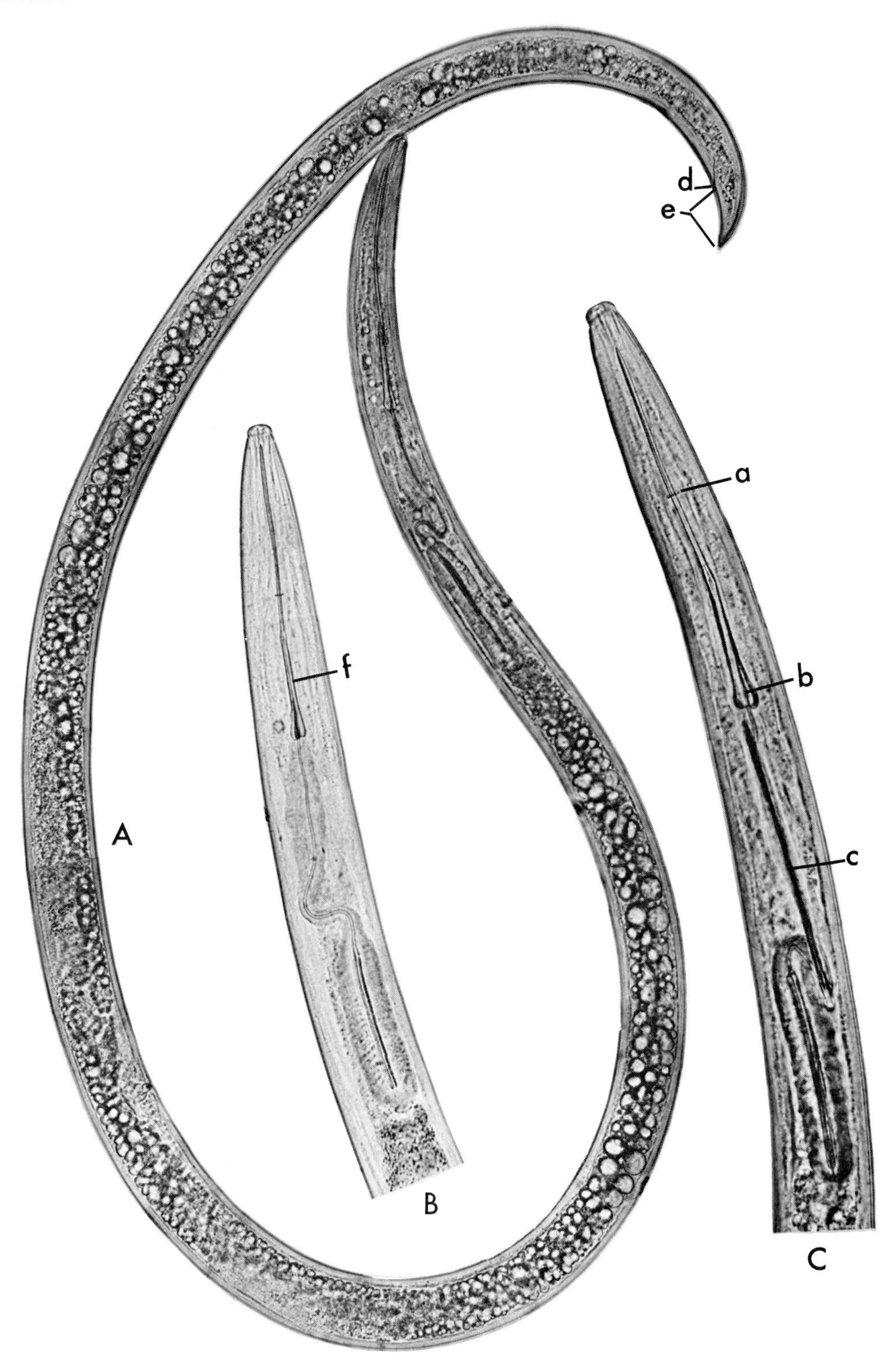
d
e
a
f
b
c
A
B
C

5. Amphid openings minute, slitlike; amphids consisting of large pouches which almost encircle the head . . . . (Plate 8) *LONGIDORUS*
Amphid openings wide, sublabial extending at least halfway across the neck at that point; amphid pouches funnel- to stirrup-shaped . . . . (Plate 9) *PARALONGIDORUS*

Longidorus (Micoletzky 1922) Thorne and Swanger 1936

Type species: *Longidorus elongatus* (de Man 1876) Thorne and Swanger 1936

Description: Longidoridae. Body greatly attenuated. Lips bearing the usual 2 circlets of 6 and 10 papillae. Lateral cords comparatively broad, the series of lateral pores in two lines. Amphid apertures minute, slitlike, exceedingly difficult to observe. Amphids consisting of abnormally large pouches which practically encircle the head. They contain coiled fibrillar terminals and small refractive bodies which probably are nervous in function. The nerve tubes can easily be traced back to the sensillae. Spear greatly attenuated. Guiding ring located near lip region. Esophagus reduced to a slender, flexible tube with an elongated basal bulb. The dorsal and the anterior pair of submedian gland nuclei are easily visible, while the posterior submedian pair is rather small and obscure. Ovaries 2 in all known species, reflexed, very short compared with the total body length. Vulva transverse. Tails of sexes similar. (From Thorne 1961.)

General Characteristics

Nematodes of this genus, the needle nematodes, are closely related to those of the genus *Xiphinema* and can be characterized as follows (Siddiqi 1959): "(1) Body long and greatly attenuated; (2) Amphids abnormally large, practically encircling the head; amphid aperture very minute and difficult to see; (3) Spear greatly attenuated, with guiding ring located near its apex; (4) Spear extension without basal swellings or flanges."

Several of the approximately 30 species are recognized as important plant pathogens. One species, *L. menthasolanum,* is a serious pest of mint in Oregon. Damage in individual mint fields ranges from a trace to large barren areas which occasionally envelop an entire field (Konicek and Jensen 1961). The feeding habits and type of injury caused by members of this genus are similar to those of the dagger nematodes (Christie 1959). Two other genera of the Adenophores, containing species pathogenic to plants, are *Xiphinema* and *Trichodorus*. Although another closely related genus, *Paralongidorus,* is found associated with plant roots, there is no experimental evidence that these nematodes are plant parasites.

Recent evidence indicates that some species transmit plant viruses.

Plate 8. Longidorus menthasolanum. A. Vulva, 500 X. B. Female tail, 500 X. C. Basal region of esophagus, 325 X. D. Anterior portion, 440 X. E. Mature female, 43 X. a. guiding ring.

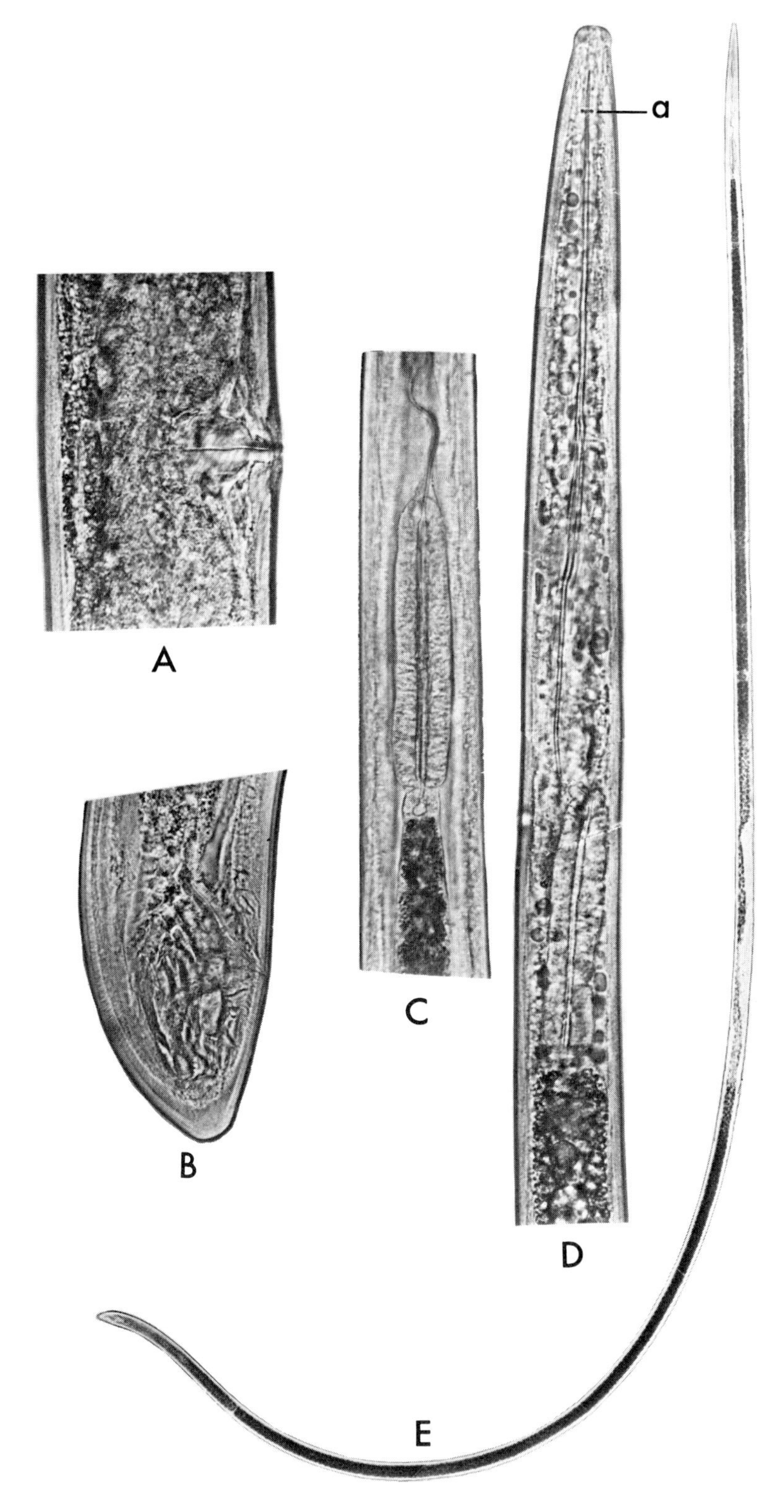
a
A
B
C
D
E

Paralongidorus Siddiqi, Hooper, and Khan 1963

Type species: *Paralongidorus sali* Siddiqi, Hooper, and Khan 1963

Description: Longidorinae Thorne 1935. Body elongate, over 2 mm long, slender ($a > 50$). Body cuticle smooth, marked by fine transverse striations. Lateral hypodermal chords are broad from which ducts ending in pores arise throughout the body length. Dorsal and ventral pores may also be present at the anterior end. Lips amalgamated, bearing 16 papillae arranged in an inner circlet of 6 and an outer circlet of 10. Amphid openings wide, sublabial, extending at least halfway across the neck at that point. Amphid pouches funnel- to stirrup-shaped in lateral view. Buccal stylet elongate, cylindroid; joins extension without pointed projections; stylet extension elongate, swelling slightly at its base, but swellings not sclerotized or flanged. Stoma ending in a conspicuous single guiding ring located around the anterior part of the stylet. Vulva transverse, near middle of the body. Gonads paired, opposed, ovaries reflexed. (From Siddiqi, Hooper, and Khan 1963.)

General Characteristics

This genus possesses some characteristics resembling those of both the genera *Xiphinema* and *Longidorus*. The elongate stylet extension with its slightly swollen nonsclerotized base and single, anteriorly placed, guiding ring are characters similar to those of *Longidorus*. The wide sublabial amphid openings and funnel- to stirrup-shaped amphid pouches somewhat resemble those of *Xiphinema*.

In an emendation of the generic description Abdoul-Eid (1970) stated that the amphid openings often extend completely across the head in lateral view, and the lip region may be offset by a constriction at the level of the amphidial opening.

Siddiqi (1964) presented a key to 5 species of *Paralongidorus*.

All species of *Paralongidorus* are associated with plant roots and probably all are plant parasitic. Edward, Misra, and Singh (1964) stated that *P. fici* caused galls on the roots of fig similar to those caused by *Xiphinema diversicaudatum* on roses.

Plate 9. Paralongidorus sali. A and B. Anterior end of female, lateral. C. Female. D. Female face view of lip region. E. Vulval region, lateral. F. Female head, dorsal. G. Vulva, ventral. H. Anterior end of female parasitized by a sporozoan. I. "Spores" from parasitized female. J-L. Larval tails, lateral: J. ? Second stage. K. ? Third stage. L. ? Fourth stage. M. Female tail, lateral. (After Siddiqi, Hooper and Khan 1963. Courtesy of Nematologica.)

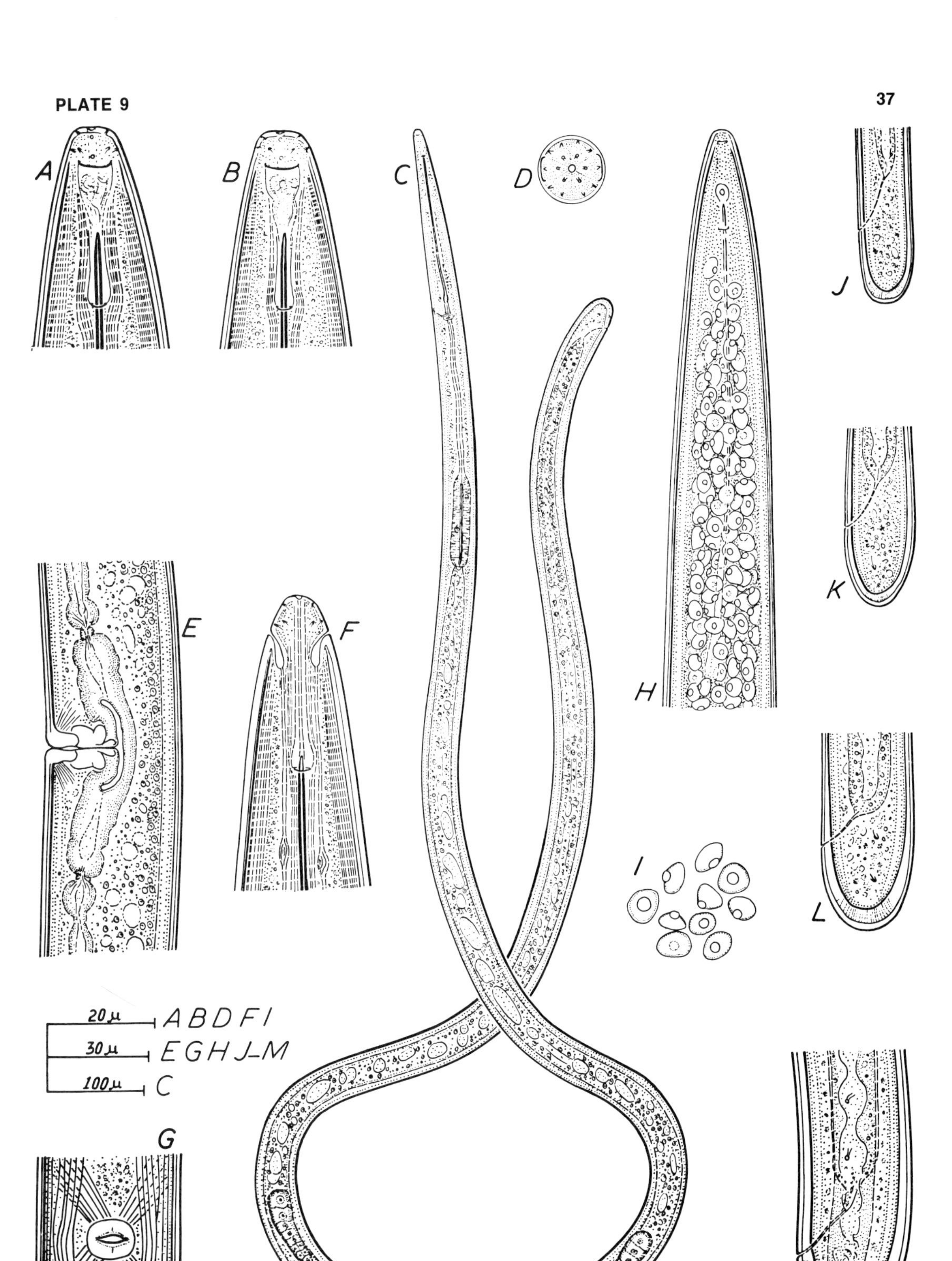
A
B
C
D
E
F
G
H
I
J
K
L
M
20 µ A B D F I
30 µ E G H J–M
100 µ C

6. Dorsal esophageal gland outlet in metacorpus, anterior to valve or in that position when median bulb absent (usually difficult to see); metacorpus very large, often appears nearly as wide as the diameter of the body . . . . (Example: Plate 11) APHELENCHOIDEA 7
   Dorsal esophageal gland outlet in procorpus (usually can be seen more readily in recently prepared water mounts than in glycerine mounts); metacorpus moderate to reduced in size (less than 3/4 body width) . . . . (Example: Plate 4B) TYLENCHOIDEA 9
7. *a* (length/greatest width) = less than 80, vulval flap absent; vagina normal . . . . (Example: Plate 8) 8
   a = around 100; vulva with wide overlapping flap; vagina curved (Plate 10) *RHADINAPHELENCHUS*

Rhadinaphelenchus Goodey 1960

Type species: *Rhadinaphelenchus cocophilus* (Cobb 1919) J. B. Goodey 1960

Description: Aphelenchoididae. Very slender nematodes (*a* = about 100); female with vulval flap; long postuterine sac, and long digitate tail. Male with terminal, bursal flap. (From Goodey 1963.)

General Characteristics

This genus contains only a single species, *R. cocophilus*, a parasite of the roots and trunk of the coconut palm. Infection by large numbers of this economically important nematode results in death of leaves, growing points, flowers, and eventually the entire tree. This species is found in the Caribbean Islands and Central and South America.

A new genus was erected for this species, formerly included in the genus *Aphelenchoides*, because of its distinct morphological characters. Concerning these characters, Thorne (1961) comments: "*Rhadinaphelenchus* is distinctive because of the very slender body, massive sclerotization of the labial arches, elongated median bulb, wide vulvar flap, unusual curved vagina, spiculum form, and sclerotized spadelike extension of the male tail. No other genus of the superfamily has such outstanding diagnostic characters."

Plate 10. Rhadinaphelenchus cocophilus. A. Female. B. Sagittal section of head end. C. Surface view of head end. D. End-on view of head. E. Sagittal section through vulva. F. Ventral view of vulva. G. Lateral view. H. Ventral view of male tail. J. Head end of preadult larva. K. Tail end of preadult larva. (After Goodey 1960. Courtesy of Nematologica.)

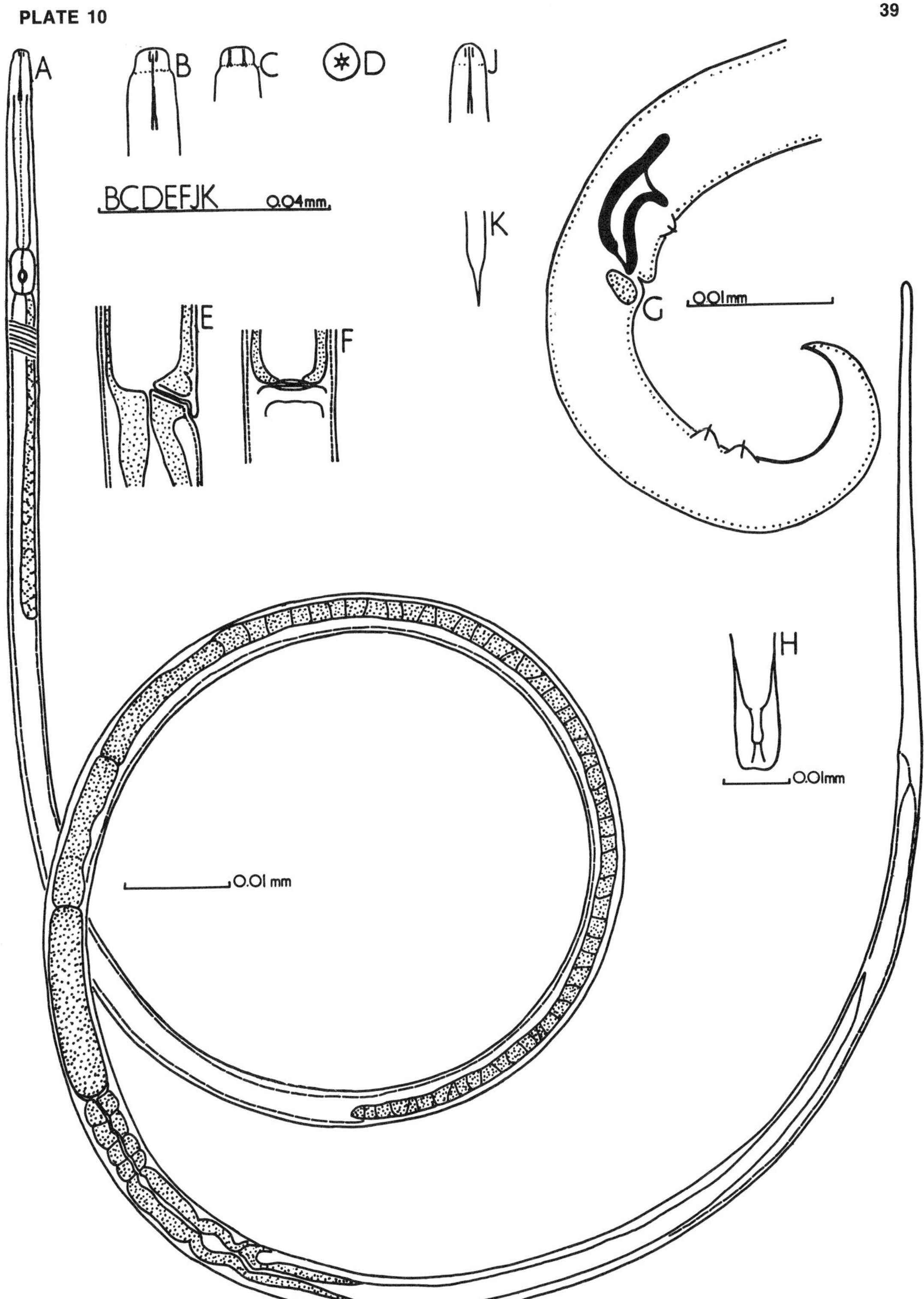
A
B
C
D
J
BCDEFJK 0.04mm
K
E
F
G
0.01mm
H
0.01mm
0.01 mm

8. Tail of female blunt; lateral field with 6 to 14 incisures; male with bursa and gubernaculum .... (Plate 11) *APHELENCHUS*
Tail of female usually conoid, often with one or more sharp points at the terminus (mucronate); lateral field with 2 to 4 incisures; male without bursa or gubernaculum ...... (Plate 12) *APHELENCHOIDES*

Aphelenchus Bastian 1865

Type species: *Aphelenchus avenae* Bastian 1865

Description: Aphelenchidae. Body tapering anteriorly. Cuticle transversely striated. Lateral field with numerous incisures. Deirids present at about level of the excretory pore. Head slightly offset. Spear shaft with slight thickenings at the base. Procorpus cylindrical, constricted slightly where it joins the ovoid median esophageal bulb which contains prominent, median, crescentic valve plates. Esophageal glands usually with, sometimes without, a lobe overlapping the intestine dorsolaterally and joining the alimentary canal where the nerve ring surrounds it just posterior to the median bulb. Excretory pore about opposite nerve ring. Intestine joined to median bulb by a short isthmus about $1\frac{1}{2}$ body-widths long. Vulva posterior, ovary outstretched, prodelphic; a postvulval sac present, rather obscure but usually reaching about halfway from vulva to anus. Vagina with thickened walls. Rectum about 1 to 2 body-widths long. Tail between 1 and 4 anal-body-widths long, cylindrical to a rounded end. Phasmids subterminal. Male with bursa usually supported by 1 preanal and about 3 postanal, subterminal pairs of ribs. Spicules paired, slender, ventrally slightly arcuate, proximally slightly cephalated. Gubernaculum about 1/3 as long as the spicules. (From Goodey and Hooper 1965.)

General Characteristics

The members of this genus have an almost cylindrical body. An outstanding characteristic is the well-developed esophageal bulb which occupies 3/4 of the body width and is usually visible under a dissecting microscope.

A. avenae is very widespread, occurring in or on decaying bulbs, tubers, rhizomes, and roots. It is not an obligate parasite of higher plants. It can be cultivated easily on a nutrient agar medium on which fungi of various species are growing. In greenhouse tests, *A. avenae* invaded lesions in corn roots infected with *Pythium arrhenomanes* but did not invade similar healthy roots in which this fungus was not present (Rhoades and Linford 1959).

Plate 11. *Aphelenchus avenae*. A. Mature female, 305 X. B. Anterior portion of mature female, 820 X. a. stylet without basal knobs. b. very large metacorpus.

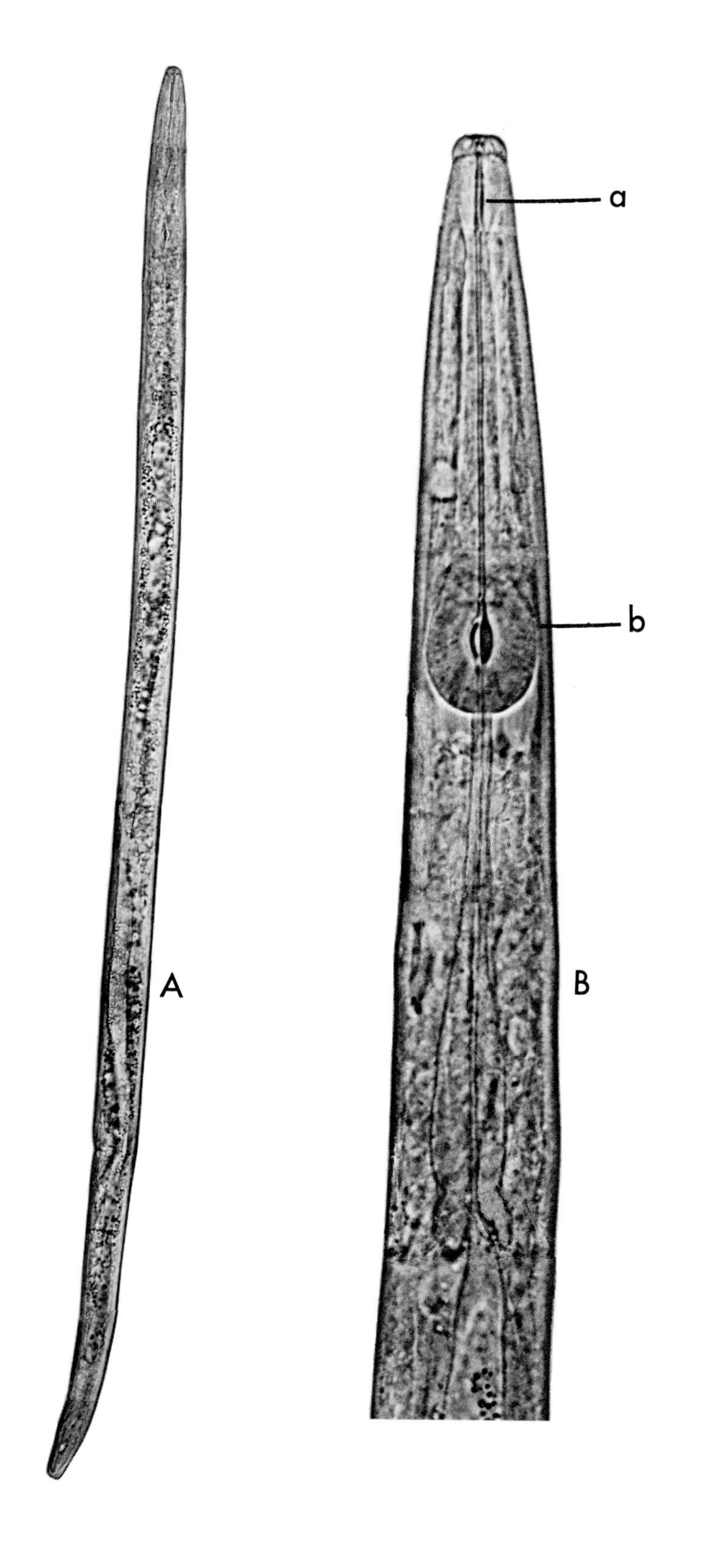
a
b
A
B

Aphelenchoides Fischer 1894

Type species: *Aphelenchoides kuehnii* Fischer 1894

Description: Aphelenchoidinae. Cuticle marked by fine transverse striae. Lateral fields with 2, 3, or 4 lines or incisures. Lip region not striated. Six amalgamated lips, slightly hexagonal when seen from a face view. Amphid apertures minute, at apex of lateral lips. A single papilla usually present on each of the 4 submedian lips. Labial framework hexaradiate, lightly sclerotized except for refractive spear guide at oral aperture. Esophageal glands in long lobes extending back over intestine. Hemizonid posterior to excretory pore. Ovary outstretched, with oocytes arranged in tandem or multiple rows. Posterior uterine branch short and collapsed or forming an elongate reservoir for spermatozoa. Males without bursa or gubernaculum. Two or 3 pairs of ventrosubmedian papillae usually present. Phasmids excessively small and difficult to observe, usually subterminal. Tails of both sexes conoid to blunt or pointed terminus, often mucronate, never filiform. (From Thorne 1961.)

General Characteristics

Goodey (1963) comments concerning the genus *Aphelenchoides*: "Several of the species of *Aphelenchoides* are important plant parasites, others are free-living, saprophagous forms occurring in soil or decaying plant material, etc. and probably feeding on fungal hyphae. Some of these species, including the plant parasite *A. fragariae*, can be successfully cultured on nutrient agar along with fungi."

Five or 6 species, called bud and leaf nematodes, live as parasites in buds and foliage of plants. Under a dissecting microscope, members of this genus appear as relatively slender nematodes with conical tails. The most important diagnostic character is the large metacorpus occupying 3/4 or more of the width of the esophagus.

Sanwal (1961) presented a key to 35 species of *Aphelenchoides*. *Paraphelenchoides,* a genus closely related to *Aphelenchoides,* and a new species *P. capsulophanus* were erected by Khak in 1967. *P. xylophilus* n. comb. and *P. limberi* n. comb. were transferred to the new genus from *Aphelenchoides*. In *P. capsuloplanus* the male and female tails are conical with rounded tips, the posterior uterus is about 4 body diameters at the level of the vulva, the spermatheca is long, and the lateral fields have 4 lines.

Plate 12. Aphelenchoides sacchari. A. Female nematode. B. Lateral view of head. C. Male tail. D. Section in midbody region showing lateral field. E. Lateral view of cuticle in midregion. a. mucro. (After Hooper 1958. Courtesy of Nematologica.)

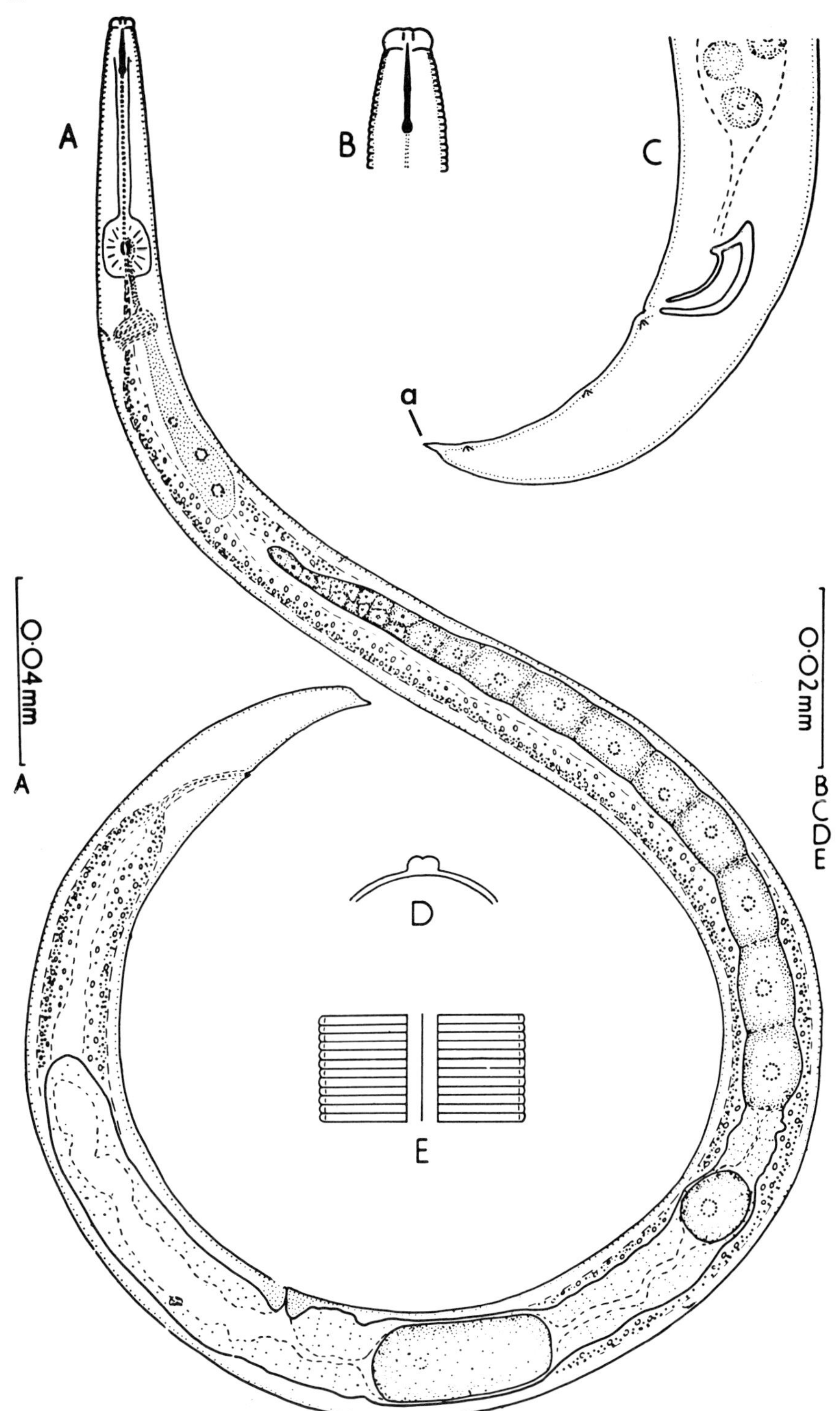
A
B
C
a
D
E
0·04mm
A
0·02mm
B
C
D
E

9. Head with setae. No plant parasites . . . . (Picture not included) *ATYLENCHUS*
. . . . (Picture not included) *EUTYLENCHUS*
Head without setae. Numerous plant parasities 10
10. Metacorpus absent or reduced; if reduced, no sclerotized valve . . . .
. . . . (Example: Plate 14) *NEOTYLENCHIDAE*
Example of family . . . . (Plate 13) *NOTHANGUINA*
Example of family . . . . (Plate 14) *NOTHOTYLENCHUS*
Metacorpus with sclerotized valves present (usually can be seen more readily in recently prepared water mounts than in glycerine mounts) . . . . (Examples: Plate 3A, B, and C) 11

Nothanguina Whitehead 1959

Type species: *Nothanguina cecidoplastes* (Goodey 1934) Whitehead 1959

Description: Nothotylenchinae. Body stout, ventrally coiled in female; head flattened, with 6 lips; small knobbed spear; esophagus without median bulb, narrowing to slender isthmus behind aperture of subventral esophageal glands; gonad single, prodelphic, reflexed anteriorly; no distinct spermatheca in female but with a postvulval uterine sac. The male has a ventral bursa not extending to the tail tip, but no gubernaculum. The tails of both sexes taper to a minute digitate, terminal process. (From Whitehead 1959.)

General Characteristics

Thorne (1961) states: "The general body characters of *Nothanguina* resemble those of *Anguina,* while the arrangement of cephalic papillae and amphid apertures are more like those of *Ditylenchus.* The two known species *N. cecidoplastes* and *N. phyllobia*, represent other divergent types. The ovary of *N. cecidoplastes* contains multiple rows of oocytes arranged about a rachis, and a gubernaculum is lacking, while in *N. phyllobia* the oocytes are arranged in single file and the gubernaculum is present."

Individuals of *N. cecidoplastes* are associated with galls on leaves, stems, and flowers of *Bothriochloa pertusa.* Goodey (1934) described the nematodes infecting these galls and the internal structure of the galls. These galls have not been reproduced by inoculating healthy plants with this nematode.

Plate 13. Nothanguina cecidoplastes. Female and anterior end of female and male tail. (After Goodey 1934. Courtesy of Journal of Helminthology.)

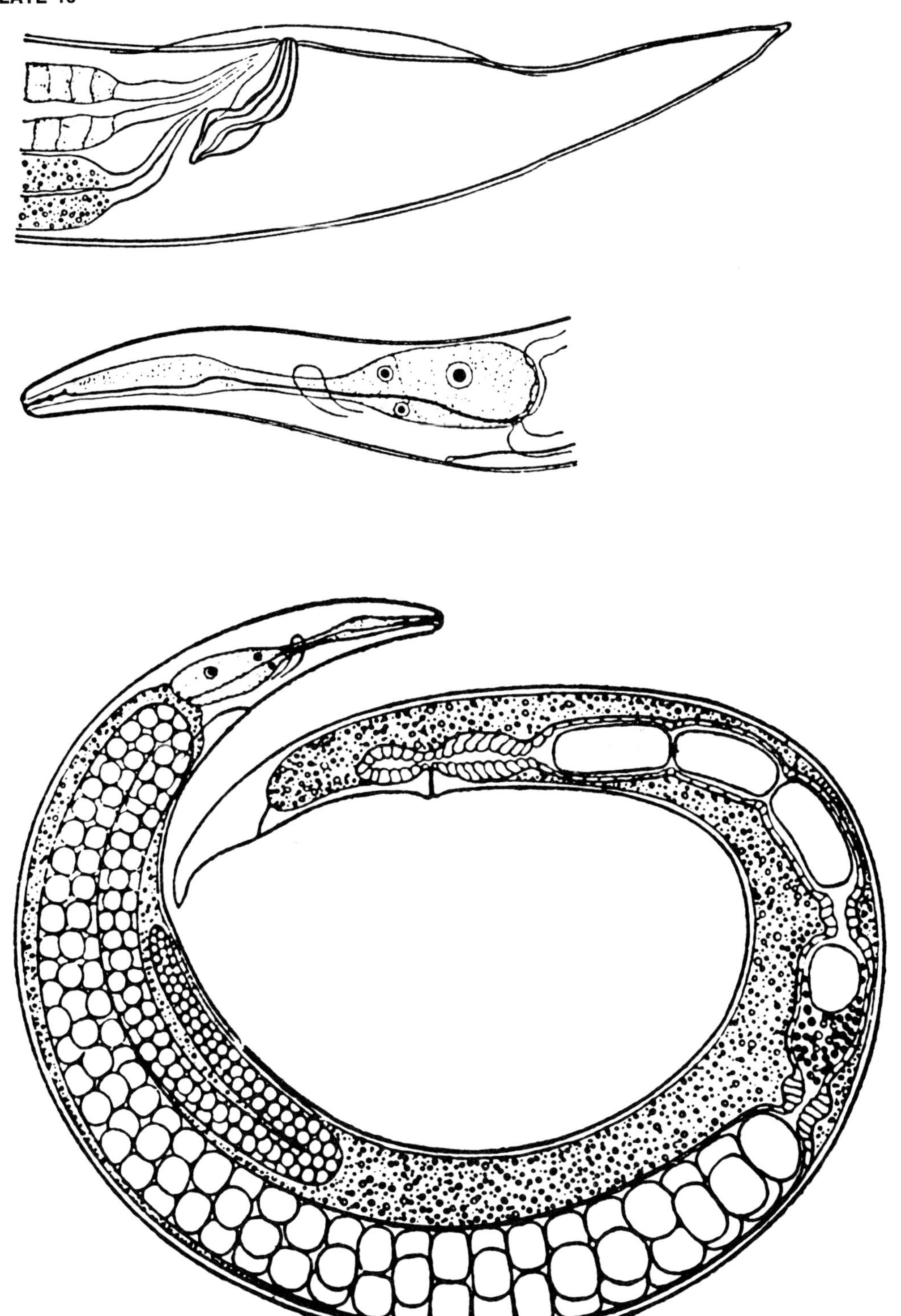

Nothotylenchus Thorne 1941

Type species: *Nothotylenchus acris* Thorne 1941

Description: Nothotylenchinae. Cuticle thin with fine transverse striae. Lateral field with 4 or more incisures. Spear with rounded knobs. Precorpus and corpus of esophagus cylindrical or fusiform; terminal bulb distinctly offset from intestine, sometimes slightly lobed. Ovary single, prodelphic, outstretched; postvulval sac present. Spicules and gubernaculum tylenchoid. Bursa extending nearly to middle of tail. (From Goodey 1963.)

General Characteristics

These nematodes are associated with roots and above-ground parts of several plant species. Nishizawa and Iyatomi (1955) produced abnormal symptoms in strawberry plants by inoculating healthy buds with individuals of *N. acris*. Symptoms included twisting of young petioles, crimping of leaves, and dwarfing of the entire plant. The nematodes were found in buds, leaflets, and axils of leaflets.

Plate 14. Nothotylenchus acris. A. Anterior portion of body. B. Posterior portion of female. C. Posterior portion of male. D. Body section showing wing area. (After Thorne 1941. Courtesy of The Great Basin Naturalist.)

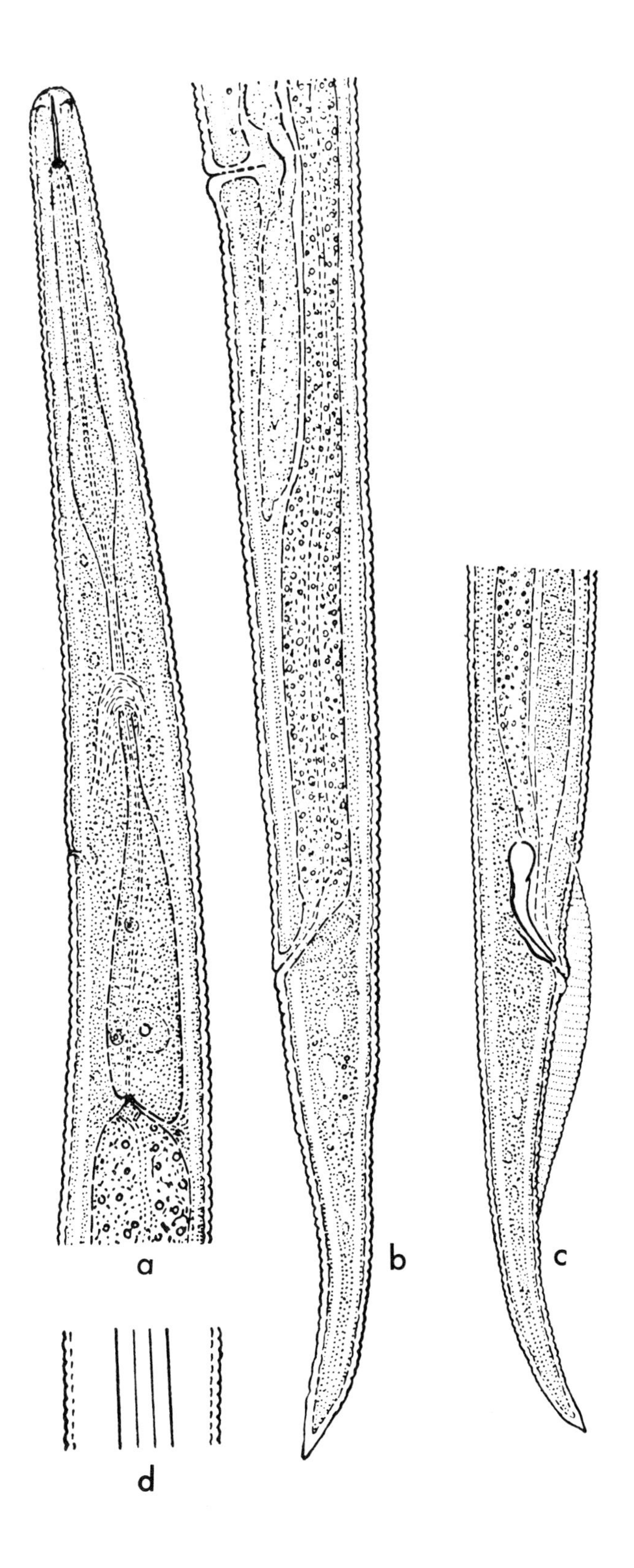
a
b
c
d

11. Mature female greatly enlarged (pear-shaped, lemon-shaped, kidney-shaped, or saccate); found in roots of plants, either embedded or attached by neck; some occur as cysts in the soil .... (Example: Plate 15A) 12
Mature females vermiform, may be slender to slightly swollen .... (Example: Plate 15B) 23
12. Mature female bodies soft; elongate-saccate, or kidney-shaped with tail; except for *Sphaeronema*, which is spherical in shape without a tail .... (Example: Plate 16) 13
Mature females becoming cysts or remaining soft-bodied; pyriform-saccate; spheroid or lemon-shaped, usually without a tail .... (Example: Plate 23) 18

Plate 15. A vermiform and a swollen female. A. Mature female of *Meloidogyne* sp., 280 X. B. Mature female of *Tylenchorhynchus* sp., 370 X.

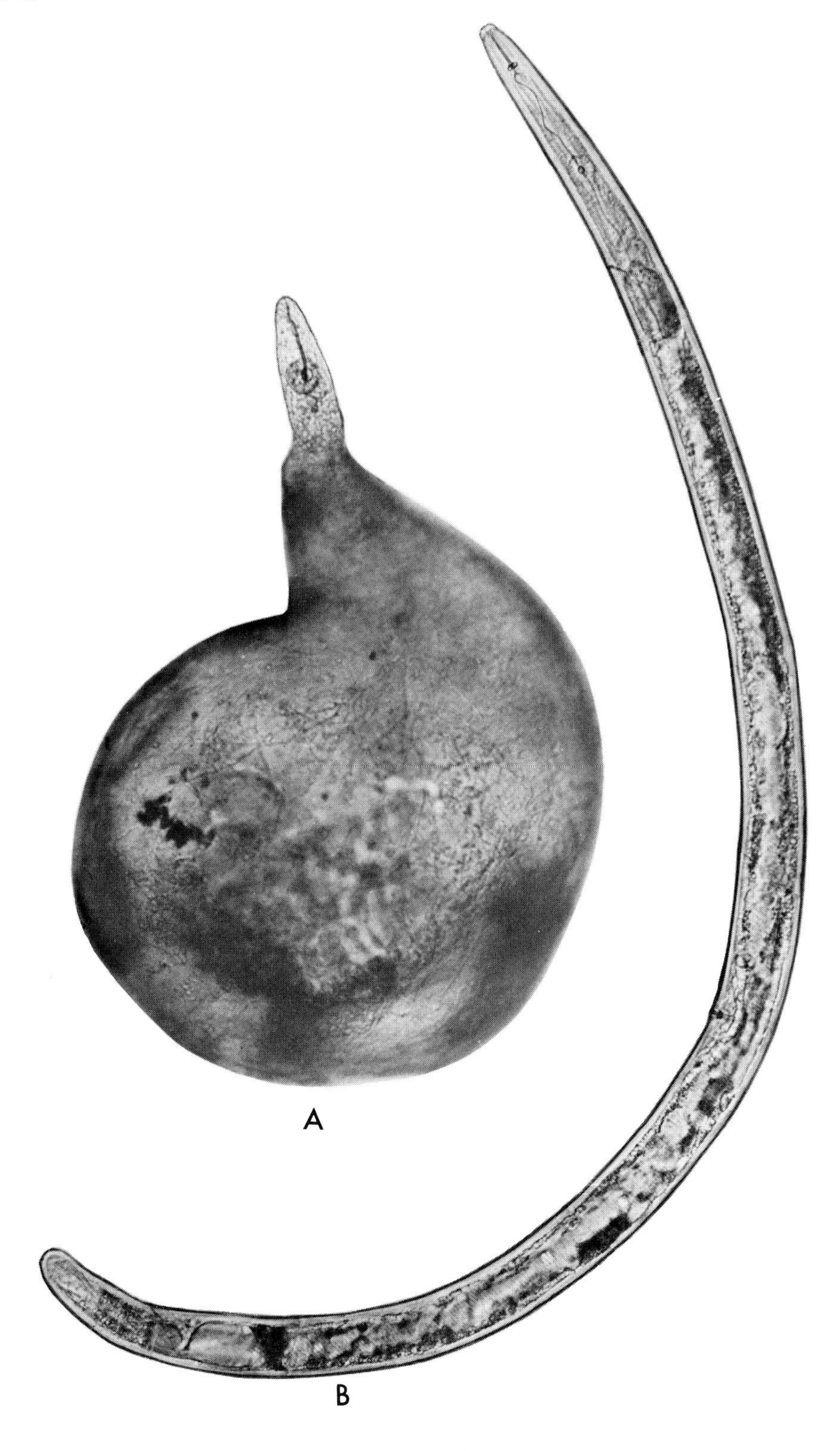
A
B

13. Mature female has 2 ovaries . . . . (Plate 16) *ROTYLENCHULUS*
    Mature female has 1 ovary . . . . (Example: Plate 18) 14

Rotylenchulus Linford and Oliveira 1940

Type species: *Rotylenchulus reniformis* Linford and Oliveira 1940

Description: Rotylenchulinae. Mature female, swollen, kidney-shaped; male vermiform. Lip region of immature female not set off, cephalic framework conspicuous. Dorsal gland orifice more than 1/2 stylet length posterior to base of stylet knobs. Esophagus long with narrow isthmus, glands overlapping intestine laterally, ventrally, more often laterally. Vulva post-median ovaries didelphic, amphidelphic with 2 flexures in immature female, highly convoluted in mature female. Female tail usually more than twice anal body diameter. Larval tail more rounded than tail of immature female. Phasmid porelike, anterior to middle of tail. Male with weak stylet and stylet knobs, reduced esophagus, indistinct median bulb and valve. Caudal alae adanal. Lateral field of male, immature female, and larvae with 4 incisures, nonareolated. Eggs deposited in gelatinous matrix. (From Dasgupta, Raski, and Sher 1968.)

General Characteristics

Rotylenchulus reniformis, the reniform nematodes, are obligate parasites with a highly specialized life history and with only the females feeding. Linford and Oliveira (1940), in studies made with observation boxes, did not observe either the males or larvae feeding. The young female is the infective stage. The females embed themselves partly, or sometimes entirely, in the root cortex and begin feeding. As the female feeds, the posterior part of the body enlarges to form the characteristic kidney shape. This species is highly specialized for a sedentary mode of life, i.e., with a superimposed series of molts without growth intervals, degeneration of males, and transformation of adult females to a reniform shape (Linford and Oliveira 1940).

Loof and Oostenbrink (1962) described a second species of this genus *R. borealis,* and Husain and Khan (1965) a third, *H. stakmani.*

The young adult females resemble members of the genus *Pratylenchus*. With the exception of the bulbous head the males are similar to the males of *Radopholus similis*. Linford and Oliviera (1940) originally placed *Rotylenchus* in the subfamily Pratylenchinae, and Thorne (1961), with a question mark, also included it in the subfamily Nacobbinae. Husain and Khan (1967) erected a new subfamily, Rotylenchulinae, to include only this genus. Allen and Sher (1967) also classified it in this new subfamily. Dasgupta, Raski, and Sher (1968) redescribed this genus.

Plate 16. Rotylenchulus reniformis. A. Male. B. Anterior portion of male. C. Posterior portion of male. D. Mature reniform female (this is semidiagrammatic in that the ovaries are represented as superimposed on the intestine). (After Linford and Oliveira 1940. Courtesy of Helminthological Society of Washington.)

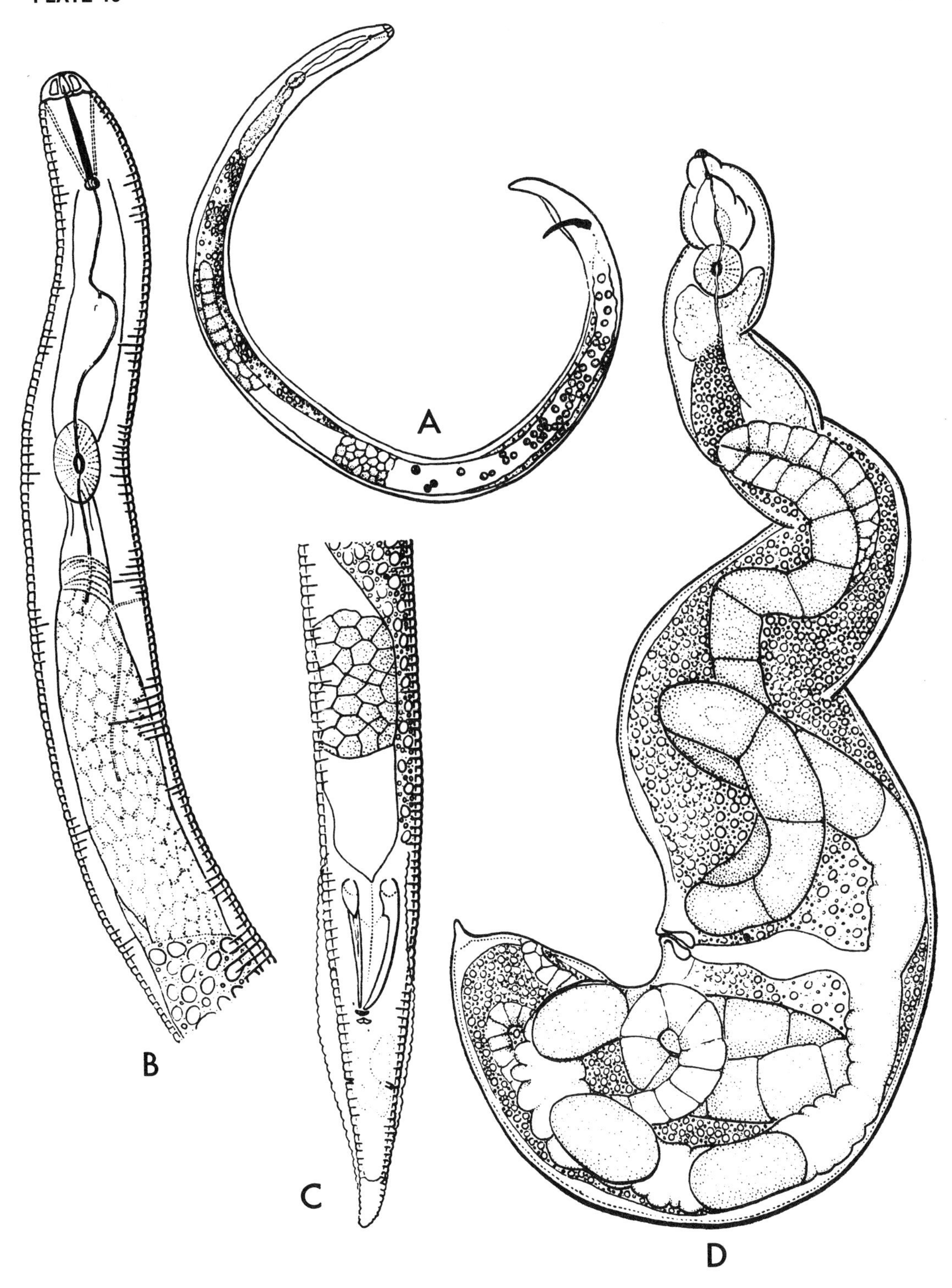
A
B
C
D

14. Excretory pore located in normal position, near nerve ring . . . . (Example: Plate 17) 15
    Excretory pore located posterior to nerve ring 16
15. Mature female subspherical; cuticle marked with coarse reticulate pattern; may have a prominently protruding vulva, subterminal in position . . . . (Plate 17) *SPHAERONEMA*
    Mature female spiral, thickened; without protruding vulva . . . . (Plate 18) *TROPHONEMA*

Sphaeronema Raski and Sher 1952

Type species: *Sphaeronema californicum* Raski and Sher 1952

Description: Sphaeronematinae. Larva provided with a strongly developed spear. Sclerotization of head with hexaradiate symmetry. Female subspherical, spear well developed, 1 or 2 very obscure annules present near lip region. Esophagus very strongly developed. Ovary single, uterus unusually large with thick muscular walls. Males slender, active, spear lacking, esophagus degenerate, bursa absent, spicule sheath present. Testis one. (From Raski and Sher 1952.)

General Characteristics

Members of this genus are mostly sedentary endoparasites. Females of *S. californicum* were found wholly embedded in roots of *Umbellularia californica* (Raski and Sher 1952). But large numbers of *S. minutissimum* were found on the roots of Citrus spp. with only the fore part of the females embedded in the root tissue. Sledge and Christie (1962) described a third species *S. whittoni* from the roots of sweet gum (*Liquidambar styraciflua*) L., and Kirjanova (1970) found a species on the roots of *Rumex confertus.*

Plate 17. Sphaeronema californicum. A. Neck region of larva. B. Face view of larva. C. Tail of larva. D. Neck region of male. E. Male tail. F. Male. G. Female head. H. Cross section of median bulb of female. I. Female. (After Raski and Sher 1952. Courtesy of Helminthological Society of Washington.)

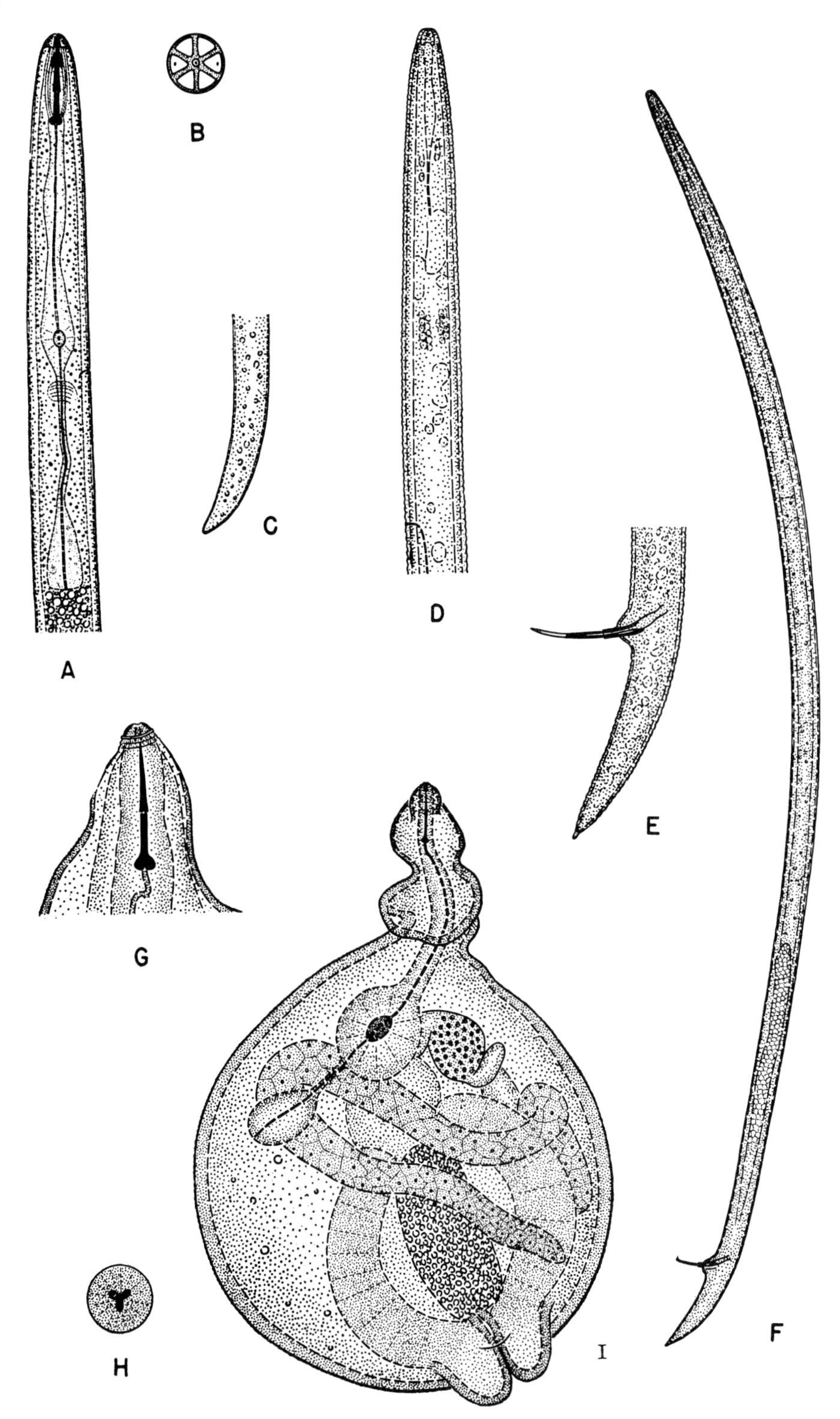
A
B
C
D
E
F
G
H
I

Trophonema Raski 1957

Type species: *Trophonema arenarium* (Raski 1956) Raski 1957

Description: Sphaeronematinae. Considerable sexual dimorphism present; females spiral, thickened; males slender; female esophagus tylenchoid, median bulb conspicuous; stylet well developed; anus indicated but apparently nonfunctional; female tail arcuate, thick; 1 ovary anterior to vulva, oviparous, 1 egg at a time in uterus deposited in a mucoid mass surrounding vulva; body greatly reduced in diameter posterior to vulva; male tail elongate conoid with sharply pointed terminus; spicules arcuate; setaceous; caudal alae absent. (From Raski 1957.)

General Characteristics

The adult females of this genus are sedentary ectoparasites. *Trophonema*, although closely related, differs from the genus *Sphaeronema* primarily in the shape of the females; in *Sphaeronema* the adult females are spherical, while in *Trophonema* the adult females are elongate-saccate and assume a spiral form. *Trophonema* is differentiated from *Trophotylenchulus* in the position of the excretory pore and in the shape of the lip region. In *Trophonema* the excretory pore is just behind the median bulb, while in *Trophotylenchulus* it is posterior to the esophagus. The females and larvae of *Trophotylenchulus* have a lip region with a distinct circumoral elevation while this character is absent in females and larvae of *Trophonema*.

T. arenarium was collected in California on the roots of Pacific rush (*Juncus leseurii*).

Plate 18. Trophonema arenarium. A. Neck region of larva. B. Tail of larva. C. Tail of female. D. Head of female. E. Female. F. Male tail. G. Neck region of male. (After Raski 1956. Courtesy of Helminthological Society of Washington.)

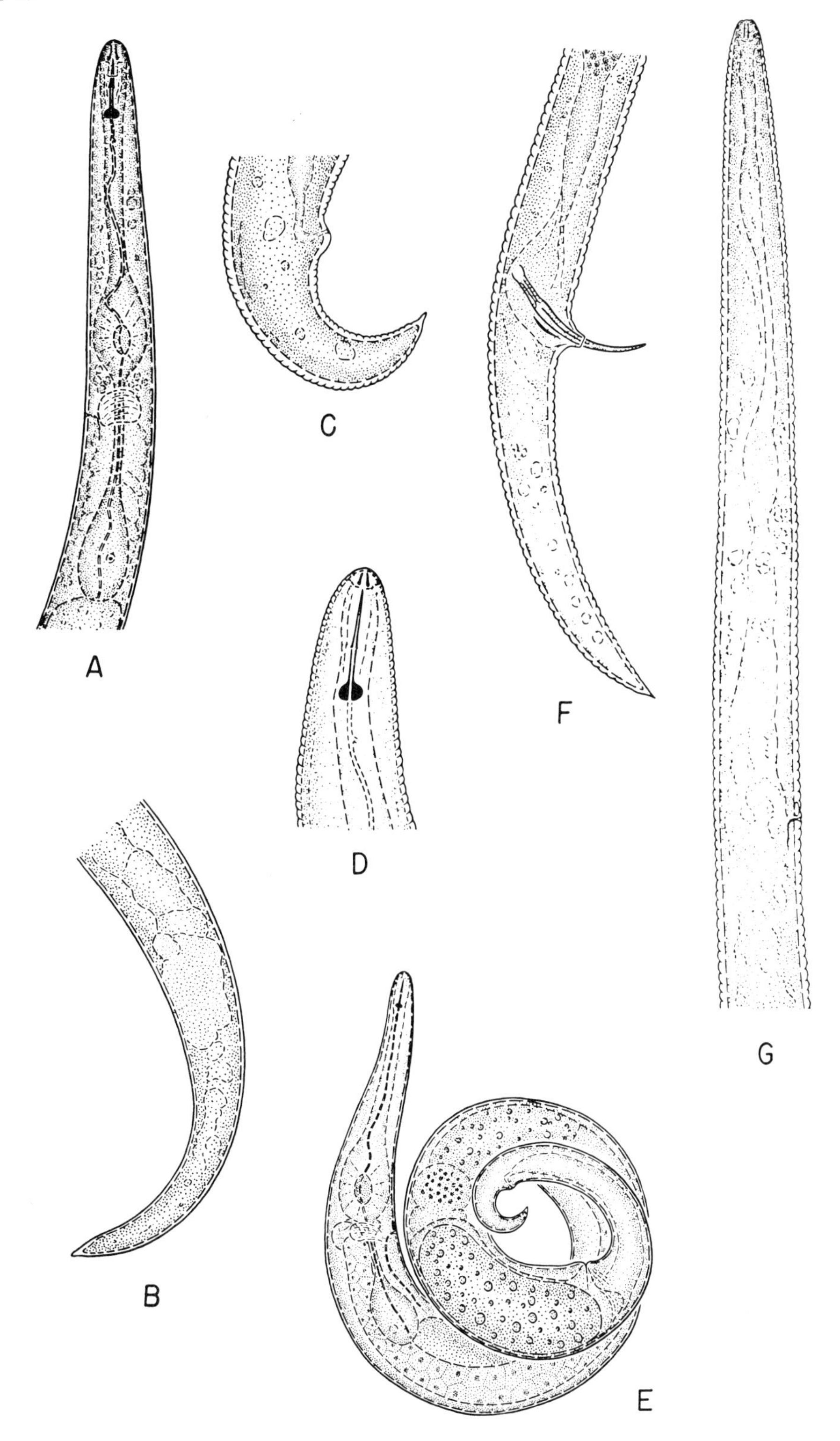
A
B
C
D
E
F
G

16. Circumoral elevation present in females and larvae . . . . (Plate 19) *TROPHOTYLENCHULUS*
    Circumoral elevation absent 17

Trophotylenchulus Raski 1957

Type species: *Trophotylenchulus floridensis* Raski 1957

Description: Tylenchulinae. Considerable sexual dimorphism present. Females spiral, thickened; males slender; lip region with distinct circumoral elevation in females and larvae; esophagus tylenchoid, median bulb conspicuous; stylet well developed; anus indicated but apparently nonfunctional; excretory pore far posterior to esophagus (33–44% from anterior end of body), even more posterior in larvae; female tail arcuate, thick; 1 ovary anterior to vulva; oviparous, 1 egg at a time in uterus, deposited in a mucoid mass surrounding vulva; body greatly reduced in diameter posterior to vulva; male tail elongate, conoid with bluntly rounded terminus; spicules arcuate, setaceous, with distinct bend near distal end; caudal alae absent. (From Raski 1957.)

General Characteristics

This genus is closely related to the genus *Tylenchulus.* Members of the genus *Trophotylenchulus* differ from *Tylenchulus* chiefly in that the lip region of females and larvae of the genus *Trophotylenchulus* is modified to form a "circumoral elevation" which does not occur in *Tylenchulus.* Another character useful in differentiating these genera is the location of the excretory pore—nearer the vulva in *Tylenchulus* than *Trophotylenchulus.*

The genus differs from *Trophonema* in the modification of the lip region and the more posterior position of the excretory pore. In this respect, location of the excretory pore in *Trophotylenchulus* is intermediate between *Trophonema* and *Tylenchulus.*

T. floridensis was collected in Florida from the roots of southern red oak (*Quercus falcata*), magnolia (*Magnolia* sp.), and persimmon (*Diospyros* sp.). On the roots of red oak, females occur individually in small darkened galls which are composed of host root material and are 2 cells or more in thickness (Raski 1957).

Goodey (1963) transferred the species *Tylenchulus mangenoti* to the genus *Trophotylenchulus.*

Plate 19. Trophotylenchulus floridensis. A. Male tail. B. Head of male. C. Head of larva. D. Tail of larva. E. Tail of female. F. Head of female. G. Female. (After Raski 1957. Courtesy of Nematologica.)

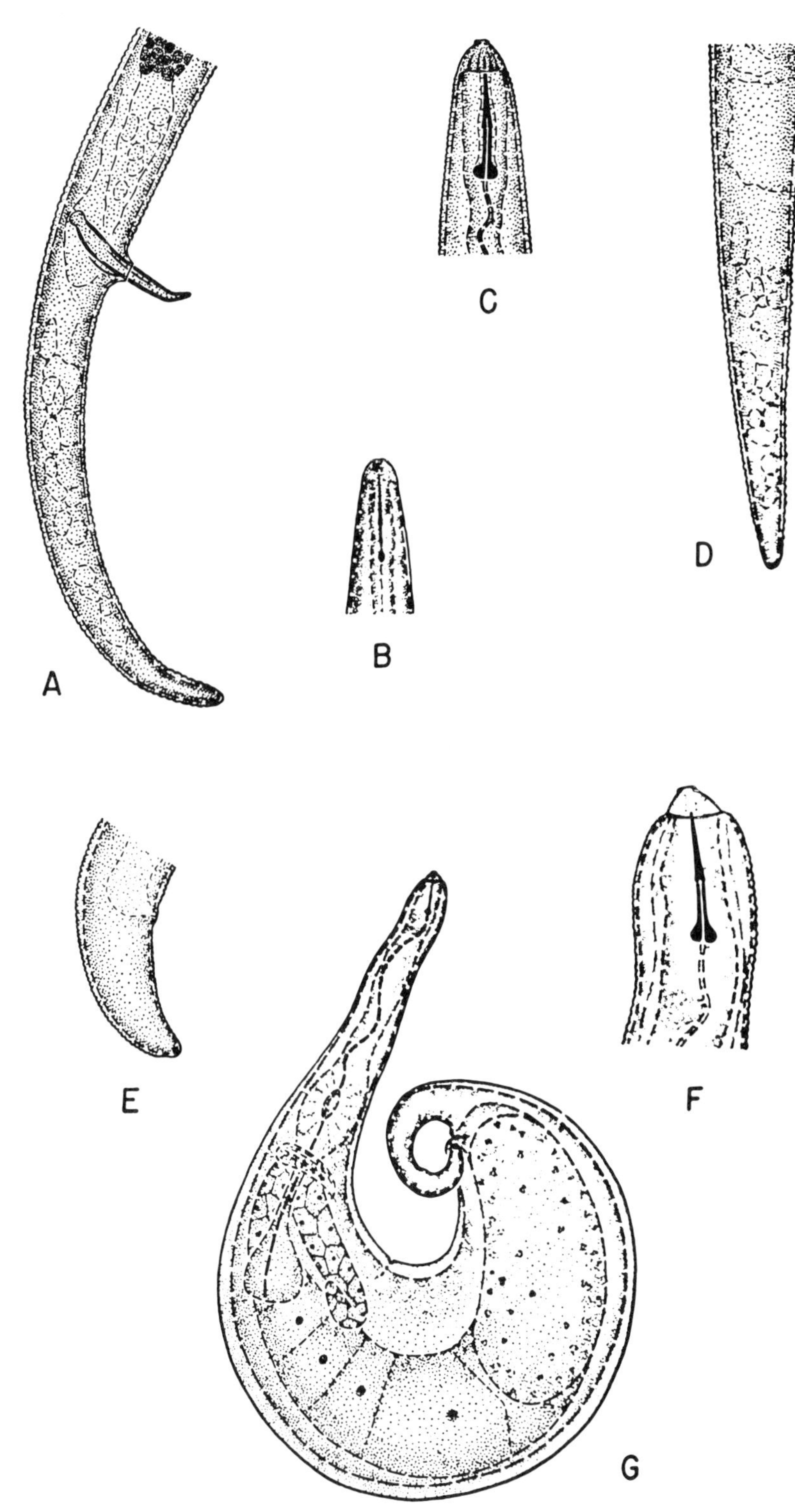
A
B
C
D
E
F
G

17. Excretory pore near vulva . . . . (Plates 20, 21) *TYLENCHULUS*
Excretory pore located near basal region of esophagus . . . . (Plate 22) *NACOBBUS*

Tylenchulus Cobb 1913

Type species: *Tylenchulus semipenetrans* Cobb 1913

Description: Tylenchulinae. Excretory pore postequatorial in position. Hind part of female swollen, elongate-saccate, with projecting tail. Anterior part variously swollen, fitting between the cortical cells of the host root, the posterior part being external. Cuticle of fore part annulated; of swollen body smooth and very thick. Head with hardly any skeleton, not offset. Spear fairly well developed, with large rounded basal knobs. Orifice of dorsal esophageal gland about a spear's length from spear base. Procorpus cylindrical, widening to a powerful, rounded, median bulb containing large valve plates. Isthmus narrow, widening to ovate, terminal bulb. Ovary single, coiled, with spermatheca. Male vermiform, spear slender, esophagus reduced; bursa absent, tail tapering to a broadly rounded tip; spicules and gubernaculum present. (From Goodey 1963.)

General Characteristics

The citrus nematode, *Tylenchulus semipenetrans*, occurs in most citrus-growing areas causing considerable economic loss. Females feed on roots during every larval stage and do not develop without host roots. After the final molt, the female becomes embedded in the root with most of her body protruding. As the female grows, the protruding part of the body enlarges becoming saccate but retaining the pointed tail. It differs from most tylenchs in the position of the excretory pore, which is equatorially or postequatorially placed.

Infected plants show much necrosis. The branch rootlets become swollen with an irregular appearance. Large numbers of females are commonly present in a small area. Christie (1959) gives a list of host plants.

Luc (1957) described another species *T. mangenoti* which was found in the Ivory Coast infecting the roots of *Dorstenia embergeri*. According to Allen (1960) *Tylenchulus mangenoti* and *Trophotylenchulus floridensis* are not readily distinguishable on the basis of morphology. Goodey (1963) transferred *T. mangenoti* to the genus *Trophotylenchulus*.

Colbran (1961) described a new species, *T. obscurus*, collected from around the roots of *Hodgkinsonia frutescens* in Queensland, Australia.

Plate 20. Tylenchulus mangenoti. (Female.) A. Full view. B. Anterior portion. (Male.) F. Full view. C. Anterior portion. E. Posterior portion. (Larva.) G. Full view. D. Anterior portion. (After Luc 1957. Courtesy of Nematologica.)

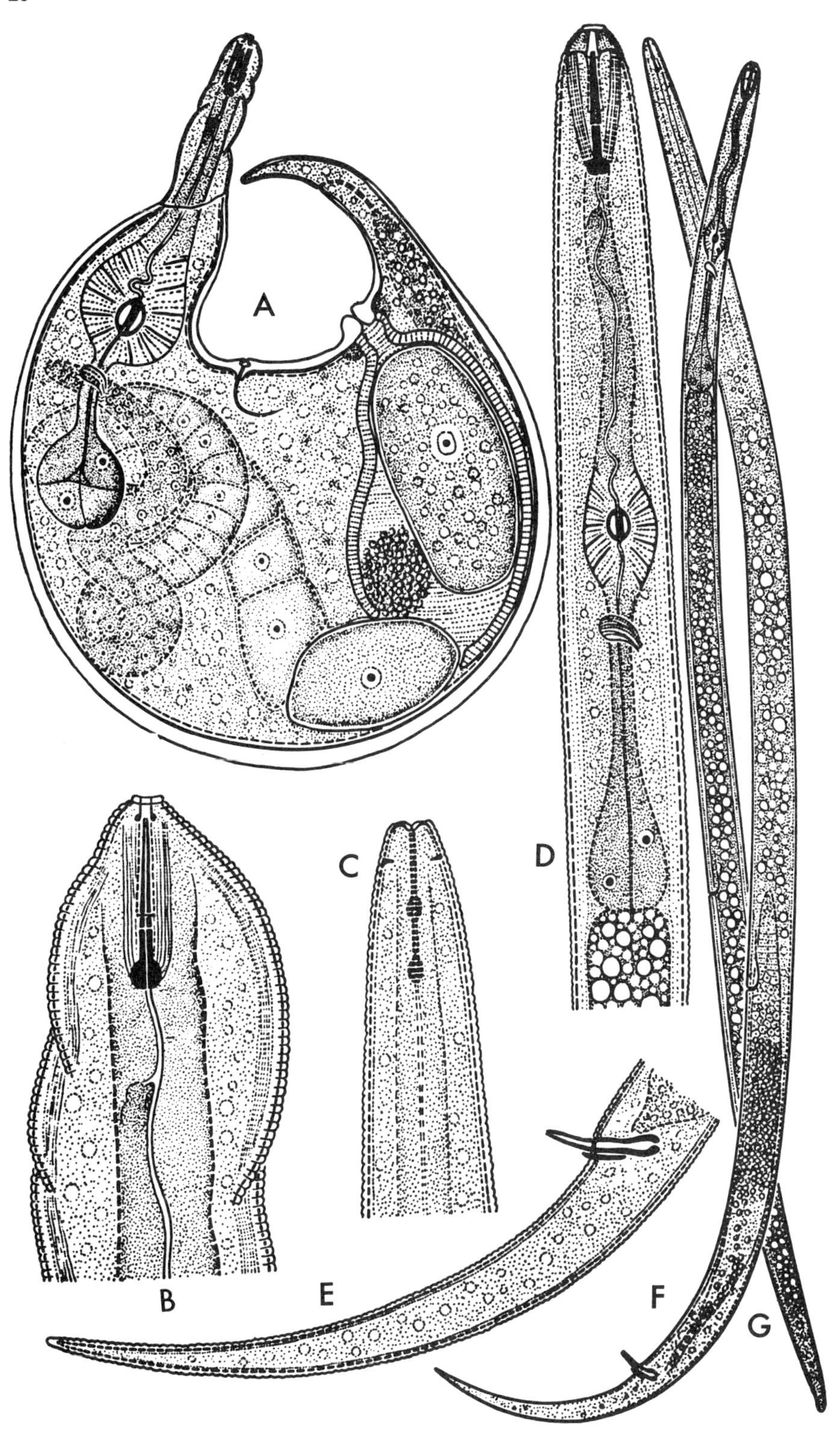
A
B
C
D
E
F
G

Plate 21. I. Swollen females of *Tylenchulus semipenetrans*. A. Head embedded in root. B. Showing position of vulva. II. Vermiform types of *T. semipenetrans*. (After Van Gundy 1958. Courtesy of Nematologica.) A. Mature male. B. Young male within fourth-molt cuticle. C. Fourth-stage male larva. D. Third-stage male larva. E. Second-stage male larva. F. Second-stage female larva.

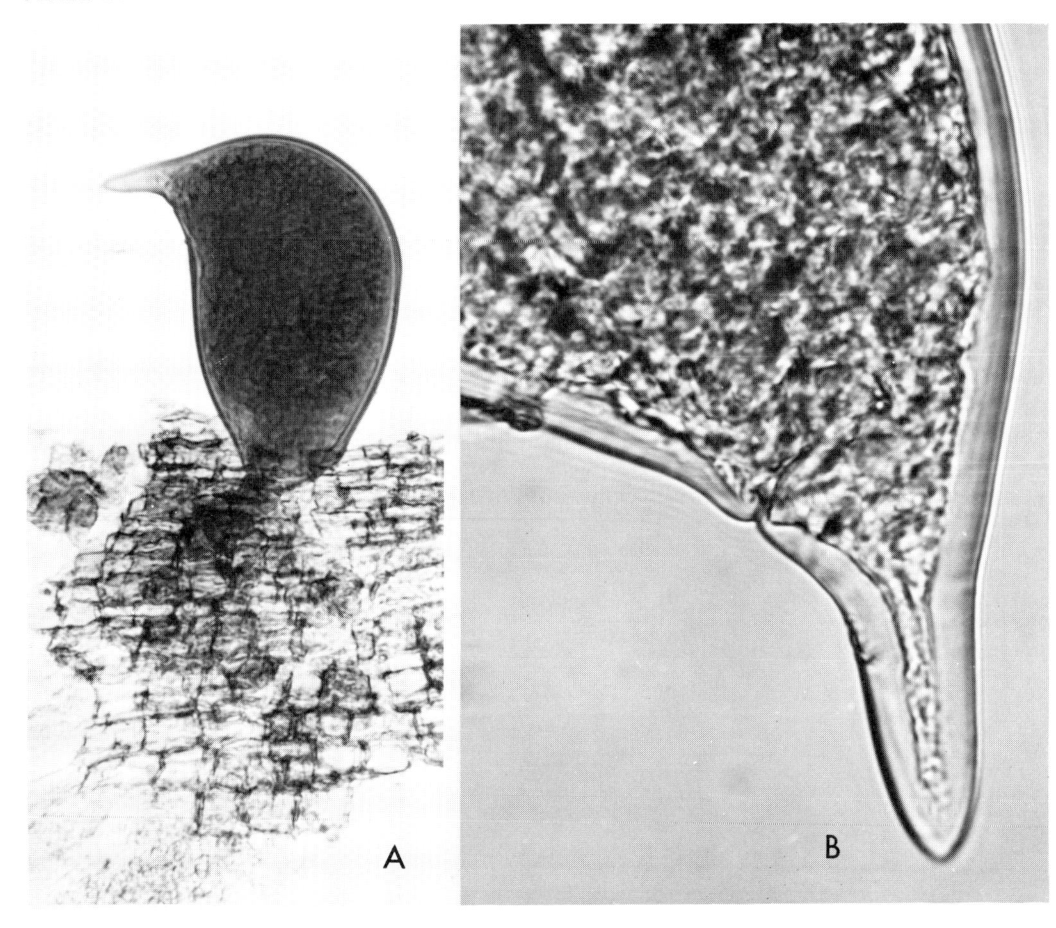
A
B

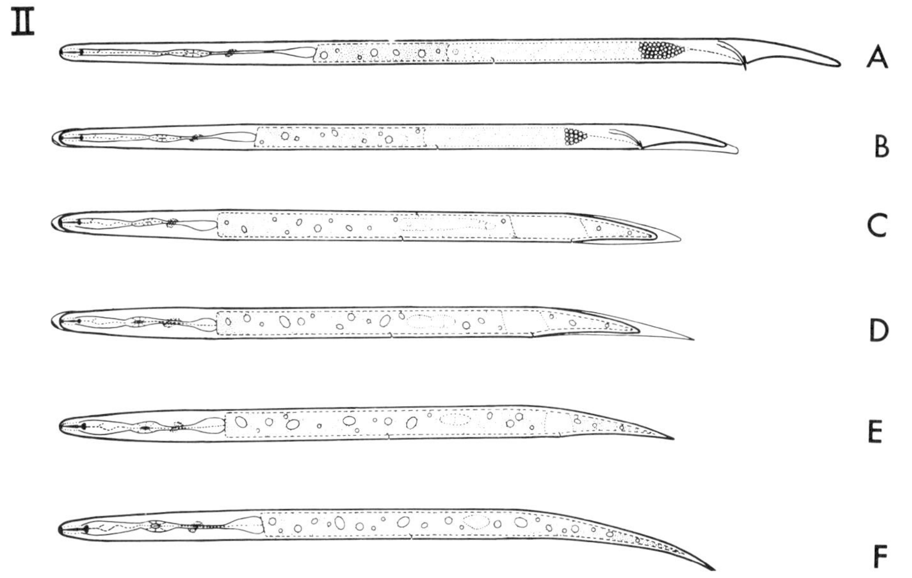
II
A
B
C
D
E
F

Nacobbus Thorne and Allen 1944

Type species: *Nacobbus dorsalis* Thorne and Allen 1944

Description: Nacobbinae. Sexual dimorphism present, mature female swollen, monodelphic; sedentary in the galls or swellings of plant roots. Male vermiform, caudal alae not well developed, enveloping tail. Cephalic framework and stylet well developed. Median bulb with strongly developed valve, except in juveniles with normal valve. Esophageal glands elongated, overlapping intestine dorsally. (From Sher 1970.)

General Characteristics

Sher (1970) considered the genus *Nacobbus* to contain only 2 closely related species, *N. aberrans* and *N. dorsalis*, that can be easily distinguished by the number of annules between the vulva and anus of the immature, vermiform female. *N. batatiformis* and *N. serendipiticus bolivianus* were proposed as synonyms of *N. aberrans*.

With respect to geographical distribution and host-parasite relationships, Sher states: "This genus appears to be native to the western part of North and South America where its members parasitize a number of native and agricultural plants causing galls on the roots that are similar to those produced by the genus *Meloidogyne*. This genus apparently has been introduced to England and the Netherlands, where it has been found infecting tomato plants under glasshouse conditions. The pathogenicity of *Nacobbus* to important agricultural crops, the similarity of its symptoms to those caused by root-knot nematodes, its wide host range and incompletely known geographical distribution, and its classification in the *Tylenchoidea* make it a most interesting and important genus to observe."

Plate 22. Nacobbus dorsalis. A. Anterior portion of male showing arrangement of esophageal glands; *gl sal dsl,* dorsal salivary gland; *sub dsl gl sal l*, left subdorsal salivary gland; *sub dsl gl sal r*, right subdorsal salivary gland. B. Posterior portion of young female; *spm*, spermatozoa in uterus. C. Head of male. D. Head of adult female. E. Posterior portion of male. F. Posterior portion of first-stage larva. G. Posterior portion of adult female. H. Adult female. I. Section through gall on root of *Erodium circutarium*. J. Basal bulb of second-stage larva from which the esophageal glands have not yet developed. (After Thorne and Allen 1944. Courtesy of Helminthological Society of Washington.)

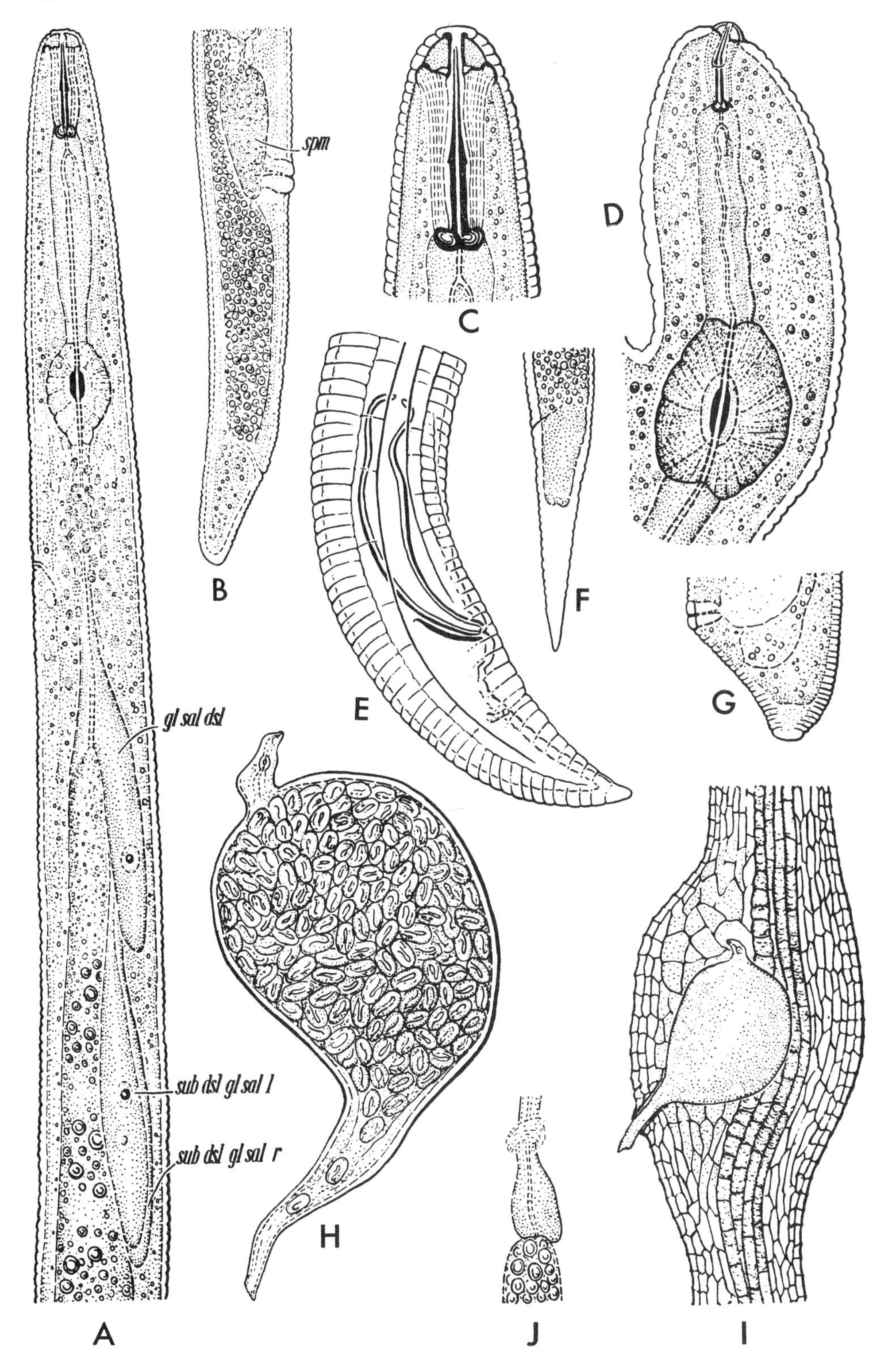
spm
gl sal dsl
sub dsl gl sal l
sub dsl gl sal r
A
B
C
D
E
F
G
H
I
J

18. Females with irregular body annules around perineum (perineal pattern); excretory pore at level with stylet or close behind it; lip region with 2 lateral lips wider than 4 sublateral lips. Second-stage larval stylet less than 20μ; weakly developed labial framework; usually marked galling of the host roots . . . . (Plate 23) *MELOIDOGYNE*
    Females with no irregular body annules around perineum; excretory pore posterior to median bulb; lip region with 2 lateral lips narrower than 4 sublateral lips. Second-stage larval stylet usually more than 20μ; well-developed labial framework. Usually no galling of the host range . . . . (Example: Plate 24) 19

Meloidogyne Goeldi 1887

Type species: *Meloidogyne exigua* Goeldi 1887

Description: Heteroderinae with marked sexual dimorphism. Adult females pear-shaped to spheroid with elongated neck. Body not transformed into cystlike structure. Six lips marked by 6 radial circumoral sclerotization. Lateral lips markedly larger than submedians. Caplike structure present on lips. Amphid apertures slitlike. Spear slender with weakly developed basal knobs. Excretory pore located anterior to median bulb, usually 12–25 annules posterior of lip region. Vulva terminal or subterminal. Anus opening on border of slight depression occupied by vulva. Cuticle of female with simple cross annulation, forming a variable more or less circular pattern in perineal region. Eggs not retained in body but deposited in a gelatinous matrix. Females usually endoparasitic, causing formation of galls or knots on roots of most hosts. Obligate plant parasites.

Males elongate, cylindrical. Lip region with or without distinct annulation, bearing a caplike structure. Amphid apertures slitlike, conspicuous, leading to broad pouches in lateral lips. Six radial circumoral cephalic framework present. Lateral lips much larger than submedians. Spear strongly developed with well-developed basal knobs. Bursa absent. Spicules and gubernaculum present. One or 2 testes, outstretched anteriorly, sometimes reflexed at distal end.

Second-stage infective larvae with well-defined lip region, plain or with 1 to 3 annulations. Amphid apertures slitlike. Lip region bearing a caplike structure. Six lips markedly larger than submedians. Spear slender with well-defined basal knobs. (From Allen 1952.)

General Characteristics

Members of this genus are known as root-knot nematodes; their feeding results in irregular, knotty enlargements of the roots. The size and shape of the gall is dependent upon the number of nematodes in the root, the species of nematode involved, and the plant species. Root-knot affects many different kinds of plants. Almost all cultivated plants are susceptible; however, some plants are susceptible to only 1 or 2 species of *Meloidogyne* and are more or less resistant to others.

Whitehead (1968) published a comprehensive treatise on the taxonomy of this genus which includes a number of illustrations and a key to 23 species.

Sledge and Golden (1964) proposed a new genus *Hypsoperine*, related to *Meloidogyne, Heterodera, Meloidodera,* and *Cryphodera.* According to them it differs from *Meloidogyne* because the body wall is "tougher," its feeding results in root swellings rather than root galls, and it completes its life cycle on grasses rather than on a wide variety of plants. *Hypsoperine* differs from *Heterodera* and *Meloidodera* because most females are embedded in roots, and a circular perineal pattern is present. One of the most important differences between *Cryphodera* and *Hypsoperine* is that the eggs of *Cryphodera* are retained within the female body while those of *Hypsoperine* are deposited in a gelatinous sac outside the body.

Because the characters of *Hypsoperine*, used to separate it from *Meloidogyne*, were present in some species of *Meloidogyne*, this genus was synonymized with *Meloidogyne* by Whitehead (1968), and Wouts and Sher (1971) concurred.

Pogosian (1966) proposed the genus *Meloidoderita* which differs from related genera in having over the entire body conical spines 3μ high and 1.5μ wide at the base with hyaline tips. Franklin (1971) described other characters of this genus.

Plate 23. Meloidogyne sp. A. Vermiform larva, 237 X. B. Anterior portion of vermiform larva, 660 X. C. Tail of second-stage larva, 660 X. D. Mature female, 265 X. E. Sausage-shaped larva from plant root, 265 X. F. Mature male, 237 X.

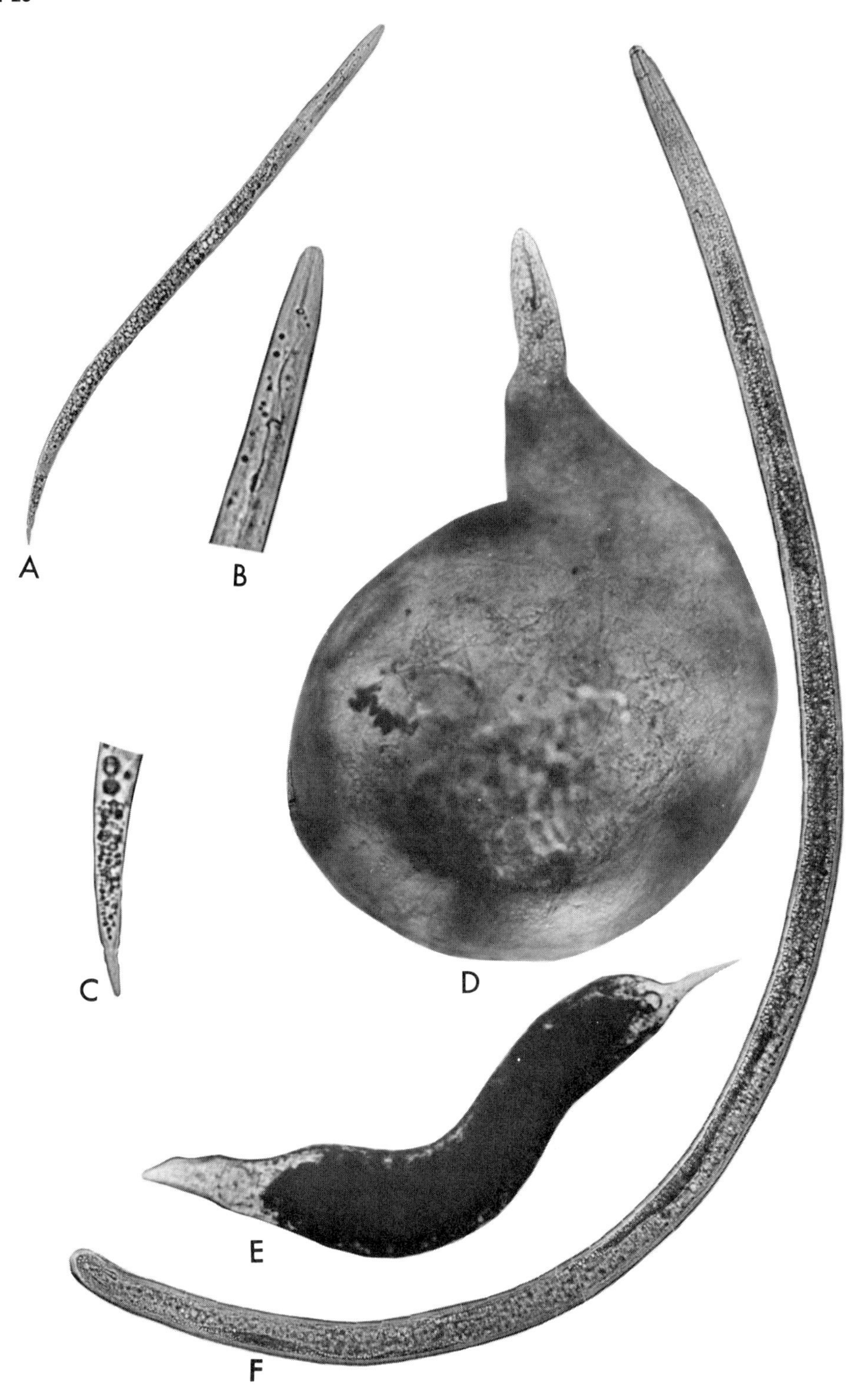
A
B
C
D
E
F

19. Vulva subequatorial, cuticle annulated . . . . (Plate 24) *MELOIDODERA*
    Vulva terminal or subterminal. Cuticle annulated or lacelike 20
20. Cuticle annulated . . . . (Plate 25) *CRYPHODERA*
    Cuticle with lacelike pattern 21

Meloidodera Chitwood, Hannon, and Esser 1956

Type species: *Meloidodera floridensis* Chitwood, Hannon, and Esser 1956

Description: Subfamily Meloidoderinae. Female: Vulva subequatorial, anus ventrally subterminal, not protruding out of body contour, phasmids terminal. Second-stage juvenile: Stylet less than 31μ long, four lateral lines, length, clear part of tail less than 25μ. Males: Not going through saccate stage, less than 600μ long, stylet shorter than stylet or second-stage juvenile, longitudinal striations on basal lip annule absent, lateral field with four incisures. (From Wouts 1973.)

General Characteristics

This genus possesses characters of both *Meloidogyne* and *Heterodera*. The females differ from those of *Meloidogyne* spp. in having a thickened cuticle, a subequatorial vulva, subcuticular dots in the cuticle, and in not forming either root galls or egg masses. They differ from those of *Heterodera* spp. in not having a cyst stage and in having well-developed annulations, a subequatorial vulva, and a smaller number of eggs contained in the body.

Cryphodera and *Zelandodera*, a genus described by Wouts in 1973, appear to be closely related. A major difference between *Meloidodera* and *Cryphodera* is that the vulva is postequatorial in the former and subterminal in the latter. *Meloidodera* and *Zelandodera* differ in that the anus is ventrally subterminal in the former and dorsal to the vulva in the dorsal body curvature in the latter.

Wouts (1973) recognized 2 previously described species, *M. floridensis* and *M. charis*, and described a new species, *M. belli*. *M. floridensis* has been found in a number of locations including soil and root samples taken in Alabama, Florida, Georgia, New Jersey, and North Carolina. It is often found associated with the roots of pine (*Pinus* spp.). The type host of *M. charis* is honey mesquite, *Prosopis juleflora* (Sev.) D.C. var *glandulosa* (Torr.) Cockerell, and the type locality is a greenhouse at the Texas Agricultural Experiment Station, College Station, Texas. The type habitat and locality of *M. belli* is soil around the roots of sage (*Salvia* sp.) growing in German Flats near Salina, Utah.

Although species of this genus are undoubtedly plant parasites, none of the 3 species is recognized as an important plant pathogen as are numerous species of *Meloidogyne* and *Heterodera*.

Plate 24. Meloidodera floridensis. A. Head of larva, lateral view. B. Head, female. C. Anal pattern, female. D. Tail of larva, lateral view. E. Esophageal region, female. F. Vulva pattern, female. (After Chitwood, Hannon, and Esser 1956. Courtesy of Phytopathology.)

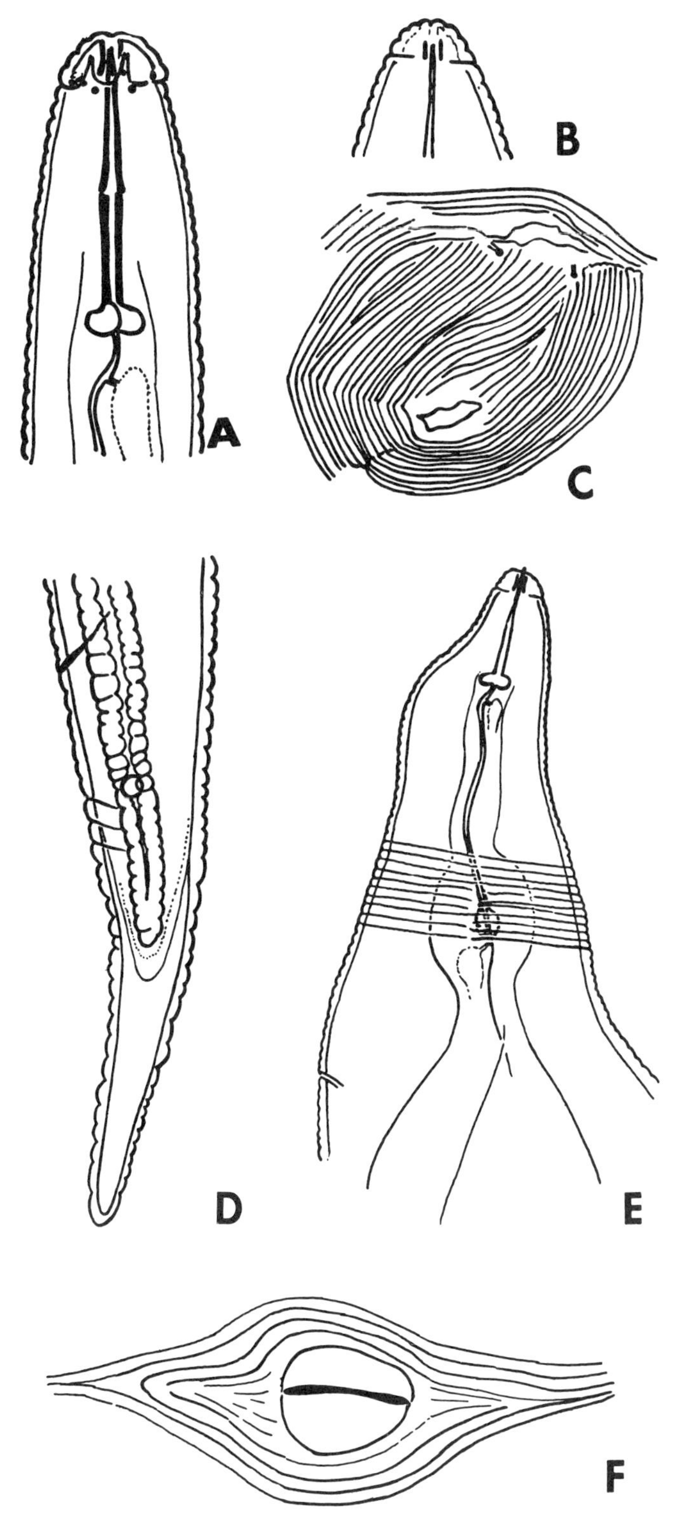
A
B
C
D
E
F

Cryphodera Colbran 1966

Type species: *Cryphodera eucalypti* Colbran 1966

Description: Heteroderidae. Females: Body ovoid with a distinct neck usually bent laterally. Cuticle thick with transverse striae and subcuticular dots. Excretory pore behind neck. Vulva 45–70μ from anus. Area between vulva and anus flat to concave. Eggs not deposited in external sac. Males: Body not constricted behind lip region. Lip cap present. Lateral lips smaller than subdorsal and subventral lips. Amphid apertures slitlike. Posterior cephalids opposite middle of stylet. Larvae: Cephalic sclerotization well developed. Lip region consists of lip cap and 3 postlabial annules. Amphid apertures slitlike. Phasmids large, a short distance behind anus. (From Colbran 1966.)

General Characteristics

C. eucalypti, the type species, is the only species of this genus. Larvae and males of this species were obtained from soil in a eucalypt forest at Long Pocket, Brisbane, and larvae from soil around eucalypts at Nambour. Females were found partially embedded in small ungalled roots of *Eucalyptus major* (a grey gum) and in roots of *E. adrewsi* (messmate). With respect to taxonomic relationships between this genus and *Meloidodera*, Colbran states: "*Cryphodera* appears to be more closely related to *Meloidodera* than to *Heterodera*. Females do not develop into a hard resistant cyst, males have a labial disc and the phasmids in second-stage larvae are large and situated a short distance behind the anus."

Zelandodera, a genus described by Wouts (1973), is closely related to *Cryphodera*. It can be distinguished from *Cryphodera* by the position of the anus which is located in the dorsal curvature of the female body, the less pronounced vulval lips, the higher number of lip annules in juveniles, and the 4 lines in the lateral fields of males.

Plate 25. Cryphodera eucalypti. Left. (Female.) A. Lateral view. B. Head. C. Pattern of striations around vulva. *Center.* (Male.) A. Lateral view. B. Head. C. *En face* view. D. Cross section of lateral field in midbody region. E. Lateral field in midbody region. F. Lateral view of tail. G. Ventral view of tail. *Right* (Second-stage larva.) A. Lateral view. B. Head. C. *En face* view. D. Cross section of lateral field in midbody region. E. Ventral view of anal region. F. Tail. (After Colbran 1966. Courtesy of Queensland Journal of Agricultural and Animal Sciences.)

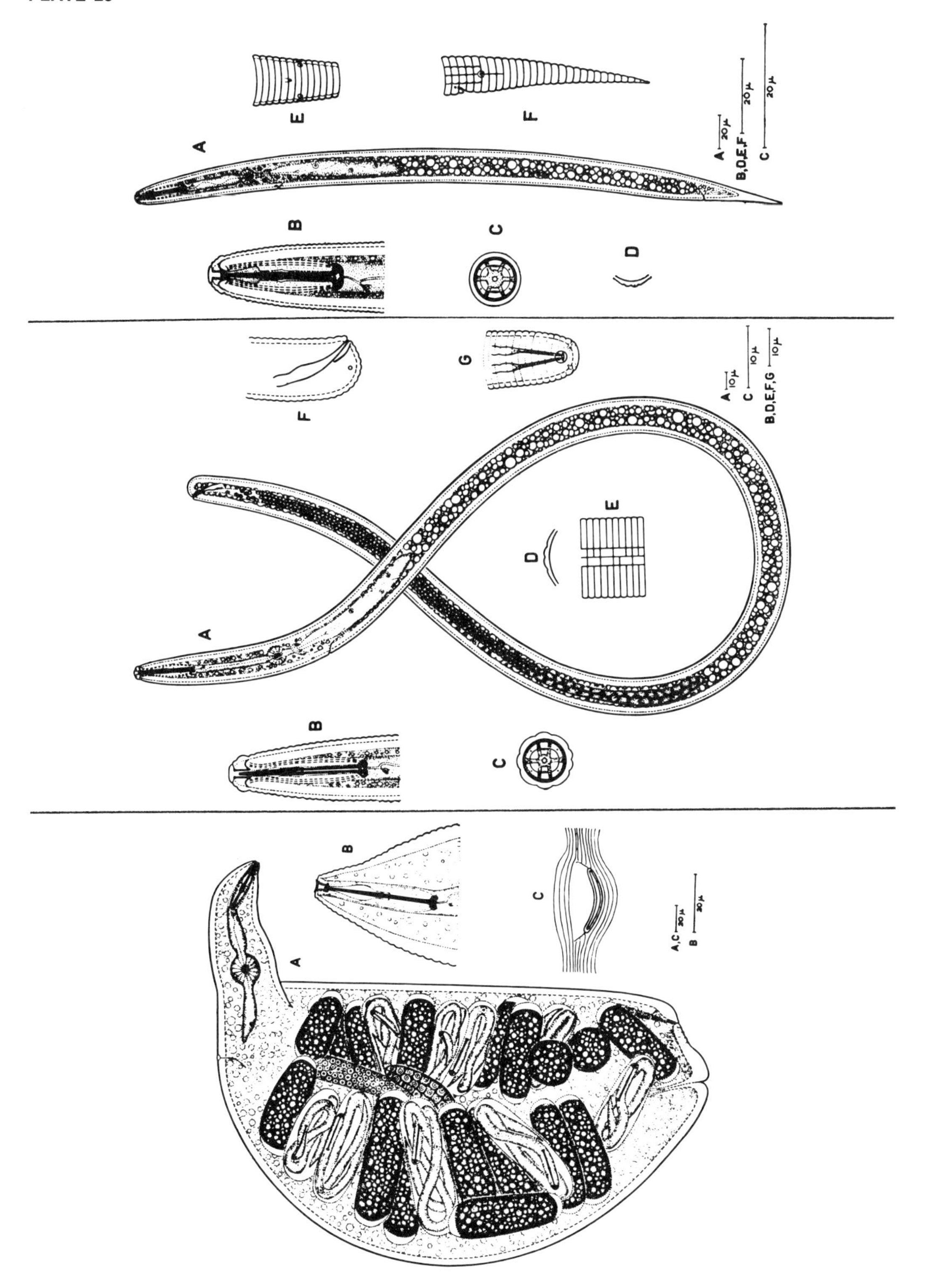
A
B
C
D
E
F
A 20 μ
B,D,E,F 20 μ
C 20 μ
A
B
C
D
E
F
G
A 10 μ
C 10 μ
B,D,E,F,G 10 μ
A
B
C
A,C 20 μ
B 20 μ

21. Cyst stage. Vulva terminal, anus dorsal, not on vulva lip; or vulva sunken into terminal vulval cone with anus on upper inside of dorsal vulva lip . . . . (Example: Plate 27) 22
No cyst stage. Vulva and anus terminal, on prominence . . . . (Plate 26) *ATALODERA*

Atalodera Wouts and Sher 1971

Type species: *Atalodera ucri* Wouts and Sher 1971

Description: Subfamily Heteroderinae. Female: No cyst stage. Cuticle with lacelike pattern. Anus and vulva terminal, on prominence. No fenestration around vulva. Second-stage juvenile: Labial disc absent. Stylet less than 30μ. Esophageal glands do not fill body width. Phasmids with lenslike structure in muscle layer. Male: To 1.5 mm. Region immediately behind lip region not constricted. Labial disc prominent. No longitudinal striations on basal lip annule. Tail present. Spicules more than 30μ. (From Wouts and Sher 1971.)

General Characteristics

Atalodera ucri is the only described species of this genus; however, Wouts and Sher (1971) mention two undescribed species. *A. ucri* was found in a number of localities and habitats in southern California.

With respect to relationships with other genera, Wouts and Sher state: "The genus *Atalodera* is closest to the genera *Heterodera* and *Cryphodera*. It can be distinguished from *Heterodera* by the absence of a cyst and the terminal anus of the female; the lenslike structure associated with the phasmids and the narrow esophageal glands of the juvenile; and the prominent labial disc of the male. It differs from the genus *Cryphodera* in that the cuticle of the adult female is not annulated, and the vulva and anus are terminal; the narrow esophageal glands in the juvenile; and the larger males with longer spicules."

Plate 26. Atalodera ucri. (A–E, juvenile.) A. Face view. B. Anterior end. C. Esophagus. D. Posterior end, lateral. E. Posterior end, dorsoventral. (F–I, male.) F. Face view. G. Anterior end. H. Posterior end, lateral. I. Posterior end, dorsoventral. (J–L, female.) J. Face view. K. Anterior end. L. Female, entire. M. Male, entire. (After Wouts and Sher 1971. Courtesy of Journal of Nematology).

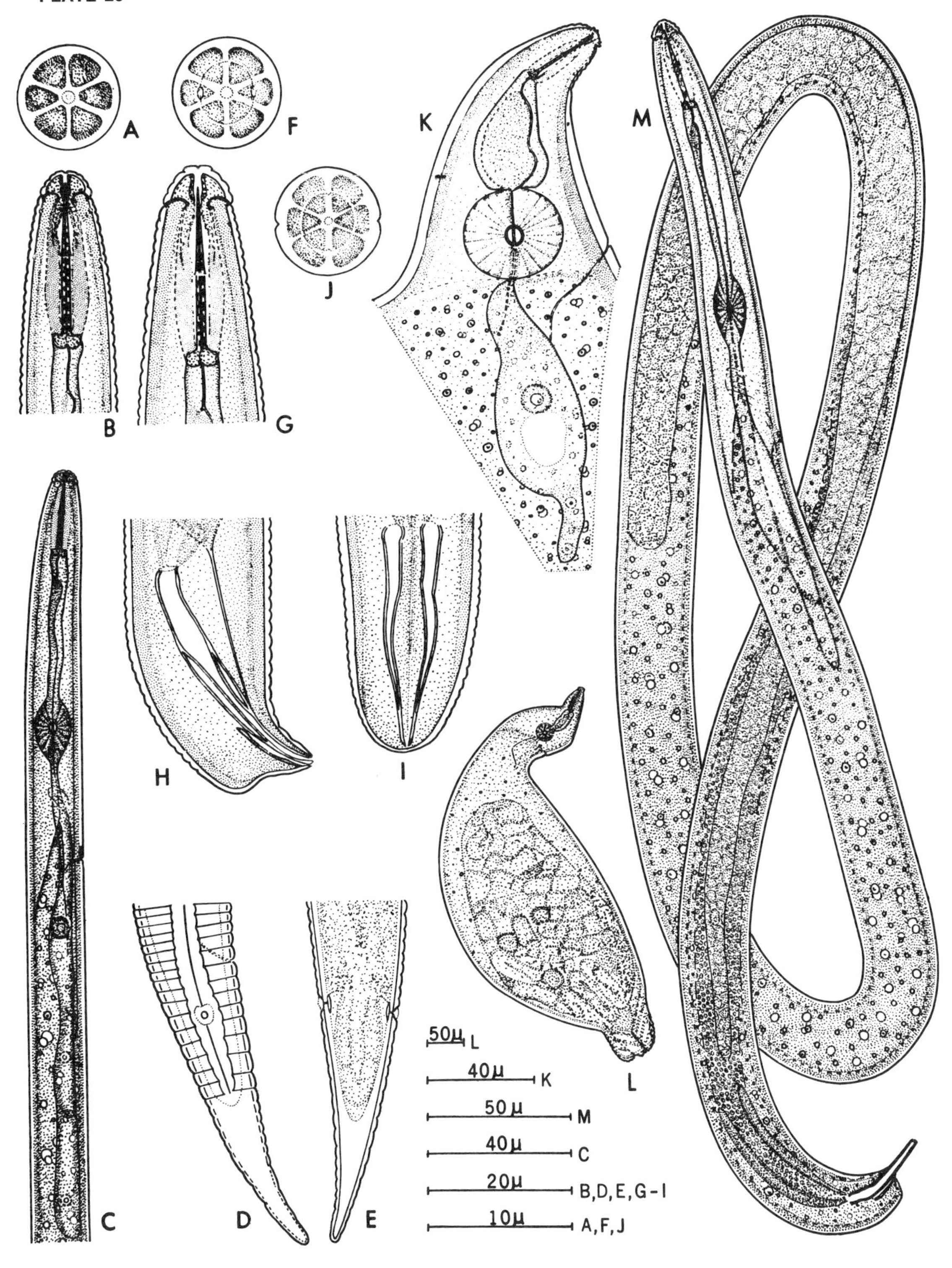

A
B
C
D
E
F
G
H
I
J
K
L
M
50μ L
40μ K
50μ M
40μ C
20μ B,D,E,G-I
10μ A,F,J

22. Vulva terminal, usually on prominence. Anus dorsal, not on vulva lip. Second-stage larvae stylet less than 30μ . . . . (Plate 27) *HETERODERA*
    Vulva sunken into terminal vulva cone. Anus on upper inside of dorsal vulva lip. Second-stage larvae stylet more than 38μ . . . . (Plate 28) *SARISODERA*

Heterodera Schmidt 1871

Type species: *Heterodera schachtii* Schmidt 1871

Description: Heteroderinae. Female body wall forms a remarkably tough durable cyst; the layered cuticle is quite thick and usually there is a lacelike pattern and punctuation. Eggs are at least partly retained in the female body which acts as a protective cyst. A mucoid mass may or may not be formed at the vulva. The male has no lateral cheeks, but the head has ridges dividing the labial region into six sectors; these ridges and sectors are also present in the hatched larva; annular striations in cephalic region are prominent in both male and larva. Stylet of larva is 20–29μ long. The anus of the females varies in position but is never situated at the edge of the posterior vulva lip. Usually does not form galls on any hosts; females tend to be located on external surface of roots at maturity. (From Chitwood 1949.)

General Characteristics

Concerning the taxonomy and morphology of this genus, A. L. Taylor in "A Manual of Plant Nematology for Experiment Station Workers in the Northeast" states:

> The genus *Heterodera* was placed in the family Heteroderidae of the Tylenchoidea by Thorne (1949) with the genus *Meloidogyne* (root-knot nematodes). Prior to 1949, species of both these genera were placed in the genus *Heterodera*. Prior to 1940, there was a strong tendency to refer all of the cyst-forming nematodes to a single species, *H. schachtii* Schmidt 1871. . . .
>
> The adult females and cysts of *Heterodera* are the forms most commonly encountered. Adult females or cysts will be found on roots of various plants if these are carefully removed from the soil and washed. The nematodes are attached to the roots by the neck only, with most of the body outside the root. The females are white or yellowish in life, and the cysts are light to dark brown. Average size is about 0.5 mm by 0.75 mm. Some species are lemon-shaped, others are pear-shaped. Cysts are very highly resistant to decay and may be found in soil in which infected plants have grown, even many years afterward. The males are slender worms shaped very much like *Meloidogyne* males. That is, they are about 1.25 to 1.75 mm long, slender (a = 35–40), taper slightly anteriorly, and have a short rounded tail (Goodey 1951). Males will be found in abundance at certain times of the year but may be very scarce at other times. The larvae have an average length of about 0.5 mm. They differ from root-knot nematode larvae in that the stylet is 20–30 microns long (*Meloidogyne*, 10–11 microns) and in the shape of the anterior end. Excellent drawings of the larvae and other stages of *H. schachtii* will be found in "The Life History and Morphology of the Sugar-Beet Nematode, *Heterodera schachtii* Schmidt," by D. J. Raski (Phytopath. 40(2): 135–152, 1950).

Members of this genus are widely distributed. Of the more than 45 described species, some, such as *H. rostochiensis*, a serious potato pathogen, and *H. schachtii*, which severely damages sugar-beet roots, are major agricultural pests. Despite the relatively narrow host ranges of some species, a long crop rotation is necessary for effective commercial control due to the resistance of encysted eggs to adverse conditions. Encysted eggs of different species vary in resistance to drying.

In the mature female little can be seen but the reproductive organs. The minute anus is just dorsal to the vulva. The ovaries are paired and the vulva is located terminally. The adult females of *Meloidogyne* differ from those of *Heterodera* in 3 ways: the anterior position of the excretory pore, absence of a cyst stage, and the presence of 6 radial sclerotized ribs dividing the lip region into 6 sectors (Allen 1952).

The short and bluntly rounded male tails are very similar to those of *Meloidogyne* males. Caudal alae are absent, and the phasmids are extremely minute.

Species with round-ended or pear-shaped cysts were assigned to *Globodera*, a separate subgenus of *Heterodera* by Skarbilovich (1959).

In an article which includes excellent illustrations, Mulvey (1972) presents keys to identify species of *Heterodera* by terminal end and cone top structures.

Plate 27. Heterodera schachtii. A. Male, 330 X. B. Larva, 330 X. C and D. Enlarged female containing eggs, 68 X.

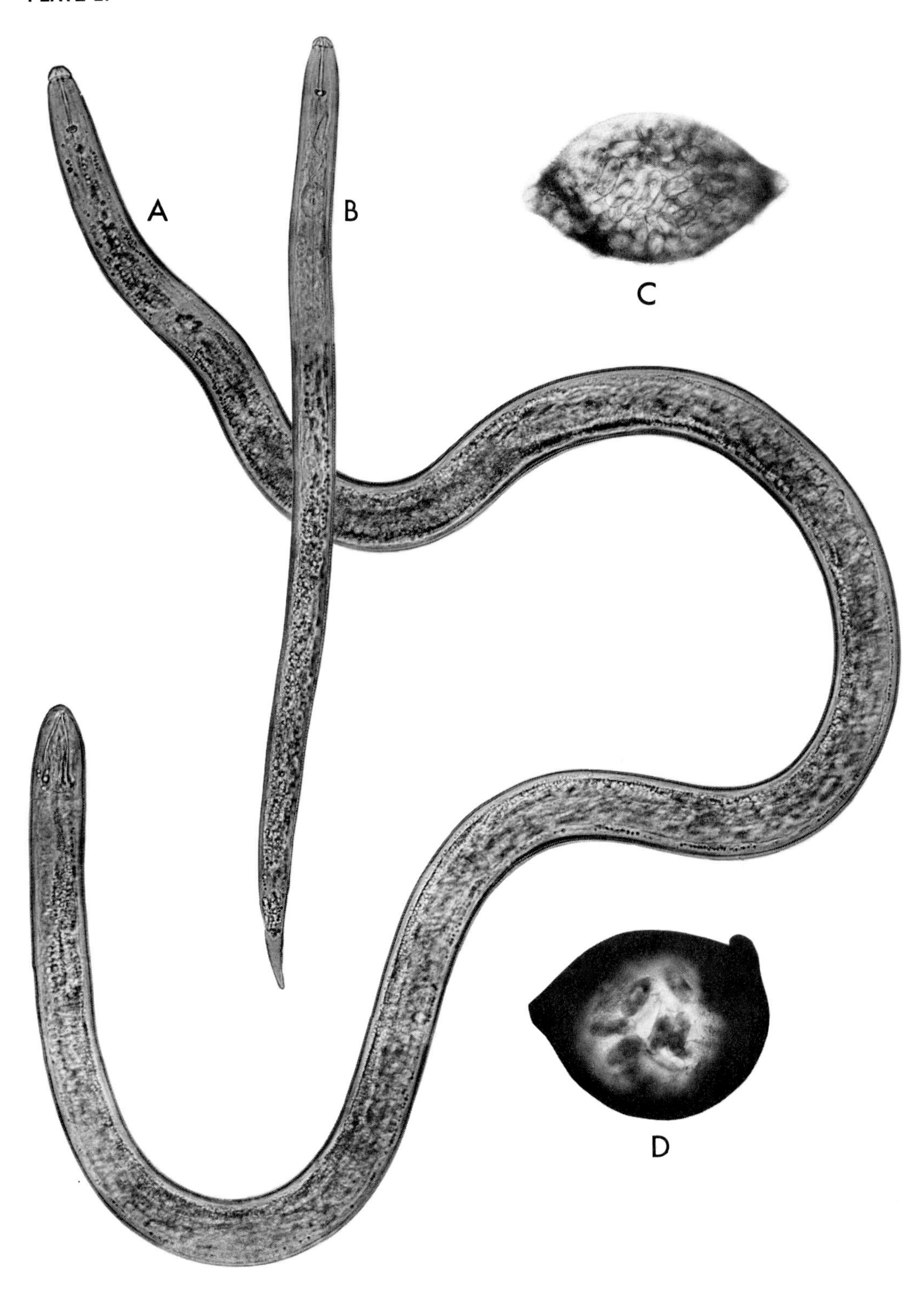
A
B
C
D

Sarisodera Wouts and Sher 1971

Type species: *Sarisodera hydrophila* Wouts and Sher 1971

Description: Subfamily Heteroderinae. Female: Cyst formed after death. Cuticle with lace-like pattern. Vulva sunken into terminal vulva lip. No fenestration around vulva. Second-stage juvenile: Labial disc absent. Stylet more than 38μ. Esophageal glands fill body width; Phasmids with lenslike structure in muscle layer. Male: To 1.5 mm. Region immediately behind lips not constricted. Labial disc absent. No longitudinal striation on basal lip annule. Tail absent. Spicules over 30μ. (From Wouts and Sher 1971.)

General Characteristics

Sarisodera hydrophila is the only described species of this genus but Wouts and Sher (1971) note that there are five undescribed species in their collection. *S. hydrophila* has been collected from a number of localities and habitats in California.

With respect to relationships to other genera, Wouts and Sher state: "The genus *Sarisodera* is closest to the genus *Heterodera* with which it shares the ability to form a cyst. It can be distinguished from all the genera in the subfamily Heteroderinae by the vulva sunken into the vulva cone, the anus located on the upper inside of the dorsal vulva lip, the long stylet and the absence of the male tail. It can further be distinguished from *Heterodera* by the lenslike structure associated with the phasmids in the juvenile."

Plate 28. Sarisodera hydrophila. (A–E, juvenile.) A. Face view. B. Anterior end. C. Esophagus. D. Posterior end. E. Tail near phasmids, ventral view. (F–I, male.) F. Anterior end. G. Esophagus. H. Posterior end, lateral. I. Posterior end, dorsoventral. (J–L, female.) J. Face view. K. Anterior end. L. Outline. (After Wouts and Sher 1971. Courtesy of Journal of Nematology.)

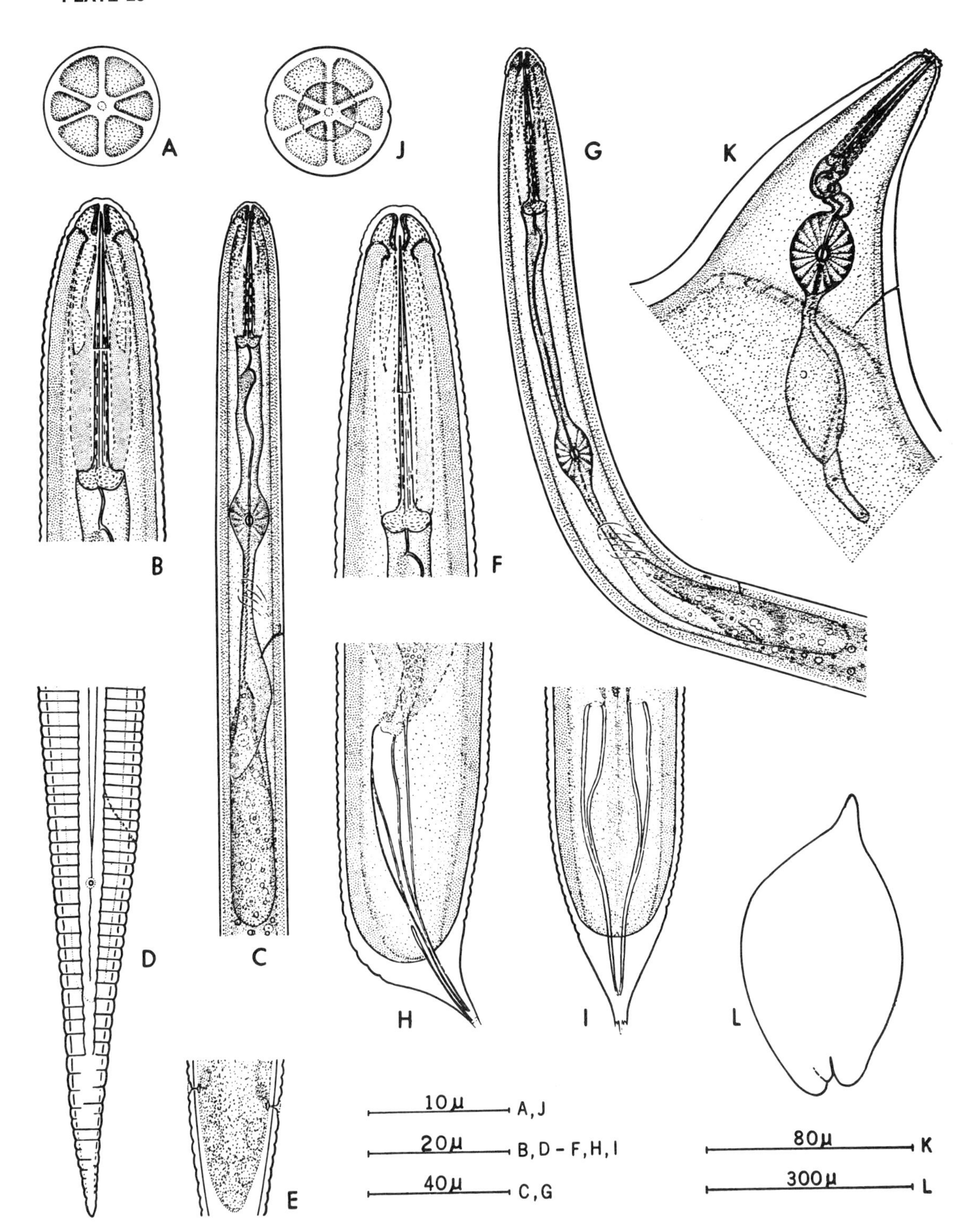
A
J
G
K
B
F
D
C
E
H
I
L
10µ A, J
20µ B, D – F, H, I
40µ C, G
80µ K
300µ L

23. Tail equal to or longer than 6 times the anal body diameter (tail filiform, with pointed or clavate terminus) . . . . (Example: Plate 29A) 24
    Tail generally less than 6 times the anal body diameter. However, if longer, tail is cylindroid rather than filiform . . . . (Example: Plate 29B–K) 28
24. Female with 2 ovaries . . . . (Example: Plate 30) 25
    Female with 1 ovary . . . . (Example: Plate 34) 26

Plate 29. Tail shapes. Entire plate, 1220 X. A. Tail long and thin. B–K. Tails not long and thin.

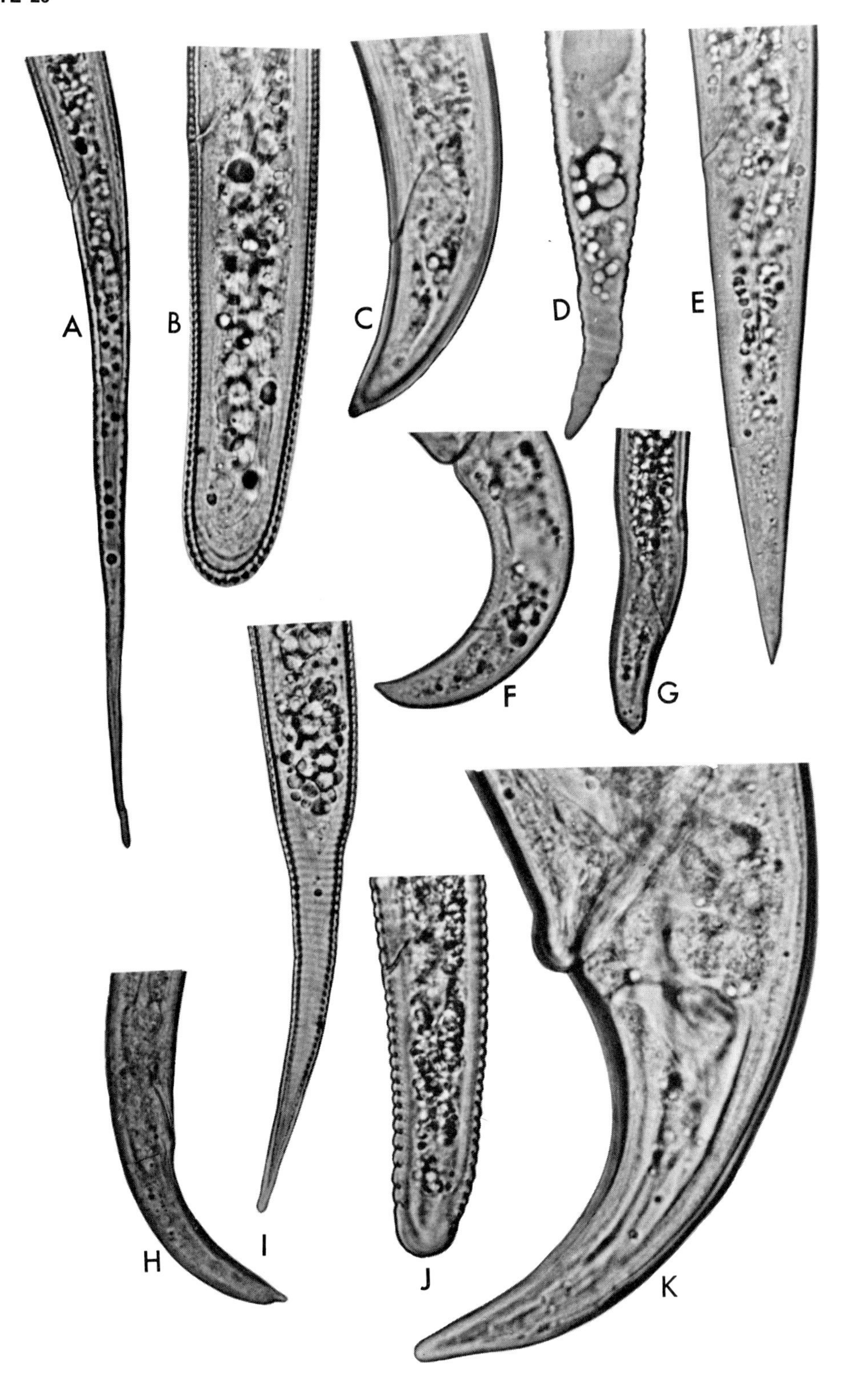
A
B
C
D
E
F
G
H
I
J
K

25. Stylet without basal knobs, no cephalic sclerotization, tail filiform with usually clavate terminus .... (Plate 30) *PSILENCHUS*
    Stylet with basal knobs; heavy cephalic sclerotization; tail filiform with pointed terminus .... (Plate 31) *BRACHYDORUS*

Psilenchus de Man 1921

Type species: *Psilenchus hilarulus* de Man 1921

Description: Cuticle and subcuticle annulated. Head framework not sclerotized. Spear slender, without basal knobs. Median esophageal bulb posterior to middle of esophagus. Basal esophageal bulb small and pyriform. Deirids and phasmids distinct. Ovaries paired, opposite and outstretched. Testis single, outstretched. Bursa adanal. Spicules equal, slightly cephalated and arcuate. Gubernaculum thin and troughlike. Tails of both sexes long, filiform, usually terminally clavate. (From Jairajpuri 1965.)

General Characteristics

The species of this genus are closely related to the genus *Tylenchus*. In 1965, Jairajpuri redefined *Psilenchus* and *Tylenchus*, retaining in *Psilenchus* only didelphic species, and placing monodelphic species in *Tylenchus*. Prior to this both didelphic and monodelphic species were included in *Psilenchus*.

Species of this genus are reported from cultivated and uncultivated areas over a widespread area. They are associated with plant roots, but feeding habits are unknown (Goodey 1963). Under a dissecting microscope the more obvious characters of nematodes of this genus are the very long tail, nonoverlapping esophagus, and delicate stylet.

Kheiri (1970) described two new species and presented a key to 11 species of *Psilenchus*.

Plate 30. Psilenchus hilarulus. A. Face view. B. Head. C. Female. D. Male tail. E. Cuticle pattern of female tail. F. Deirid region. G. Variations in female terminus. a. Bard, California. b. Reno, Nevada. c. Lewiston, Utah. (After Thorne 1949. Courtesy of Helminthological Society of Washington.)

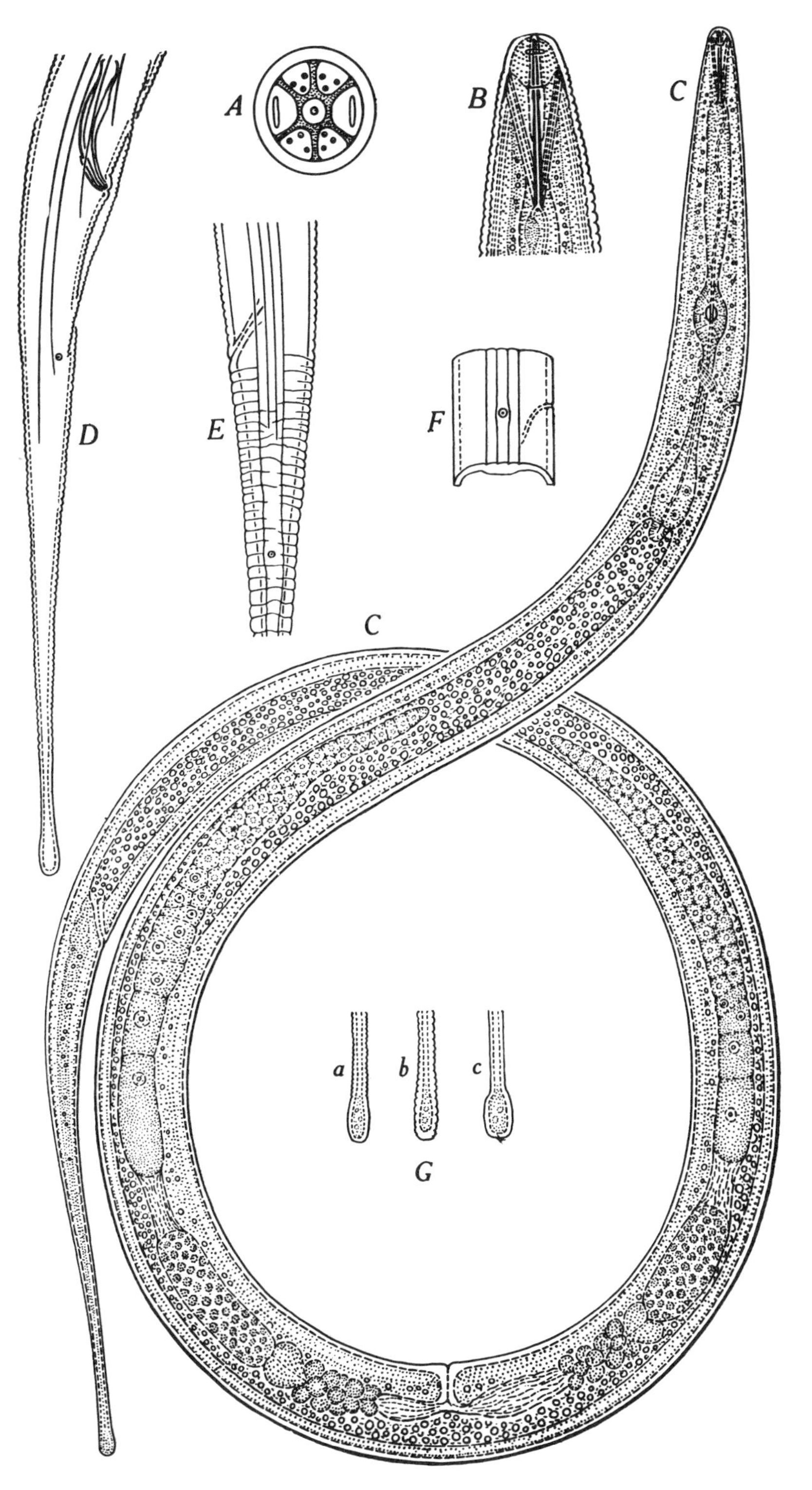
A
B
C
D
E
F
C
a
b
c
G

Brachydorus de Guiran and Germani 1968

Type species: *Brachydorus tenuis* de Guiran and Germani 1968

Description: Dolichodorinae: Moderate size, slender forms. Head rather small, lightly offset, with pronounced sclerotization. Stylet in both sexes strong but short, with basal knobs. Lateral fields present. Ovaries amphidelphic, outstretched. Female tail elongate, attenuated. Bursa trilobed, enveloping tail. (From Golden 1971.)

General Characteristics

The only described species, *Brachydorus tenuis*, was extracted from soil around the roots of *Ravenala madagascariensis* growing in Madagascar. *Brachydorus* differs from *Dolichodorus* by its small size, the short stylet, smooth lips only slightly set off from the body and not divided into 4 lobes, and by the length of the attenuated tail. It differs from *Tylodorus* by the body length and the short stylet, the shape of the lips, the heavy cephalic sclerotization, the male tail, and the didelphic gonads of the female.

Plate 31. *Brachydorus tenuis*. (Female.) A. Entire animal. B. Anterior part. C. Head. D. Cross section of the labial region at the apex and at the base. E. Posterior part, lateral view. (Male.) F. Head. G. Posterior part, lateral view. H. Posterior part, ventral view. (After DeGuiran and Germani 1968. Courtesy of Nematologica.)

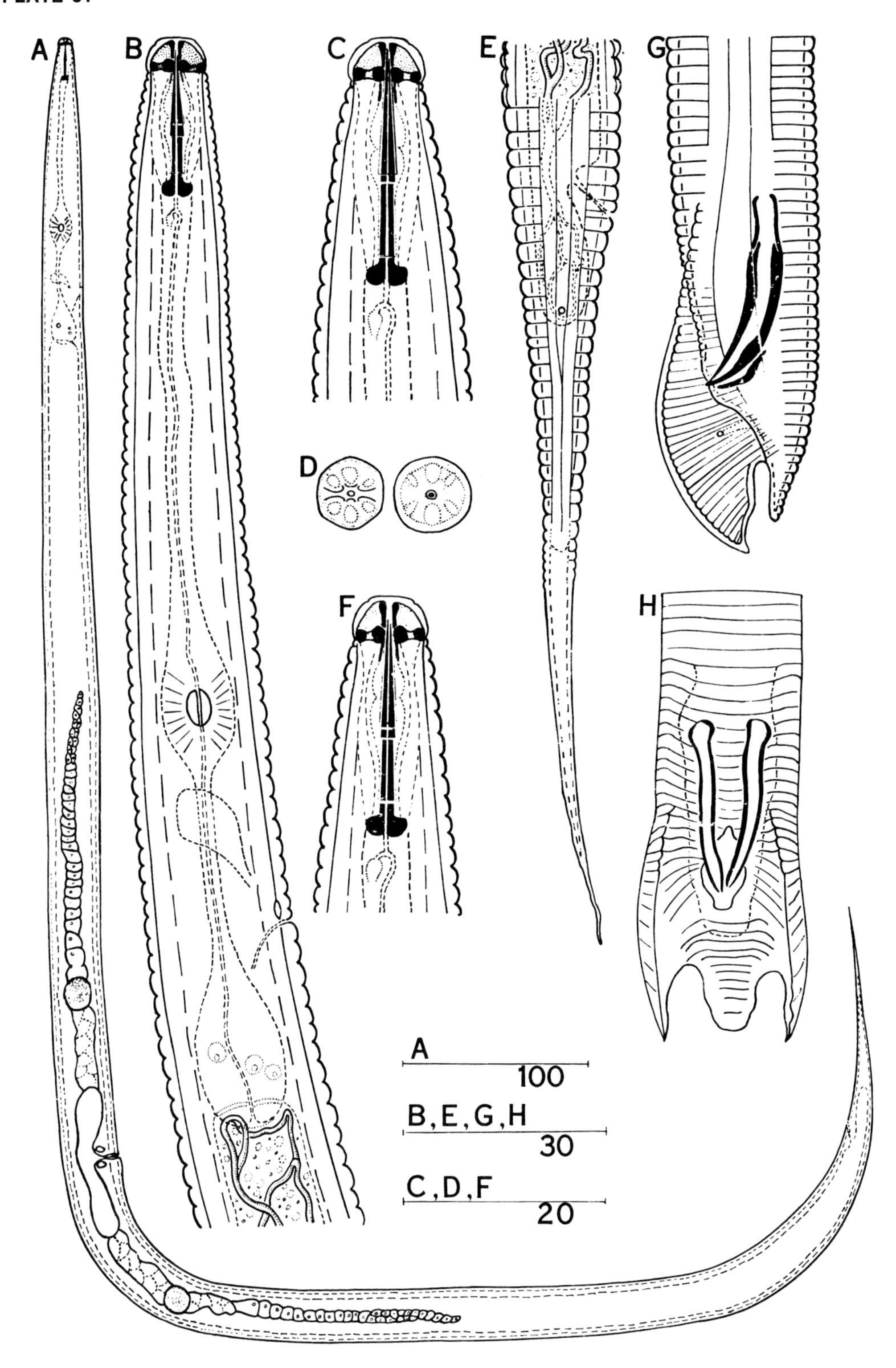
A
B
C
D
E
F
G
H
A
100
B, E, G, H
30
C, D, F
20

26. Esophagus criconematoid; thick cuticle, coarsely annulated . . . . (Plate 32) *CALOOSIA*
 Esophagus tylenchoid, thin cuticle, not coarsely annulated . . . . (Example: Plate 34) 27

Caloosia Siddiqi and Goodey 1963

Type species: *Caloosia longicaudata* (Loos 1948) Siddiqi and Goodey 1963

Description: Hemicycliophorinae. Basal knobs of spear spheroid. Female elongate-cylindrical, C-shaped at death. Cuticle thick, single, annules rounded, moderately wide. Lateral fields absent. Excretory pore behind the terminal esophageal bulb. Isthmus and terminal bulb forming a cylinder, slightly swollen posteriorly. Anterior part of intestine overlaps terminal bulb. Ovary single, prodelphic; vulva posterior; vagina sloping forward from the outside; spermatheca present. Tails of both sexes similar, elongate filiform. Male bursa large with crenate edge, ending about midtail length. Spicules large, slightly ventrally arcuate, protruding through a sheath at the cloaca. A short gubernaculum present. Male spear absent; esophagus degenerate; excretory pore well behind esophagus. Testis single, outstretched. (From Siddiqi and Goodey 1963.)

General Characteristics

The genus *Caloosia* was erected to include nematodes formerly placed in *Hemicycliophora longicaudata*, but which lack a sheath in larvae and adult females and have almost straight spicules in males.

Loos (1948) described *Hemicycliophora longicaudata* found in jungle or patna soils in Ceylon. The adult female lacks the cuticular sheath commonly associated with *Hemicycliophora*; the male exhibits almost straight spicules as opposed to the hooked or sickle-shaped usually found in this genus.

H. longicaudata (Loos 1948) was used as the type species of *Caloosia*, while the *H. longicaudata* of Siddiqi (1961) was described as a separate species, *Caloosia paralongicaudata*, primarily because of differences in the head region.

Mathur et al (1969) described two new species, *C. paxi* and *C. exilis*, and provided a key to the species of *Caloosia*.

Plate 32. Caloosia paxi, n. sp. A. Entire female. B. Vulval region. C. Anterior end of female. D. Female tail. E. Head of female.

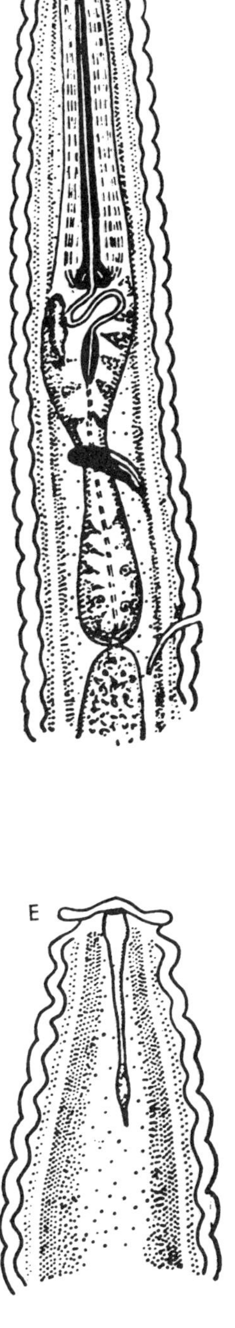

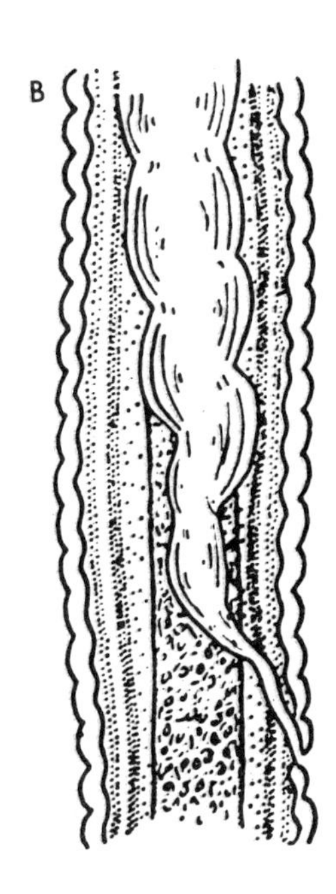

10 μ E
20 μ B, C, D
100 μ A

27. Long stylet (s[stylet length/body diameter at base of stylet] = 2.5 or more) .... (Plate 33) *TYLODORUS*
Short stylet (s = 2.5) .... (Plate 34) *TYLENCHUS*

Tylodorus Meagher 1963

Type species: *Tylodorus acuminatus* Meager 1963

Description: Tylenchinae: nematodes of moderate size (1–2 mm). Bodies of male and female slender. Tails filiform. Cuticle with distinct transverse striae irregularly crossing the outer lateral incisures. Deirids distinct. Phasmids not seen. Head continuous with body contour. Lip cap present. Lip region unstriated, 4-lobed. Cephalic sclerotization absent. Amphid openings porelike at lateral margins of lip cap. Amphids distinct. Stylet very long with basal knobs. Opening of dorsal esophageal gland a short distance behind spear base. Median bulb elongate-ovate, set off from enlarged procorpus. Isthmus well defined. Basal bulb pyriform. Gonad single, outstretched anteriorly. Oocytes in single file. Vulva situated in posterior region of body. Postvulval uterine branch longer than body diameter at vulva. Testis single, outstretched. Caudal alae adanal and well developed. Spicules tylenchoid, dorsal flanges meeting medially. Gubernaculum simple, trough-shaped. (From Meagher 1963.)

General Characteristics

Placed in the Tylenchinae, *Tylodorus* bears characters similar to those found in a number of genera. The long filiform tail resembles those of *Tylenchus* and *Psilenchus*. The esophageal region is somewhat similar to those of *Dolichodorus* and *Macrotrophurus*.

Tylodorus may be separated from the other monodelphic genera in the Tylenchinae by the elongate stylet, presence of a lip cap, and transverse striae which irregularly interrupt the lateral field.

Meagher states: "*Tylodorus* may provide a link to support the classification of *Dolichodorus* in Tylenchinae, the case for which was argued by Loof (1958). *Tylodorus* seems to occupy a systematic position between *Tylenchus* and *Dolichodorus*."

The single species included in this genus was found in Australia associated with *Eucalyptus* sp. roots growing in a grey silty-clay soil.

Plate 33. Tylodorus acuminatus. A. Adult female. B. Adult male. C. Lateral view of female esophageal region. D. Lateral surface view of female anterior end. E. End-on view of head. F. Dorsoventral view of head. (After Meagher 1963. Courtesy of Nematologica.)

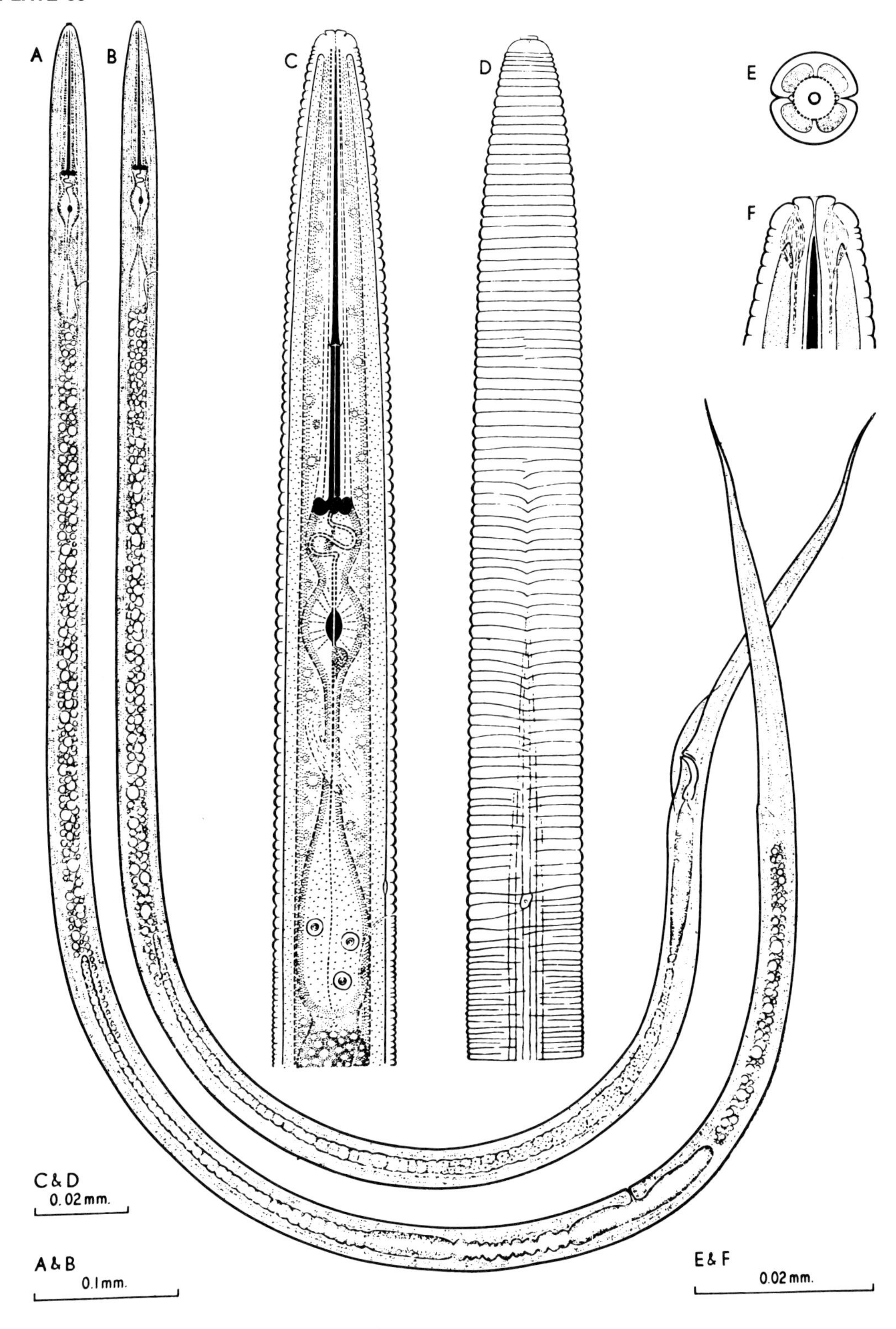
A
B
C
D
E
F
C & D
0.02 mm.
A & B
0.1 mm.
E & F
0.02 mm.

Tylenchus Bastian 1865

Type species: *Tylenchus davainei* Bastian 1865 (Filipjev 1934)

Description: Tylenchinae. Tails filiform. Lip region striated. Vulva well behind middle of body. Anterior ovary outstretched. Posterior uterine branch short, rudimentary. Bursa short, adanal. Developing oocytes and spermatocytes usually arranged in single file. Deirids generally prominent, located near the latitude of the conspicuous excretory pore. Phasmids not observed.

Cuticle striated, lateral fields marked by incisures. No sclerotized cephalic framework present. Spear well developed with basal knobs; the protrudor muscles anchored to the cephalic walls. Median esophageal bulb ovate with refractive valvular apparatus. Isthmus long, slender, ending in a somewhat pyriform basal bulb containing the usual 3 nuclei. Cardia present. Intestinal cells usually packed with coarse granules which obscure details of the cell nuclei. (From Thorne 1949.)

General Characteristics

A member of the genus *Tylenchus* is recognized by its long pointed tail, the nonoverlapping esophagus, and delicate, well-developed stylet. The species of this genus are closely related to the genus *Psilenchus*. Jairajpuri (1965) redefined *Psilenchus* to include only didelphic species and placed in *Tylenchus* all monodelphic species formerly included in *Psilenchus*.

On the basis of morphological characteristics and phylogeny, Andrassy (1954) divided the genus into 4 species groups, which he indicated were well defined and deserved subgeneric rank. Four new subgenera were proposed: *Tylenchus*, *Aglenchus*, *Filenchus*, and *Lelenchus*. Meyl (1961) raised the 4 subgenera to generic rank. J. B. Goodey (1963) recognized them as subgenera and added *Miculenchus* (Andrassy 1959).

Jairajpuri (1965) erected another subgenus *Clavilenchus* in the genus *Tylenchus* to include a monodelphic species having a clavate tail. This species, *Tylenchus* (*Clavilenchus*) *tumidus*, was formerly included in the genus *Psilenchus* and differs from other species in *Tylenchus* by having a clavate tail. He also placed the genus *Basiria* in synonomy with *Tylenchus* (*Filenchus*) as proposed also by Goodey (1963). Golden (1971) recognized *Basiria* as a separate genus. Thorne and Malek (1968) erected *Basiroides*, a genus closely related to *Basiria*.

According to Golden (1971) genera closely related to *Tylenchus* and that have a long pointed tail, a nonoverlapping esophagus, a single ovary, and a delicate well-developed stylet include: *Aglenchus*, *Malenchus*, *Miculenchus*, *Neopsilenchus*, *Basiria*, *Basiroides*, *Clavilenchus*, *Dactylotylenchus*, and *Cephalenchus*. He synonymized the subgenera *Filenchus* and *Lelenchus* with the genus *Tylenchus*. Husain and Khan (1967) described *Ottolenchus* as a subgenus of the genus *Tylenchus*. Golden (1971) synonymized this subgenus with the genus *Aglenchus*.

Szczygiel (1969) described a new genus, *Pleurotylenchus*, from strawberry soil in Poland. The new genus is closely related to *Aglenchus*.

Dactylotylenchus, described by Wu (1968), is similar to the genus *Tylenchus* but differs in that the labial framework consists of a number of pairs of fingerlike sclerotized structures and in that the cuticle is much thicker.

Andrassy (1968) described a new genus *Malenchus*, related to *Tylenchus*, in which the body cuticle is heavily annulated and the body narrows markedly immediately posterior to the vulva.

Large numbers of these nematodes are found in cultivated fields around the roots of many plant species. Populations of *T. hexalineatus* were repeatedly found associated with roots of sugar maple and silver maple, and this nematode reproduced on sugar-maple root callus cultures grown on modified White's medium (Savage and Fisher 1966). They also observed feeding and reproduction of this nematode on alfalfa, moosewood, plantage, and popular tissue cultures.

Steiner and Albin (1946), who revived 5 specimens of *Tylenchus* from a leaf of a rye seedling after 39 years of dormancy, remark, "To our knowledge this resuscitation after almost 39 years is the longest period of dormancy as yet observed for a nematode."

Plate 34. *Tylenchus* sp. A. Male tail, 475 X. B. Mature female, 475 X.

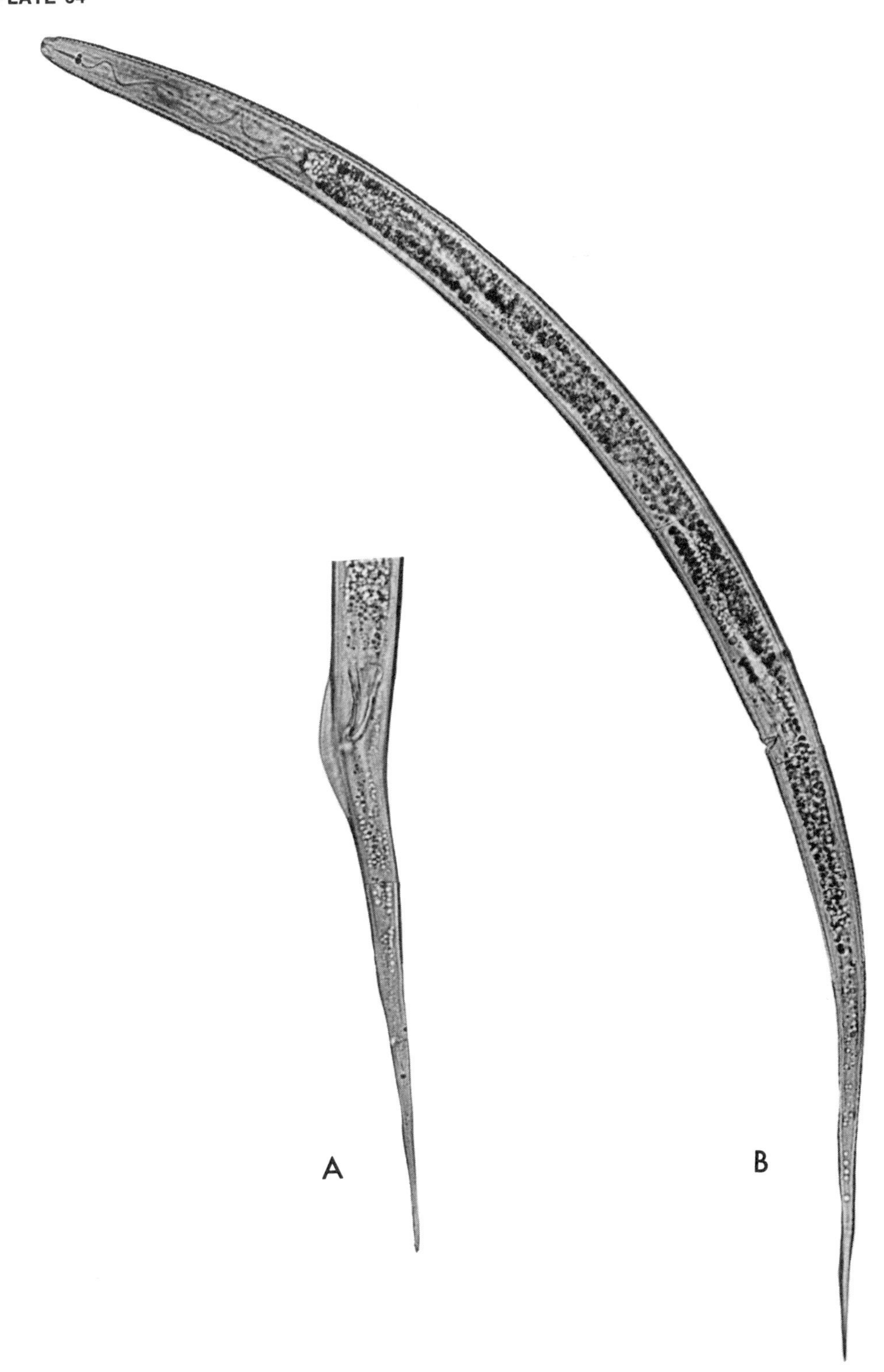
A
B

Plate 35. A. Female of *Paratylenchus* sp. with 1 ovary, 770 X. B. Female of *Tylenchorhynchus* sp. with 2 ovaries, 250 X. a. vulva at more than 65% of body length; b. vulva at less than 60% of body length.

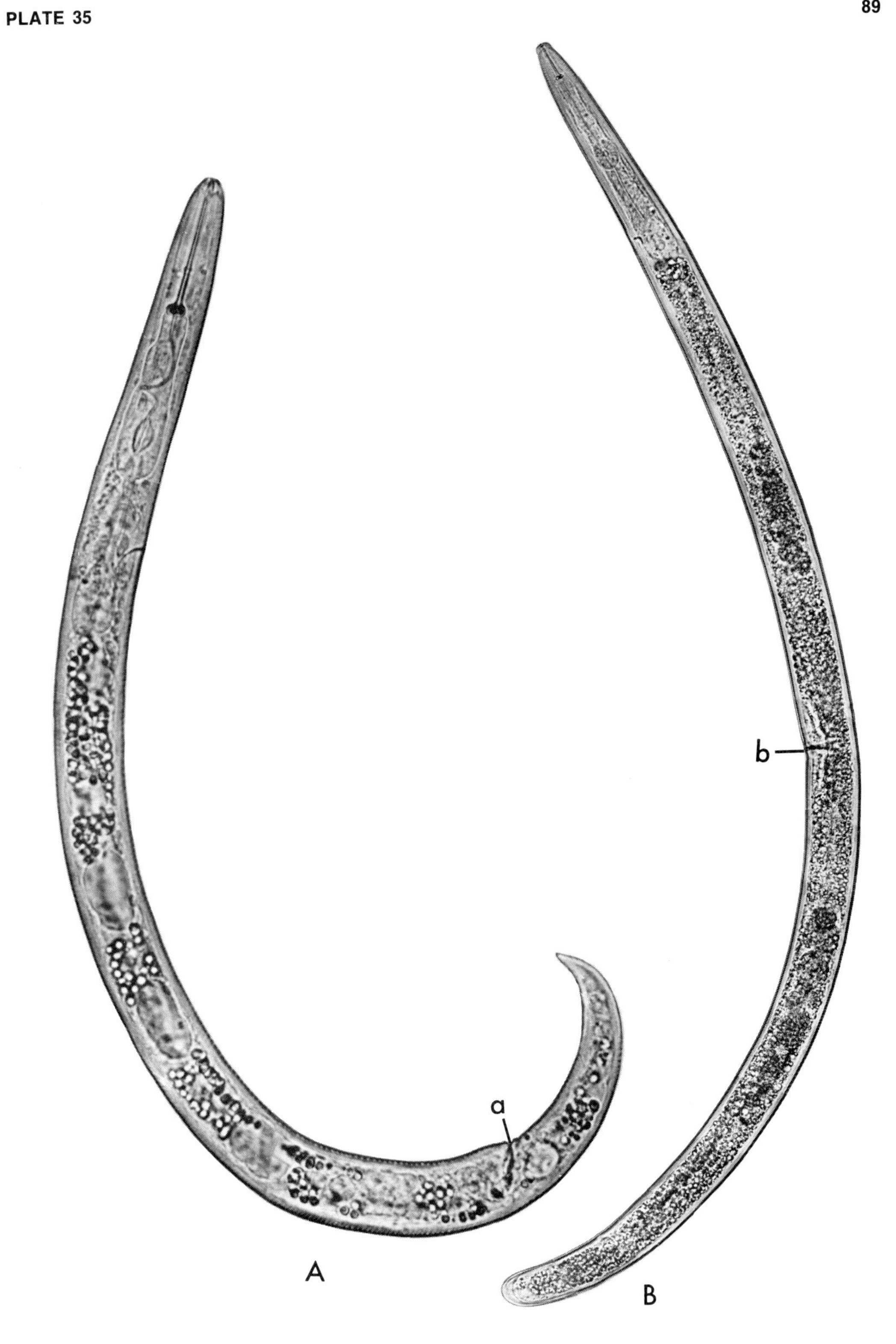
a
b
A
B

28. One ovary (vulva usually located in posterior third of body) . . . . (Example: Plate 35A) 29
    One ovary (vulva located near center of body); lip region conical, not annulated; female tail tip rounded, cuticle of tail swollen . . . . (Plate 36) *TROPHURUS*
    Two ovaries (vulva located near center of body) . . . . (Example: Plate 35B) 44

Trophurus Loof 1956

Type species: *Trophurus imperialis* Loof 1956

Description: Tylenchinae. Body moderately slender, slightly curved when killed by gentle heat; cylindrical tapering anteriorly; tail ending in a rounded tip which is not annulated. Female tail slightly clavate, terminus hemispherical. Lip region conical, not annulated, continuous with body contour, and weakly sclerotized. Stylet slender, with small but distinct basal knobs. Esophagus with an oval median and an oblong terminal bulb; esophago-intestinal cells seem to be present. The distance between the bulbs is much less than that between the stylet base and the median bulb. Female gonad single, anterior; a very short posterior uterine branch is present. Vulva somewhat behind the middle of the body. Phasmids conspicuous, located near middle of tail. Spicula and gubernaculum tylenchoid. (Mostly from Loof 1956 with additions from Caveness 1958.)

General Characteristics

Members of the genus *Trophurus* resemble the genus *Tylenchorhynchus* in general appearance. The position of the vulva and the anus, the shape of the male tail in lateral view, the clearly visible phasmids, and the broad lateral fields are similar. The chief differences are the stylet, the shape of the lip region, and the number of ovaries. Despite the position of the vulva, approximately halfway between head and tail, only 1 ovary is present in the genus *Trophurus* (Loof 1956). Loof (1956) described 2 species, *T. imperialis* and *T. sculptus*, and Caveness in 1958 described *T. minnesotensis*. These nematodes are rare, but they are found occasionally in soil around plant roots. Roman (1962) described *T. longimarginatus* obtained from soil around roots of West Indian mahogany growing in Puerto Rico.

Plate 36. *Trophurus minnesotensis*. A. Posterior portion of female. B. Posterior portion of male. C. Anterior portion of female. D. Adult female. (After Caveness 1958. Courtesy of Helminthological Society of Washington.)

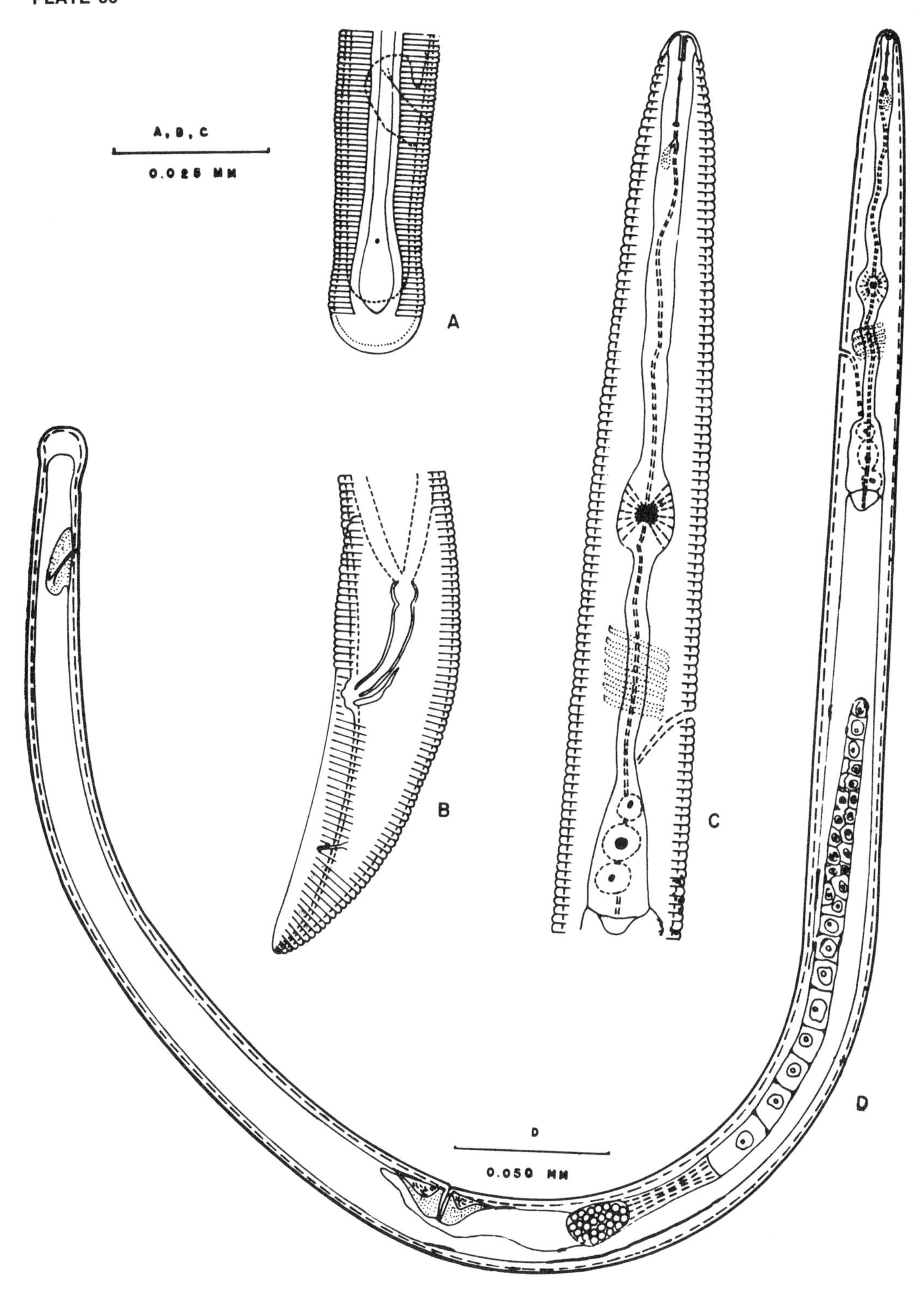
A, B, C
0.025 MM
A
B
C
D
D
0.050 MM

29. Procorpus and metacorpus not swollen and combined into a large valvular bulb . . . . (Example: Plate 40) 30
    Procorpus and metacorpus swollen and combined into a large valvular bulb . . . . (Example: Plate 45) 37
30. Stylet delicate (15μ or less in length); tail acute or subacute . . . . (Example: Plate 37) 31
    Stylet strong (generally more than 15μ in length); tail tapering or bluntly rounded . . . . (Example: Plate 40) 33
31. Ovary with oocytes in 1 or 2 lines, not arranged around a rachis. Mature female slender or stout . . . . (Example: Plate 38) 32
    Ovary with multiple rows of oocytes arranged around a rachis. Mature female mostly obese. Found in galls in leaves or flower parts . . . . (Plate 37) *ANGUINA*

Anguina Scopoli 1777

Type species: *Anguina tritici* (Steinbuch 1799) Chitwood 1935

Description: Tylenchinae. Robust, stout species with female body generally arcuate or spiral in form. Cuticle marked by fine striae which frequently are visible only on the neck and lip region. Lateral fields appearing as plain bands or as bands bearing 4 or more minute incisures. Deirids and phasmids not observed. Lip region distinctly set off with amphid apertures appearing as minute refractive elements at the apices of the lateral lips. Six minute papillae surround the oral opening with 4 submedian ones located on the outer margins of the submedial lips. Spear small with well-developed basal knobs. Median bulb of esophagus with distinct valvular apparatus; basal bulb made up of glandular tissues which frequently may become greatly swollen and irregular in form. Ovary extending forward, generally with 1 or 2 flexures. Oocytes in multiple series, arranged about a rachis. Posterior uterine branch rudimentary. Spicula joined together, arcuate, without definite cephalation, the blades generally as wide, or wider than the haft. Gubernaculum troughlike, slightly curved. Testis with spermatocytes developing in multiple rows about a rachis. Bursa enveloping tail or nearly so. Typical parasites of the seeds and stems of plants. (Mostly from Thorne 1949.)

General Characteristics

The genus *Anguina* is distinguished from the closely related genus *Ditylenchus* by the caudal bursa; by multiple rows of oocytes and spermatocytes arranged about the rachis; by the robust, amalgamated, wide spicula; and by the distended, almost immobile bodies of the females (Thorne 1949). Wu (1967) described *A. calamagrostis* n. sp. in which the females are slender, more like *Ditylenchus*, and the bursa does not quite envelop the tail terminus. *Subanguina* is a closely related genus in which the mature females are moderately stout and the single ovary extends far anteriorly, commonly with 1 or more flexures and oocytes in 2 lines. These nematodes are found in root galls of Gramineae. Another closely related genus, *Paranguina*, which causes galls on the rhizomes of *Agropyrum repens* in the U.S.S.R., differs from *Anguina* in the number and arrangement of the esophageal glands (Kirjanova 1955).

In general *Anguina* spp. cause stem and seed galls on species of the Gramineae: some species of *Anguina*, however, produce galls on dicotyledonous plants. Large numbers of larvae may be present in a single gall; counts of *A. tritici* ranging from several hundred to 90,000 per wheat gall are reported. Each female of *A. stolonifera* lays approximately 1,000 eggs during a period of about 2 weeks. Quiescent larvae, usually second-stage larvae, may remain viable in a dry condition for several years.

Plate 37. Anguina tritici. A. Neck of female with greatly developed esophageal glands. B. Spicula and cross section of gubernaculum. C. Head. D. Male tail, ventral view. E. Face view. F. End of ovary showing cap cell. G. End of testis showing cap cell. H. Male tail. I. Adult female. J. Cross section of ovary. (After Thorne 1949. Courtesy of Helminthological Society of Washington.)

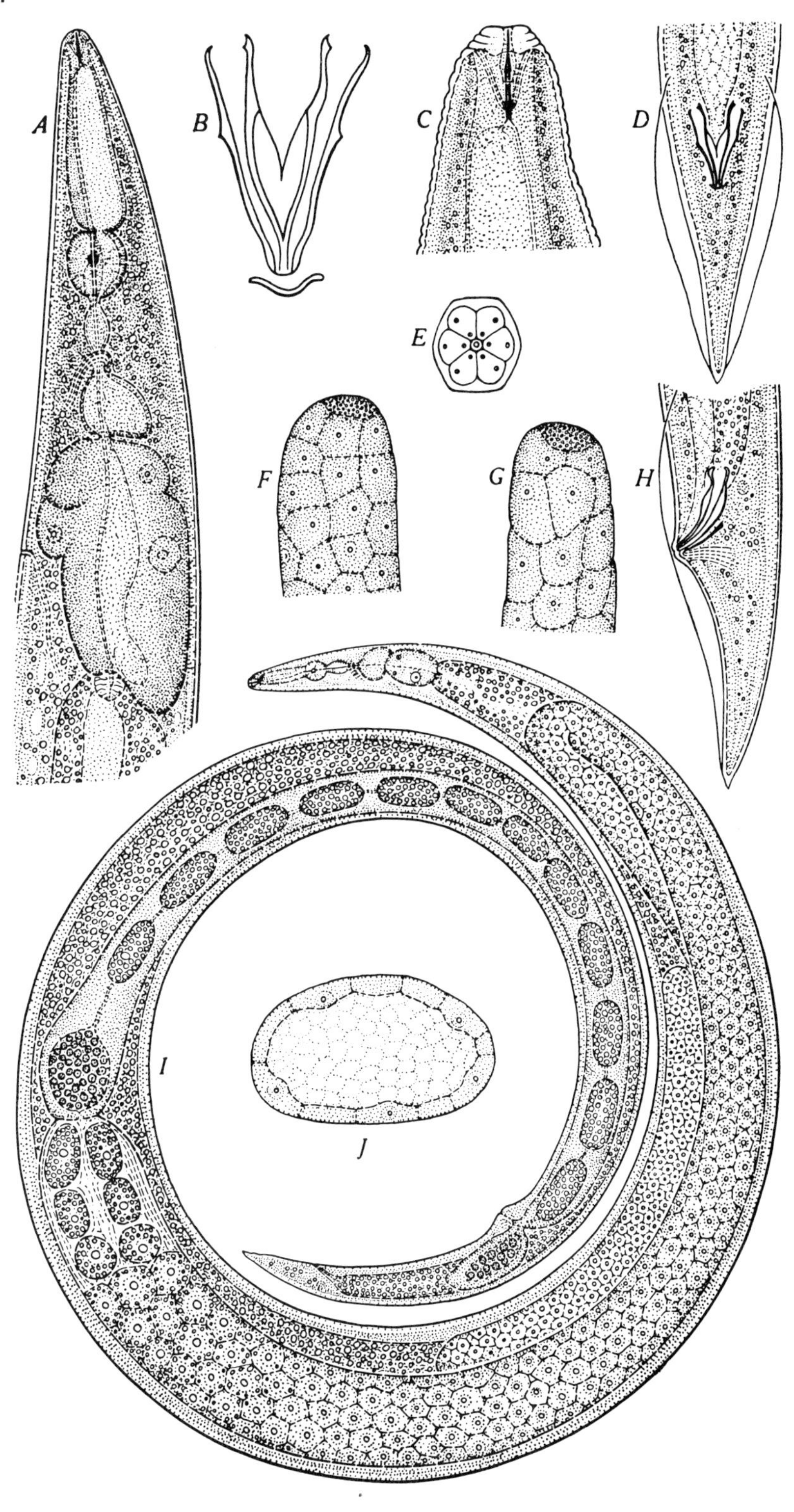
A
B
C
D
E
F
G
H
I
J

32. Ovary with 1 or more flexures, moderately stout forms. Found in root galls of Gramineae . . . . (Plate 38) *SUBANGUINA*
Ovary outstretched, slender forms. Found in bulbs, stems, leaves and tubers . . . . (Plate 39) *DITYLENCHUS*

Subanguina Parmonov 1967

Type species: *Subanguina radicicola* (Greeff 1872) Paramonov 1967.

Description: Anguininae: Moderately stout forms, tapering at extremities, with finely striated cuticle. Esophageal glands 3, enclosed in a rather swollen, broadly pyriform basal bulb. Lateral field present. Ovary extending far anteriorly, commonly with 1 or more flexures and oocytes in 2 lines. Testis also with flexures. Bursa subterminal, enveloping up to about 75% of tail. Causing galls on underground parts of grasses. (From Golden 1971.)

General Characteristics

Greeff (1872) reported that in Germany as early as 1864 he observed galls on the roots of annual meadow grass, *Poa annua*, and couch grass *Agropyron repens*, from which he obtained large numbers of nematodes which he described as a new species, *Anguillula radicicola*. Since then this nematode has been found attacking these and additional hosts in numerous locations and its species name was changed several times. Thorne (1961), assigned this nematode to the genus *Ditylenchus* but stated that it is probably an *Anguina*, not a *Ditylenchus*. Parmonov (1967) erected a new genus *Subanguina* to include the type and only species, *D. radicicola*. According to Golden (1971) this species is related to both *Ditylenchus* and *Anguina* and it might well be shown eventually that *S. radicicola* represents a complex of closely related species. In *Subanguina* the oocytes are in 2 rows, not on a radius, while in *Anguina* the oocytes occur in multiple rows arranged about a radius. *Subanguina* occurs in root galls in Gramineae while *Anguina* occurs in seeds or galls on above-ground parts of various plants.

According to Golden (1971), *Subanguina* can be separated from *Ditylenchus* because the ovary of *Subanguina* has 1 or more flexures and the ovary of *Ditylenchus* is outstretched. Also *Subanguina* females are considered to be moderately stout while those of *Ditylenchus* are slender. Individuals of *Subanguina* are found in roots, while those of *Ditylenchus* are found in bulbs, stems, leaves, and tubers.

Plate 38. Anguillulina radicicola (now named *Subanguina radicicola*). A. Adult male. B. Adult female. C. Esophagus region. D. Male tail in lateral view. (After Goodey 1932. Courtesy of Journal of Helminthology.)

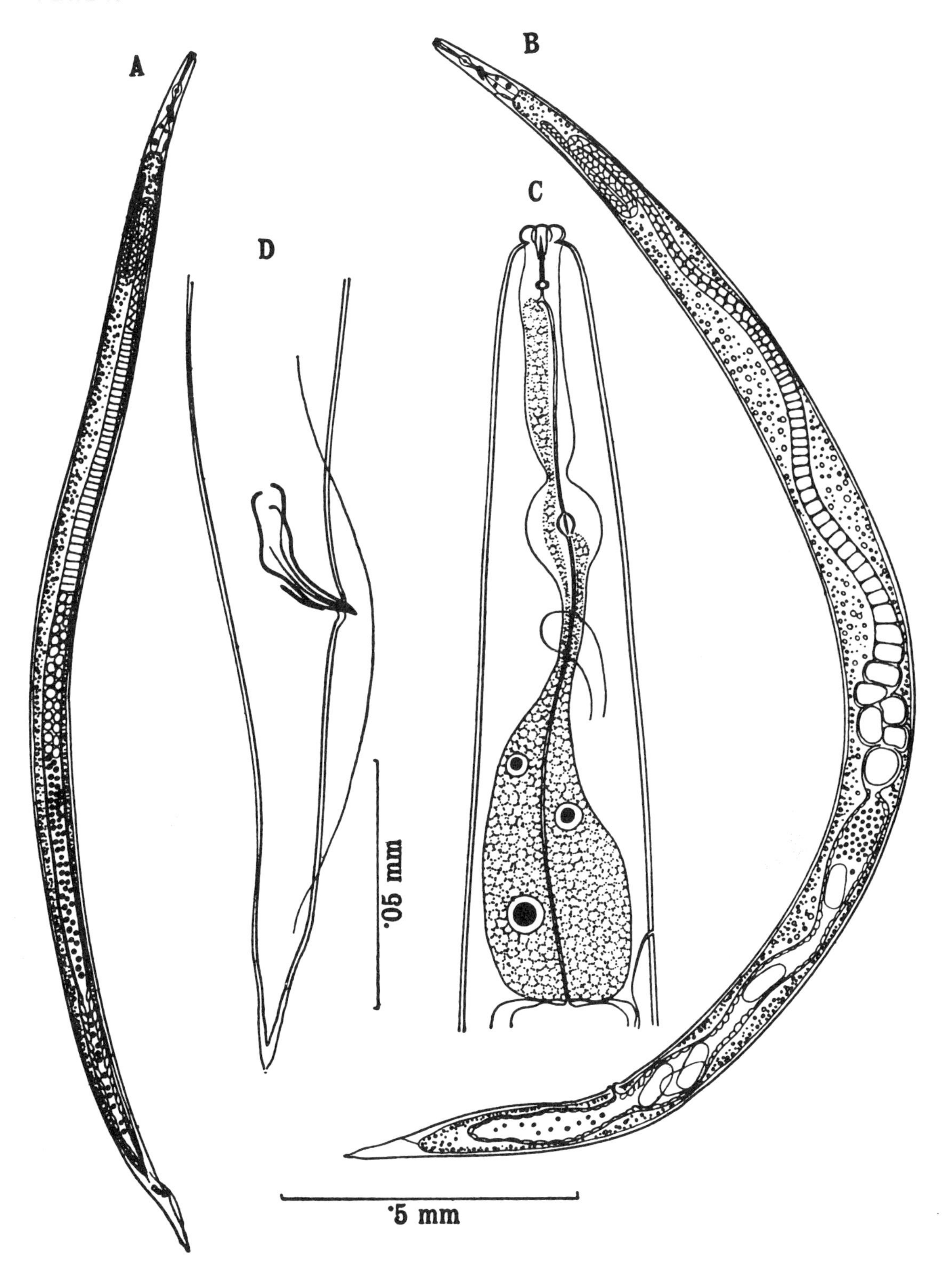
A
B
C
D
·05 mm
·5 mm

Ditylenchus Filipjev 1934

Type species: *Ditylenchus dipsaci* (Kuhn 1857) Filipjev 1936

Description: Tylenchinae. Ovary single. Rudimentary posterior uterine branch present. Lip region plain, not annulated. Gonad cells in 1 or 2 lines, not arranged about a rachis. Lateral fields marked by 4 or 6 incisures. Basal portion of esophagus a distinct bulb, occasionally with a short lobe extending back over anterior end of intestine. Deirids small but usually visible. Phasmids exceedingly minute if visible at all. Bursa enveloping 1/4 to 3/4 of the tail. Tails elongate-conoid to an acute or subacute terminus. (From Thorne 1949.)

General Characteristics

Christie (1959) comments concerning the stem nematode, *Ditylenchus* spp., and related genera: "The genus *Ditylenchus* impinges upon the genus *Tylenchus* at one extreme and upon the genus *Anguina* at the other. This had led to slight confusion between *Ditylenchus* and *Anguina*, certain species having been placed sometimes in the 1 genus, sometimes in the other. At least some and perhaps all of the species of these 3 genera have 1 feature in common: certain of the larval stages will persist in a dry condition and remain viable for many years. Nowhere else among the plant parasites has this feature of anabiosis become developed to anywhere near an equal extent. The stem nematode, *Ditylenchus dipsaci*, and the potato rot nematode, *D. destructor*, are the 2 species of this genus that are of greatest economic importance. Both are composed of what will be referred to herein as biological races. These are, in effect, host races because they are differentiated primarily by their host relationships."

Paramonov (1967) erected the genus *Subanguina* to include the species *D. radicicola*. According to Golden (1971), *Subanguina* can be separated from *Ditylenchus* because the ovary of *Subanguina* has 1 or more flexures and the ovary of *Ditylenchus* is outstretched. Also *Subanguina* females are considered to be moderately stout while those of *Ditylenchus* are slender. Individuals of *Subanguina* are found in roots while those of *Ditylenchus* are found in bulbs, stems, leaves, and tubers.

Most larvae and adults of this genus are long and slender. The stylets are short and are often difficult to observe clearly under the dissecting microscope.

Sher (1970) proposed the genus *Chitinotylenchus* as a synonym of *Ditylenchus*.

Tarjan (1958) erected *Pseudhalenchus* a genus which resembles *Ditylenchus*, but differs in that the esophageal glands lie free in the body rather than enclosed in a bulb as in *Ditylenchus*.

Plate 39. Ditylenchus dipsaci. A. Mature female, 115 X. B. Mature female, 255 X. C. Anterior portion of female, 2230 X. D. Posterior portion of male, 820 X.

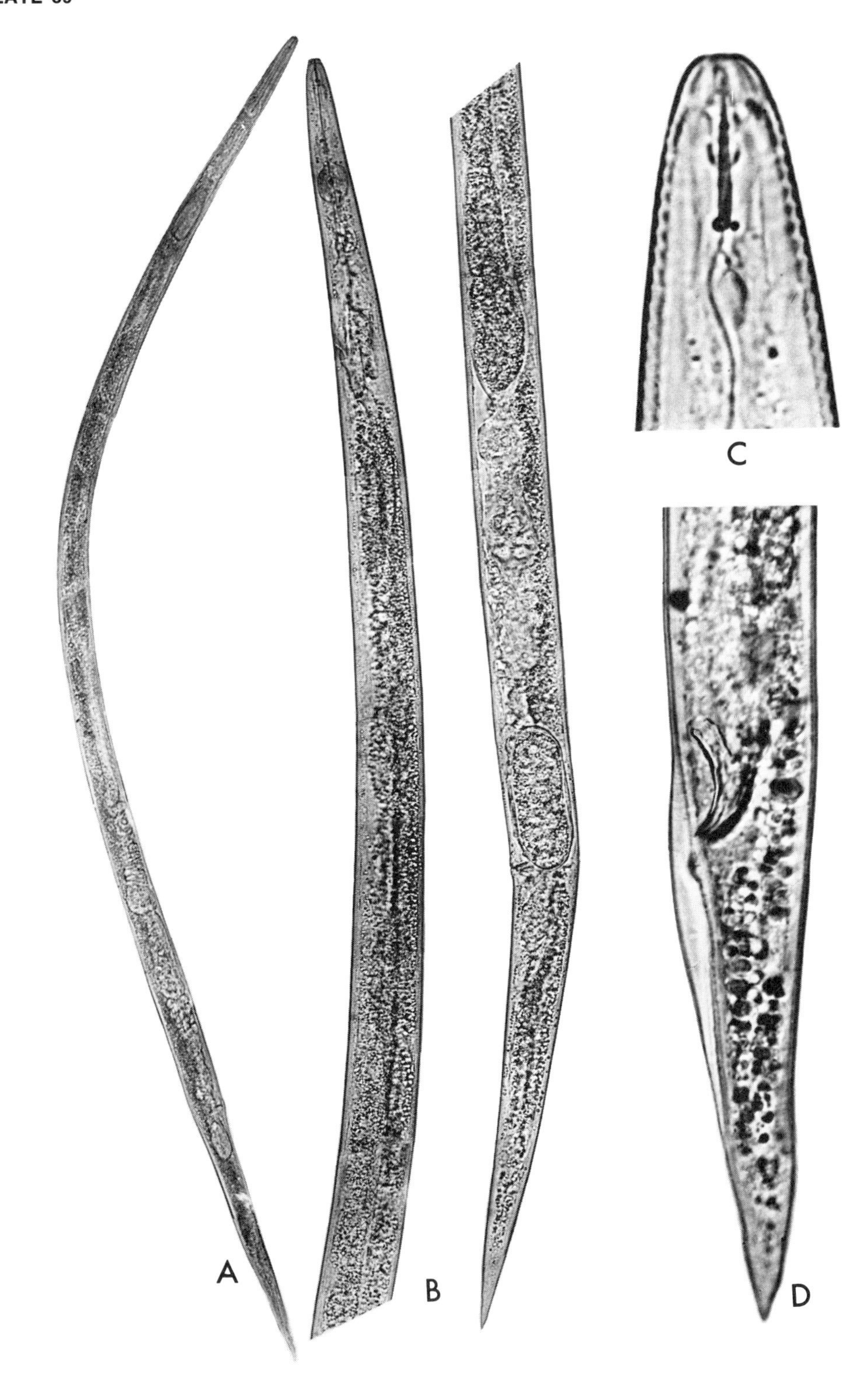
A
B
C
D

33. *s* (stylet length/body diameter at base of stylet) = 1.5 or more; tail generally 1.5 times anal body diameter or shorter . . . . (Plate 40) *ROTYLENCHOIDES*
s = less than 1.5; tail generally longer than 1.5 times anal body diameter . . . . (Example: Plate 41) 34

Rotylenchoides Whitehead 1958

Type species: *Rotylenchoides brevis* Whitehead 1958

Description: Tylenchidae. Head with 6 lips, annulated, not offset by constriction; head framework strongly developed; body cuticle striated, lateral field marked by incisures; spear long, strong, with well-developed basal knobs, esophagus with ovate, valvate median bulb and lobelike posterior esophageal region, extending over anterior end of intestine; phasmids at about level of anus; female with single prodelphic ovary, vulva posterior; tails of both sexes not longer than anal body diameter, male tail enveloped by bursa; gubernaculum present; a telamon may be present. (From Whitehead 1958.)

General Characteristics

R. brevis Whitehead, 1958, the type species, is found as an endoparasite of banana in Tanganyika. Luc (1960) described 3 other species occurring in the Ivory Coast; *R. affinis* about coffee roots; *R. intermedius* in roots of *Drypetes aylmeri* and *R. variocaudatus* about roots of *Piper nigrum* and *Abbizzia* sp. Whitehead (1958) comments concerning the taxonomic position of this new subfamily and genus:

> *R. brevis* occupies an interesting taxonomic position. In common with Hoplolaiminae the genus *Rotylenchoides* has a heavily sclerotized head framework, a massive spear at least 3 times as long as the width of the head base, with well-developed basal knobs, lobelike posterior esophageal region and a tail not longer than the anal body width. However, unlike Hoplolaiminae, *Rotylenchoides* has but a single gonad and a posterior vulva, in which characters the new genus resembles *Pratylenchus*, a genus of the subfamily Pratylenchinae. *Rotylenchoides* differs from Pratylenchinae in that the latter has a spear not more than twice as long as the width of the head base, the tails of the vermiform species are at least twice the length of the anal body diameter; the phasmids are located behind the level of the anus and the median bulb is spheroid.
>
> To include *Rotylenchoides* with Hoplolaiminae it would be necessary to emend the diagnosis of the subfamily as regards the possession of 2 ovaries and a vulva at the middle of the body, in this group. In view of the fact that the subfamily Hoplolaiminae as it stands at present, constitutes a compact group in the present classification of the order Tylenchida Thorne, 1949, it is felt that separation of *Rotylenchoides* to a new subfamily is justified.

Golden (1971) included this genus in the Rotylenchoidinae, a subfamily of the Hoplolaimidae.

Plate 40. A. General illustration of female *Rotylenchoides brevis*. B. General illustration of male *R. brevis*. C. Lateral view of cuticle of head and anterior end of body, showing origin of lateral fields of female worm. D. Detailed illustration of female head. E. Tail region of male. F. Lateral view of cuticle of female posterior body region. (After Whitehead 1958. Courtesy of Nematologica.)

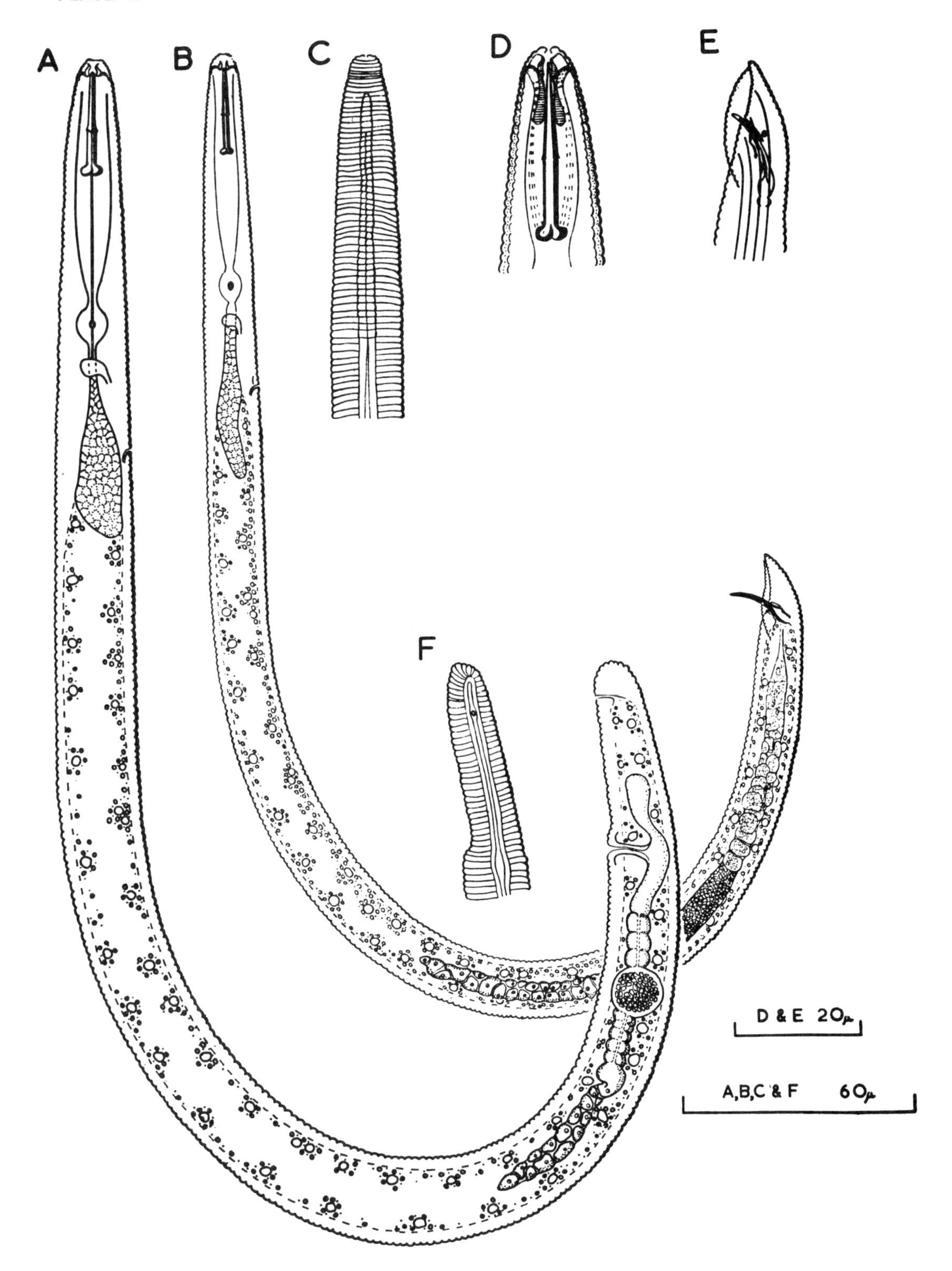
A
B
C
D
E
F
D & E 20μ
A,B,C & F 60μ

34. Esophagus overlaps intestine ventrally . . . . (Plate 41) *PRATYLENCHUS*
Esophagus overlaps intestine dorsally . . . . (Example: Plate 42) 35

Pratylenchus Filipjev 1934

Type species: *Pratylenchus pratensis* (de Man 1880) Filipjev 1936

Description: Pratylenchinae. Stout, cylindroid nemas less than 1.0 mm in length, with relatively broad heads and bluntly rounded tails. Lateral field marked by 4 incisures, except 6 in *Pratylenchus hexincisus* and variable in rare specimens of certain other species. Incisures of adult females frequently absent owing to stretching of the cuticle. Deirids rarely visible, located near latitude of excretory pore. Phasmids located 1/3 of tail length or more behind latitude of anus. Lip region bearing 2 to 4 annules (1 to 3 striae) set off by a narrowing of the head. Cephalic framework sclerotized, refractive. Spear strong, 14 to 19μ long, with massive basal knobs. Median esophageal bulb spheroid, more than half as wide as neck. Basal bulb extending back over intestine, usually in a lateroventral position. Three prominent esophageal gland nuclei. Esophageal lumen and intestine joined by an obscure muscular valve. Excretory pore prominent, about opposite nerve ring. Hemizonid slightly anterior to excretory pore. Intestine packed with numerous dark granules. Slender, muscular rectum ending in a transverse, slitlike anus. Vulva a depressed transverse slit, vagina extending in and slightly forward. Anterior ovary outstretched, with oocytes arranged in a single file except for a short region of multiplication. Posterior uterine branch rudimentary.

Males known in about half the species. Bursa enveloping tail, with phasmids located near its base. Spicula slightly arcuate, resting on a thin, troughlike gubernaculum. Testis outstretched, with spermatocytes irregularly arranged, especially in the region of multiplication. (From Thorne 1961.)

General Characteristics

Members of this genus are called lesion nematodes (because of the lesions they cause on feeder roots and occasionally on other underground parts of plants), or meadow nematodes (because of their frequent occurrence in meadows). They are commonly found in large numbers inside roots and sometimes in tubers, peanut shells, etc.

The distribution of members of this genus is worldwide. Some species appear to be adapted to cool regions and others to warm regions. Available information indicates that, in general, they are somewhat more numerous in the warmer parts of the temperate zone than they are in the tropics and subtropics (Christie 1959).

Each of the approximately 37 species has a wide host range. Apparently, all plants infected by a given species are not equally suitable hosts; on some hosts there is a substantial build-up of nematode numbers without obvious damage to roots; on other hosts the build-up is slight with no appreciable damage; on still others there is marked root damage. When damage occurs, a substantial reduction in nematodes results due to food depletion. A single plant species may be susceptible to more than 1 species of *Pratylenchus*.

Under a dissecting microscope, some diagnostic characteristics are the overlapping esophagus, the flat head, and the relatively slow graceful movement. When at rest or dead, members of this genus tend to lie in a straight line.

Plate 41. *Pratylenchus* sp. A. Mature female, 460 X. B. Anterior portion of female, 1220 X. C and D. Lateral view of male tail, 825 X. E. Mature female, 460 X. a. overlapping esophagus.

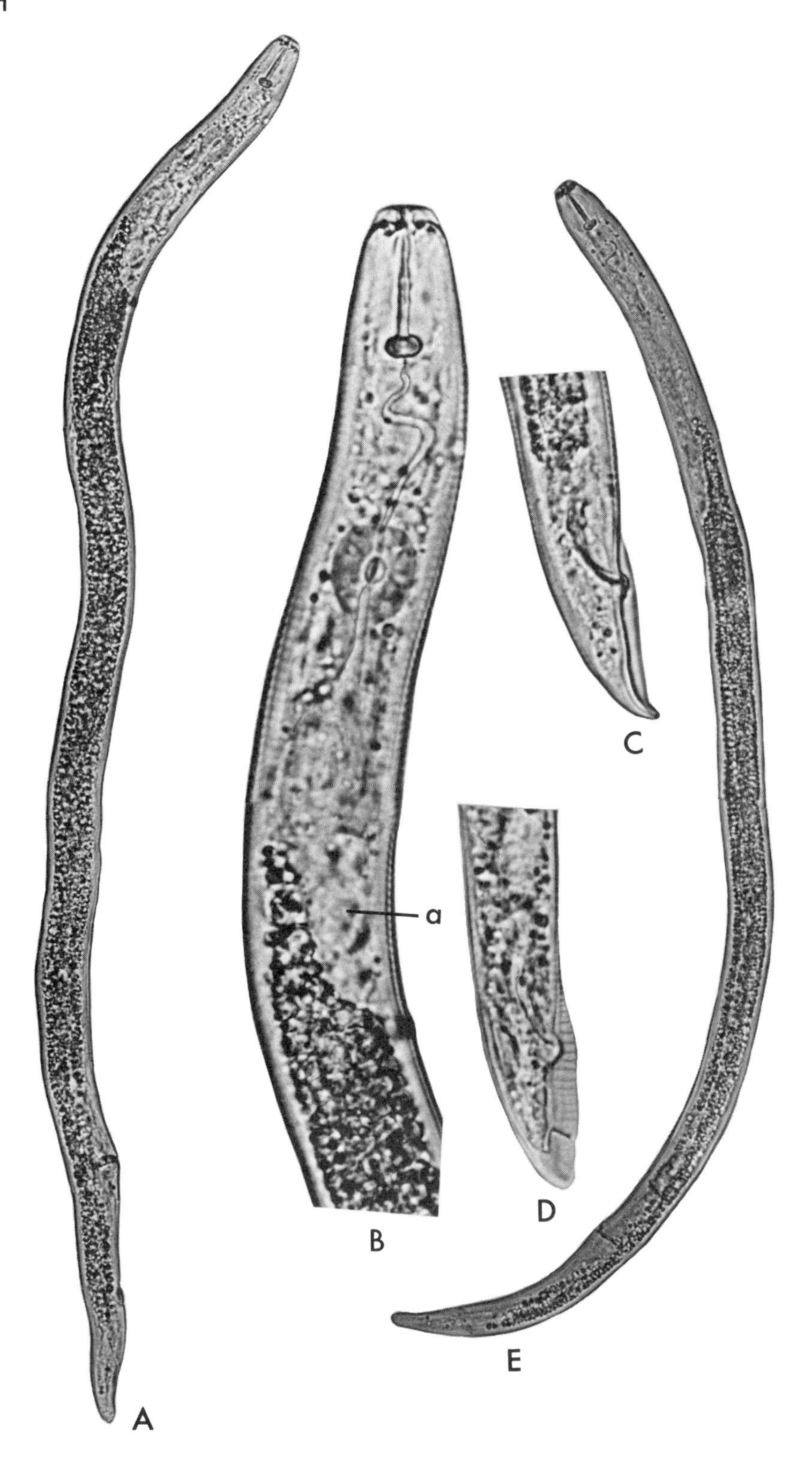
a
A
B
C
D
E

35. Lip region low, generally rounded; stylet knobs flattened anteriorly; marked sexual dimorphism . . . . (Plate 42) *RADOPHOLOIDES*
Lip region, high conoid; stylet knobs sloping anteriorly or indented; males present or absent 36

Radopholoides De Guiran 1967

Type species: *Radopholoides litoralis* De Guiran 1967

Description: Pratylenchinae. Marked sexual dimorphism. Female: 1 anterior ovary; vulva at the beginning of the last third of the body length. Lip region rounded, slightly offset; head heavily sclerotized; massive stylet; basal gland of the esophagus long and dorsally overlapping the intestine; tail, conical; posterior end blunt, irregularly striated.

Male: lip region globular, slightly sclerotized, separated from rest of body by a constriction; reduced stylet; degenerate esophagus; single testis; spicules and gubernaculum slightly bent; tail tapering, completely enveloped by the bursa. Phasmid in both sexes situated at 1/4 of the tail length. (From De Guiran 1967.)

General Characteristics

Radopholoides litoralis was collected by De Guiran (1967) from a sandy seashore in the Antalaha region of western Madagascar where it was associated with the vanilla plant.

Concerning the taxonomic position of this new genus, De Guiran states the following:

Among the genera of the subfamily Pratylenchinae, *Radopholoides* n. g. is near, as its name indicates, *Radopholus* Thorne 1949 because of the general organization and the pronounced sexual dimorphism. It is different from this genus because of the presence of only 1 ovary in the female.

Radopholoides n. g. differs from *Pratylenchus* Filipjev 1934 by the dorsal overlapping of the esophagus over the intestine and the sexual dimorphism.

Hoplotylus s'Jacob 1959 and *Radopholoides* both have a single ovary and a dorsally overlapping esophagus. But, as Siddiqi (1963) has emphasized, *Hoplotylus* fits better into the Hoplolaiminae Filipjev 1934 because of the shape and structure of the head and because of the length and the shape of the stylet. The last 2 characters separate precisely these 2 genera.

Radopholoides could therefore be defined simply as a genus near *Radopholus*, but whose female has a single ovary.

Plate 42. Radopholoides litoralis. (Female.) A. Entire animal. B. Anterior portion. C. Head. D and E. Tail. (Male.) F. Entire animal. G. Anterior portion. H. Head. I. Tail. (A and F: natural position.) (After De Guiran 1967. Courtesy of Nematologica.)

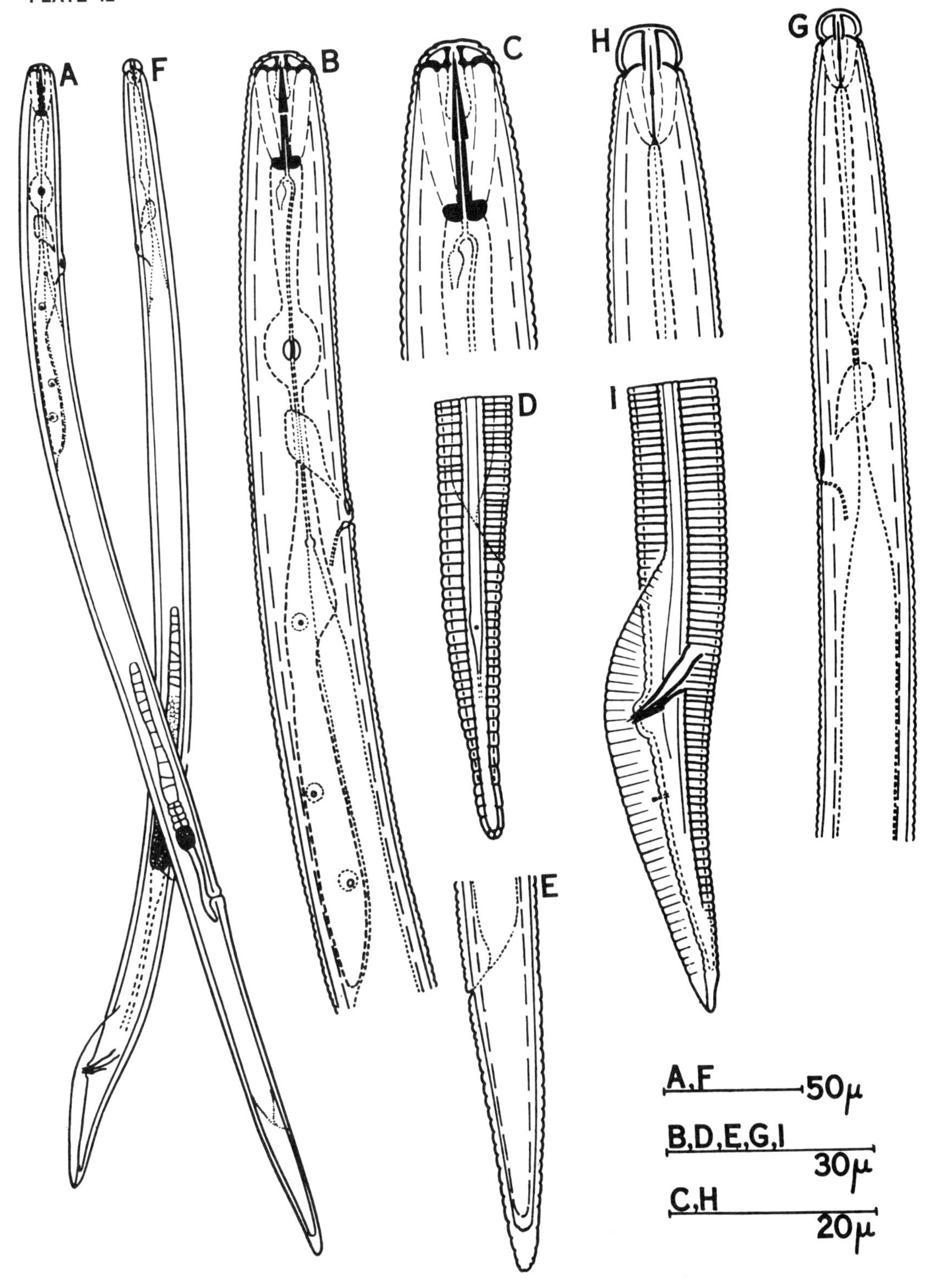
A
F
B
C
H
G
D
I
E
A,F 50μ
B,D,E,G,I 30μ
C,H 20μ

36. Female body swollen; posterior part of stylet knobs sloping anteriorly; marked sexual dimorphism .... (Plate 43) *ACONTYLUS*
Female body slender; each stylet knob tapering anteriorly to a dentate tip; males unknown .... (Plate 44) *HOPLOTYLUS*

Acontylus Meagher 1968

Types species: *Acontylus vipriensis* Meagher 1968

Description: Hoplolaimidae: Esophageal glands in long lobe overlapping intestine dorsally. Opening of dorsal esophageal gland at least half stylet length from stylet base. Body cuticle finely annulated. Phasmids porelike. Marked sexual dimorphism. Female: Body partly swollen. Single anterior ovary with postvulval sac. Head slightly offset, with sclerotized framework. Lip region high, conoid. Stylet well developed. Tail length less than twice anal body diameter. Tail annulated. Male: Body vermiform and slender. Head distinctly offset with reduced sclerotization. Lip region high, rounded. Stylet and esophagus reduced. Bursa surrounding tail tip. (From Meagher 1968.)

General Characteristics

A. vipriensis, the only species of this genus, were found attached to the roots of Eucalyptus sp. L'Heritier by Meagher (1968). They were found clustered in "colonies" of 20 to 30 nematodes which included females, males, young females, and larvae. Eggs were found singly, along with a small quantity of gelatinous material by which they were frequently attached to plant roots. Heads of mature females were embedded in the root to about the latitude of the excretory pore, but males and larvae were not observed feeding.

One of the chief distinguishing characteristics of *Acontylus* is that the bodies of mature females are irregularly swollen; slightly swollen in the anterior portion, but body diameter is almost doubled in the region of the oviduct-uterus. When relaxed by gentle heat the body becomes straight or slightly curved ventrally.

The single species is sexually dimorphic. Males differ from females in that stylet and esophagus are reduced, lip region is higher and more rounded, and the head is more distinctly offset and its sclerotization is reduced.

With respect to relationships with other genera, Meagher states:

> *Acontylus* shares some characters of the genus *Hoplotylus* s'Jacob 1959 and the genera *Radopholus* Thorne 1949, and *Radopholoides* de Guiran 1967 in Pratylenchinae. These three genera possess an elongate esophageal lobe overlapping the intestine dorsally. The shape of the male head and reduced esophagus of the male resemble *Radopholus* but the female of *Acontylus* has only one gonad and the lip region is conoid. Though *Radopholoides* has one gonad, the conoid lip region, as well as the tail and general body shape, serve to distinguish *Acontylus* from this genus. *Acontylus* is close to *Hoplotylus*, but males of *Hoplotylus* have not been found. *Acontylus* also differs from *Hoplotylus* by the swelling of the female, the distance of the openings of the dorsal esophageal gland from the stylet base, the shape of the stylet knobs, and the shape and annulation of the female tail.
>
> In Nacobbinae there are 2 genera characterized by swollen females. *Nacobbus* Thorne and Allen 1944 has an esophageal lobe overlapping the intestine dorsally but the lip region is low and rounded, and similar in the 2 sexes, and the opening of the dorsal esophageal gland is a short distance from the base of the stylet. The opening of the dorsal gland is well removed from the stylet base in *Rotylenchulus* Linford and Oliveira 1940, but this genus has 2 gonads, a short ventral overlap of the esophageal lobe and also differs from *Acontylus* in the shape of the male tail and lip region.
>
> The genus *Rotylenchoides* Whitehead 1958, the only representative of Rotylenchoidinae, does not exhibit sexual dimorphism of the lip region, and, according to Luc (1960), the esophageal lobe overlaps laterally or ventrolaterally. Nevertheless, in the possession of a single anterior gonad and in the shape of the female tail and lip region, *Acontylus* bears a similarity to this genus.

Plate 43. Acontylus vipriensis. A. Mature female. B. Egg. C. Larva. D. Larval tail. E. Anterior end of male. F. Male tail. G. Anterior end of young female. H. Young female tail. I. Young female. J. Face view, female. K. Face view, male. (After Meagher 1968. Courtesy of Nematologica.)

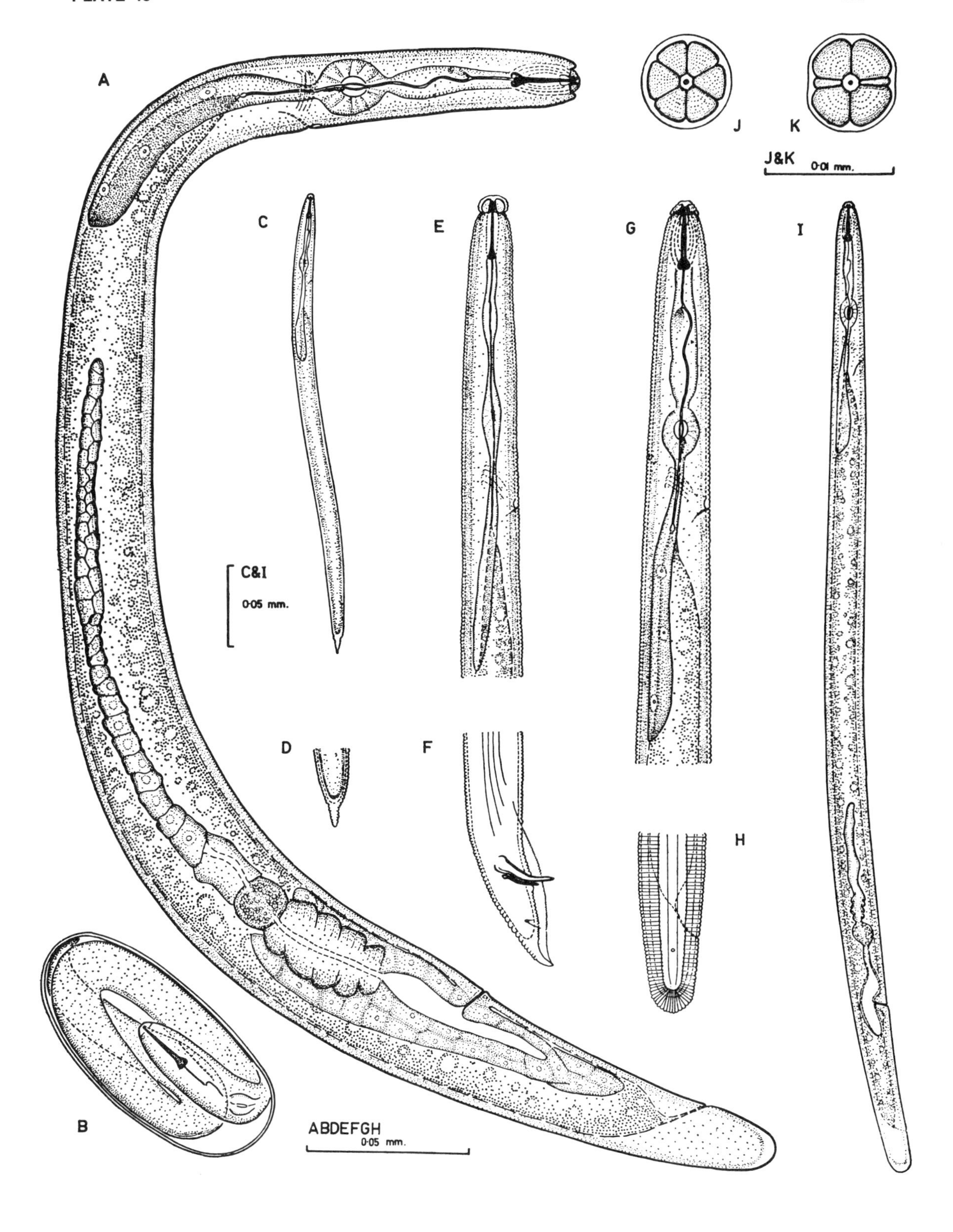
A
B
C
D
E
F
G
H
I
J
K
J&K 0·01 mm.
C&I 0·05 mm.
ABDEFGH 0·05 mm.

Hoplotylus s'Jacob 1959

Type species: *Hoplotylus femina* s'Jacob 1959

Description: Pratylenchinae. Body cylindrical, tapering to tail end. Head slightly offset with heavily sclerotized framework. Lip region high, conoid, striated. Body cuticle finely annulated, except on tail tip, where striation is coarser. Stylet well developed with 3 oval knobs, close together; the anterior margin of each knob forms teeth. Behind median bulb the esophagus narrows to isthmus and then widens into a lobelike glandular region, the posterior part of which overlaps intestine dorsally. Nerve ring present. Ovary single, outstretched; vulva situated ventrally in posterior part of body. Postvulval sac present. Phasmid apertures porelike. Tail distinctly longer than anal body width. (From s'Jacob 1959).

General Characteristics

H. femina, the only species of this genus described, was found by s'Jacob (1959) around and in the roots of English Oak (*Quercus robur* L.) and Lawson Cypress (*Chamaecyparis lawsoniana* Parl.) growing in sandy soils near Wageningen, The Netherlands.

Plate 44. Hoplotylus femina. A. Female nematode; variation in female terminus, 405 X. B. Female head, 620 X. C. Lateral view of cuticle near phasmid region, 810 X. (After s'Jacob 1959. Courtesy of Nematologica.)

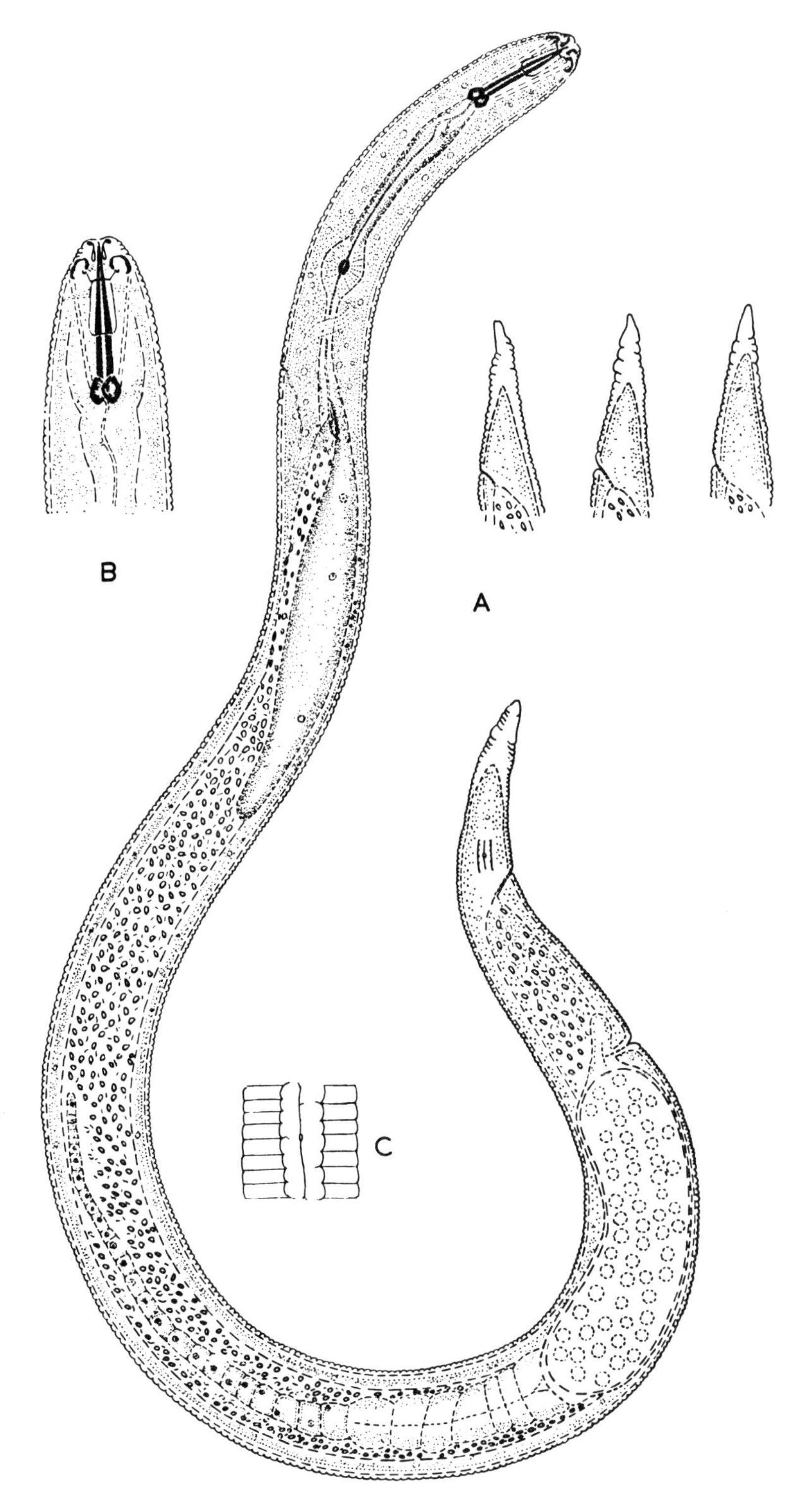
B
A
C

37. Mature female without extra cuticle or sheath . . . . (Example: Plate 45) 38
 Mature female with extra cuticle or sheath . . . . (Example: Plate 47) 40
38. Cuticle with prominent retrorse annules . . . . (Example: Plate 45) 39
 Cuticle without prominent retrorse annules . . . . (Example: Plate 47) 41
39. Annules of female with spines, scales, plates, or stalked appendages on posterior margins . . . . (Plate 45) *CRICONEMA*
 Annules of female with smooth or crenate posterior margins . . . . (Plate 46) *CRICONEMOIDES*

Criconema Hofmanner and Menzel 1914

Type species: *Criconema guernei* (Certes 1889) Menzel in Hofmanner and, Menzel 1914

Description: Criconematinae. Body stout, fusiform. Length of females 6–12 times greatest body diameter. Cuticle thick, divided into large transverse annules, most of which have spine or scalelike appendages on the posterior edge. Annules 50–150. Head apparently of 2 annules, with or without spines. Mouth armed with long axial stylet, the base of which is made up of 3 distinct lobes, each lobe with a forward-projecting process to which the muscles are attached. Terminal bulb of esophagus more or less rounded. Vulva located in posterior $\frac{1}{5}$ of body, often covered by a scale. Ovary single prodelphic. Male (known only in one species, *C. squamosum* Cobb) more slender than female, loses stylet at final molt. Spicula 2, equal. No distinct gubernaculum and no bursa. (From Taylor 1936.)

General Characteristics

Members of this genus closely resemble individuals of the genus *Criconemoides* in general structure, size, stylet, esophagus, and structure of gonad; they differ by having annules possessing backwardly directed spines or scalelike extensions or even stalked appendages on their posterior edges.

The species of this genus have not been proven to be plant parasites. Their association with plant roots and their well-developed stylets indicate that they feed on roots. Goodey (1963) recognized 21 species.

In a revision of the genus *Criconema* and related genera by Mehta and Raski (1971), *Blandicephalanema, Crossonema, Neolobocriconema*, and *Pateracephalanema* are proposed as new genera. Two subgenera (*Criconema* and *Variasquomata*) of the genus *Criconema*, and 2 subgenera of *Crossonema* (*Crossonema* and *Seriespinula*) are proposed. Six new species belonging to these genera are described and keys are given to genera of the subfamily Criconematinae and to species of the above genera.

Plate 45. Criconema celetum. 1. A mature specimen. 2. Anterior end, annules 4–12, optical section. 3. *En face* view, showing first head annule and first body annule. 4. a. spines on annules 12–14. b. spines on annule more posterior. 5. Posterior end, ventral view. 6. Annules 11–13, ventral view. 7. Posterior end, optical section. 8. Reproductive system of young specimen with portion of body wall, optical section. (After Wu 1960. Courtesy of Canadian Journal of Zoology.)

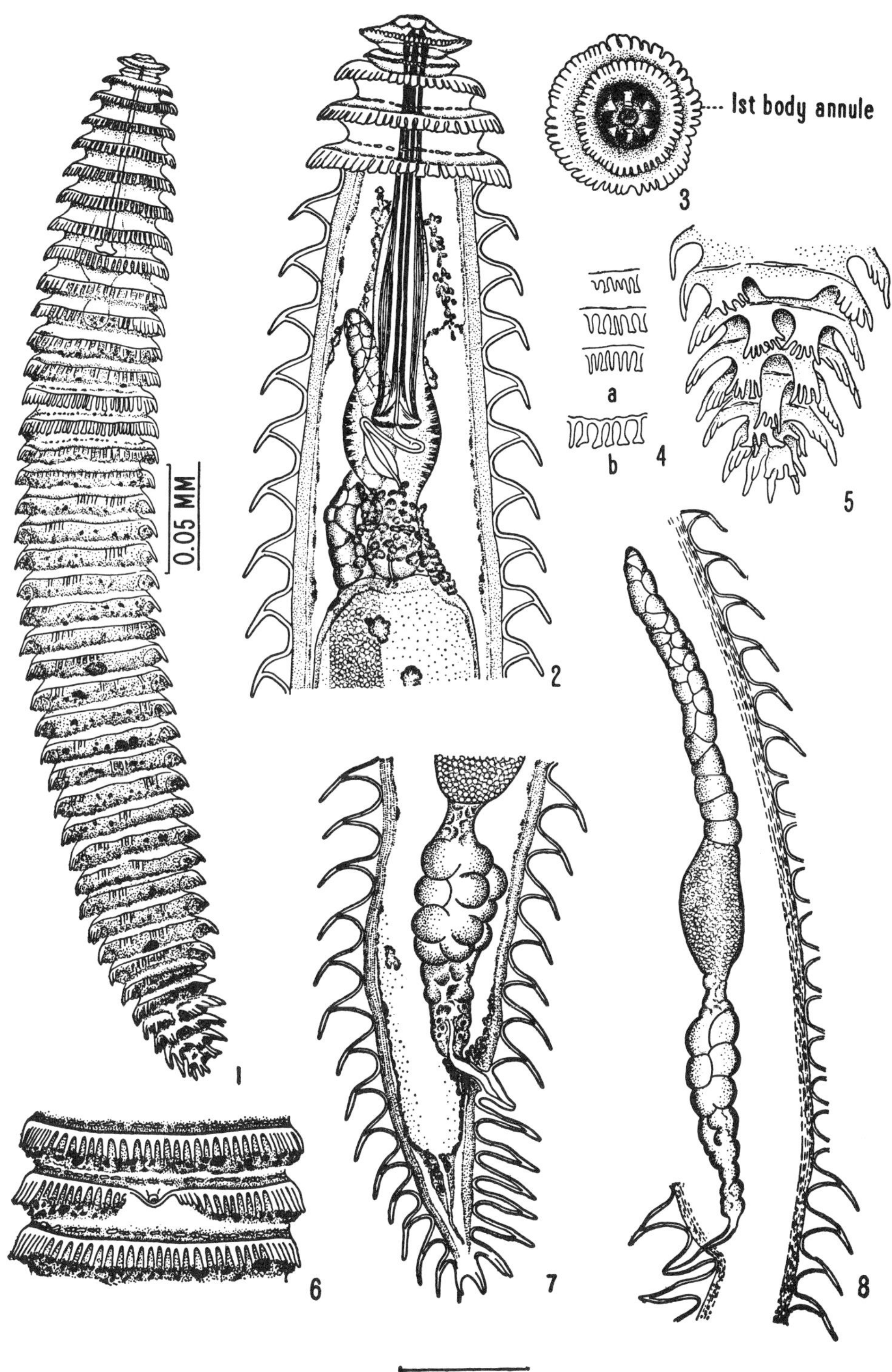
1st body annule
0.05 MM
1
2
3
a
b
4
5
6
7
8
0.03 MM

Criconemoides Taylor 1936

Type species: *Criconemoides morgense* (Hofmanner in Hofmanner and Menzel 1914) Taylor 1936

Description: Criconematinae. Body fusiform with 33–194 annules, generally coarse and retrorse with plain, irregular, or finely serrated margins. Stylet knobs with forwardly directed processes (except *microdorum*). Lateral area of female sometimes marked by union of anastomosing annules but without longitudinal lines or incisures. Tail short and conical or broadly rounded. Males with 2, 3 or 4 incisures in the lateral fields that extend on to the tail on narrow caudal alae that reach almost to the terminus. Development through juveniles that have smooth or crenated annules or with rows of scalelike cuticular protrusions. (Mostly from Raski and Golden 1965.)

General Characteristics

Members of this genus, widely distributed geographically, are commonly associated with roots of numerous plant species growing in cultivated and uncultivated soils. Experimental data and observations indicate that *Criconemoides* spp. are ectoparasites of plant roots.

The movements of individuals of this genus are sluggish. The locomotion is not serpentine, like that of most nematodes, but is earthwormlike, the body being alternately lengthened and contracted by turgor or relaxation of the muscles (Winslow 1960). Only a small percentage move through tissue paper in a Baermann funnel. All stages of females are fusiform in shape with heavy annulation. Males, known to occur in only a few species, have no stylet and indistinct esophagus and intestine. Under a dissecting microscope, identifying characteristics are lack of movement, "stout" appearance, and general rough appearance, indicating the heavy annulation.

Diab and Jenkins (1965) erected the genus *Neocriconema* to include a new species *N. adamsi* and 10 species transferred from either *Criconemoides* or *Criconema*. According to the authors, this genus is closely related to *Criconemoides* and *Criconema* and is characterized by crenations on the posterior margins of body annules of the adult female.

De Grisse and Loof (1965) declared *Criconemoides* genus inquirendum and the species of this genus species inquirenda. The species of this genus were divided into 6 genera: *Macroposthonia, Nothocriconema, Lobocriconema, Criconemella, Xenocriconemella*, and *Discocriconemella*. Loof and De Grisse (1967) re-established the genus *Criconemoides* but also recognized the 6 genera proposed by De Grisse and Loof (1966).

Neither the genus *Neocriconema* nor the genera erected by De Grisse and Loof (1966) were recognized by Raski and Golden (1965), Tarjan (1966), Allen and Sher (1967), or Jenkins and Taylor (1967). Raski and Golden (1965) published a key to 86 species of the genus *Criconemoides*.

Plate 46. *Criconemoides* sp. A. Mature female, 556 X. B. Ventral view of posterior end of female, 960 X. C. Lateral view of posterior end of female, 960 X. a. vulva.

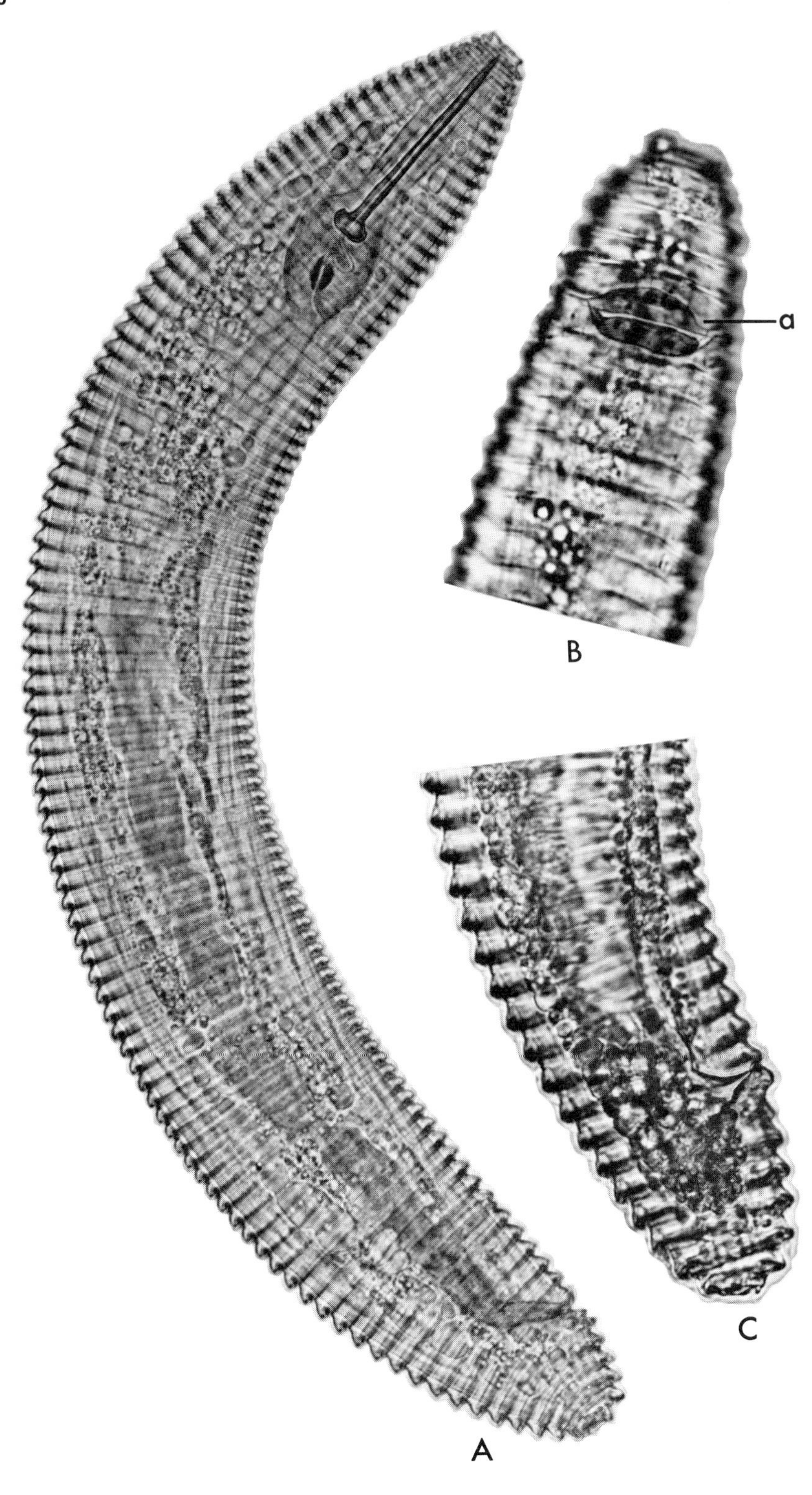
a
B
C
A

40. Stylet knobs rounded, sloping posteriorly; cuticle usually with more than 200 annules .... (Plate 47) *HEMICYCLIOPHORA*
Stylet knobs anchor-shaped, with anterior projections; cuticle usually with less than 200 annules .... (Plate 48) *HEMICRICONEMOIDES*

Hemicycliophora de Man 1921

Type species: *Hemicycliophora typica* de Man 1921

Description: Hemicycliophorinae. Basal knobs or spear spheroid, enclosing a cavity at the base. Cuticle double in adult female, the inner one apparently being a fifth produced after the fourth moult. Sheath attached at the head, vulva, and sometimes at the tail end. Annules on both cuticles plain, rounded, not retrorse. Lateral fields sometimes present on sheath and not completely interrupting the transverse striae. Longitudinal striae sometimes present on sheath. Head flat, not offset, with a few annules and a labial disc. Isthmus and terminal bulb forming a cylinder that is slightly swollen posteriorly. Tails elongate, pointed or rounded, of varying lengths. Males rare, without spear, without sheath, cuticular annulation finer. Esophagus degenerate, but nerve ring, excretory pore, and hemizonid visible. Spicules usually sickle-shaped. Bursa and gubernaculum present. Spicules protrude through a prominent cloacal sheath. Development via juveniles possessing a sheath. (From Siddiqi and Goodey 1963.)

General Characteristics

Hemicycliophora spp. are ectoparasites widely distributed geographically. These nematodes are found in a wide range of habitats, including moist locations (especially in forests and along stream banks), sandy soils of cultivated fields, and greenhouse soils.

Members of species of this genus are associated with a number of crops including celery, sugar beets, alfalfa, peach, rose, rice, and corn. The feeding of *H. arenaria* causes an enlargement of the lateral and terminal root tips and a reduction in the growth of rough lemon (*Citrus limonia* Osbeek) (Van Gundy 1958).

The double cuticle and the long stylet are distinguishing characteristics of larvae and females of this genus. The lack of a stylet and the indistinct esophagus aid in identifying the males.

Van Gundy (1959) made an excellent study of the life history of *H. arenaria*. Wolff Schoemaker (1968) presented a key to 45 species of this genus.

Plate 47. Hemicycliophora sp. A. Posterior portion of female, 841 X. B. Anterior portion of female, 841 X. C. Mature female, 215 X. D. Larva, 215 X.

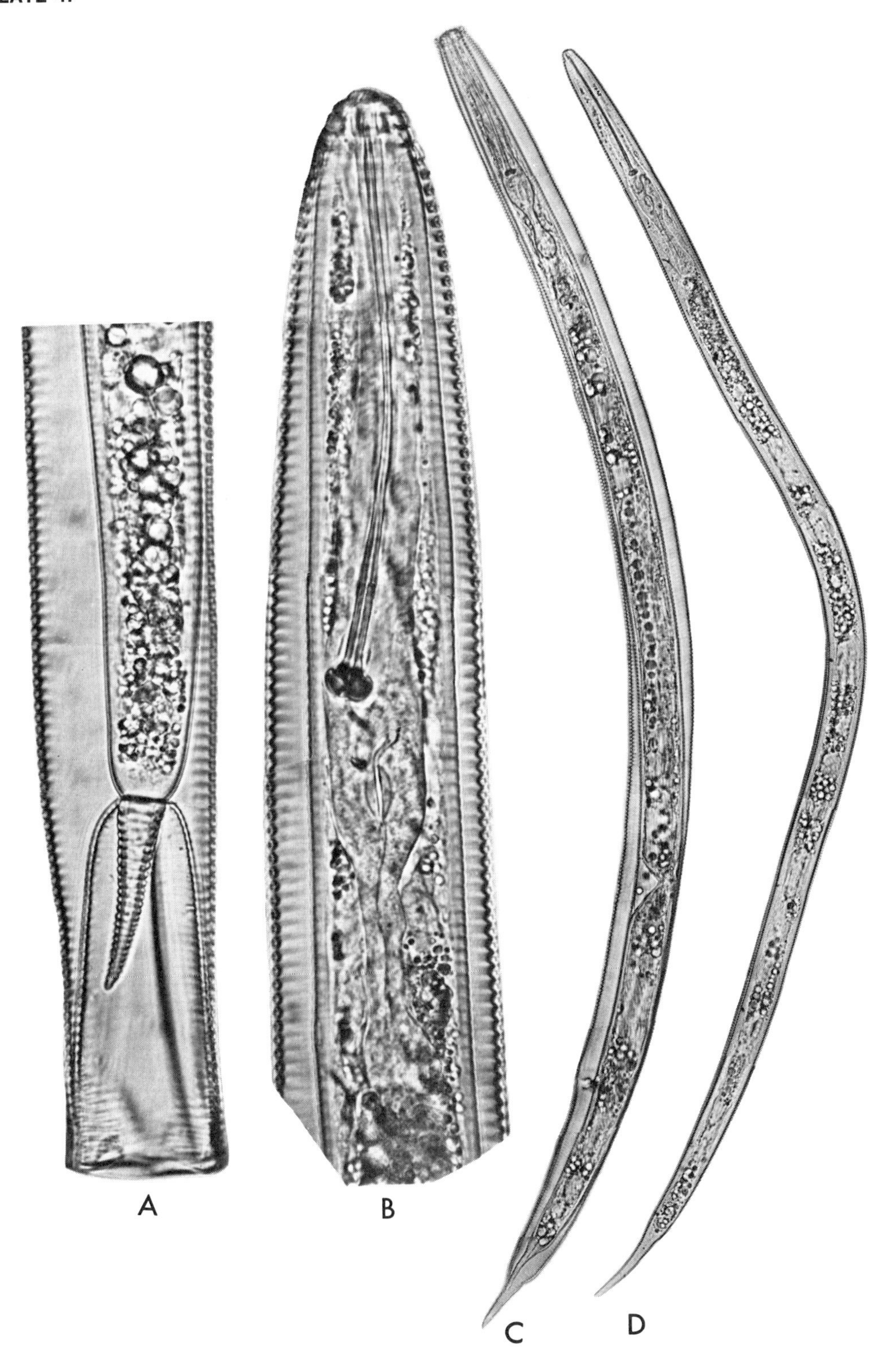
A
B
C
D

Hemicriconemoides Chitwood and Birchfield 1957

Type species: *Hemicriconemoides wessoni* Chitwood and Birchfield 1957

Description: Criconematinae. Spear knobs anchor-shaped, with prominent forward-directed processes, rarely rounded but not sloping backward. Female body elongate-cylindrical, enclosed in a cuticular sheath which is attached to the body at the head, vulva, and sometimes at the tail. Annules of sheath and body similar, coarse, smoothly rounded, not retrorse, usually less than 150 in number. Head with 2 annules, variable in shape. Vulva sometimes with lateral cuticular membranes. Males with lateral fields usually with 4 incisures. Spicules elongate-slender, slightly ventrally arcuate. Gubernaculum simple, troughlike. Bursa, when present, narrow with crenate margins almost enveloping the tail. Development through larvae that have longitudinal rows of spines or scales on the body cuticle and lack a sheath. Larval spear well developed with anchorlike base. (Mostly from Siddiqi and Goodey 1963 and Dasgupta, Raski, and Van Gundy 1969.)

General Characteristics

According to Chitwood and Birchfield (1957): "The genus *Hemicycliophora* is presently conceived as a genus in which the females have a sheath with more than 200 annules; the cuticle is rather simply marked; the males lack a sheath and a stylet, have very hooked spicules and prominent caudal alae. *Criconemoides*, on the contrary, is conceived as without a sheath in the female, less than 200 (usually 60–120 annules) and while the males lack a stylet, the spicules are nearly straight and lateral alae project only slightly as caudal alae. Our forms (placed in *Hemicriconemoides*) are distinctly intermediate, since the female has a sheath and the number of annules is from about 70–200. The males lack a stylet and also lack any distinct indication of caudal alae, since the lateral alae extend the length of the body and the spicules are arcuate or nearly straight but not hooked."

Goodey (1963) synonymized *Hemicriconemoides* with *Hemicycliophora* due to lack of evidence for separating the females of the 2 groups, and because male characters of 2 *Hemicycliophora* species resemble those of *Hemicriconemoides*. Siddiqi and Goodey (1963), after a more detailed study of the Criconematidae, reinstated *Hemicriconemoides* as a valid genus and erected *Caloosia* to include those species of *Hemicycliophora* in which adult females lack the sheath commonly found in *Hemicycliophora* and males possess slightly curved spicules.

Plate 48. *Hemicriconemoides* sp. A. Mature female, 360 X. B. Eggs, 805 X. C. Anterior portion of female, 570 X. D. Female tail, 570 X.

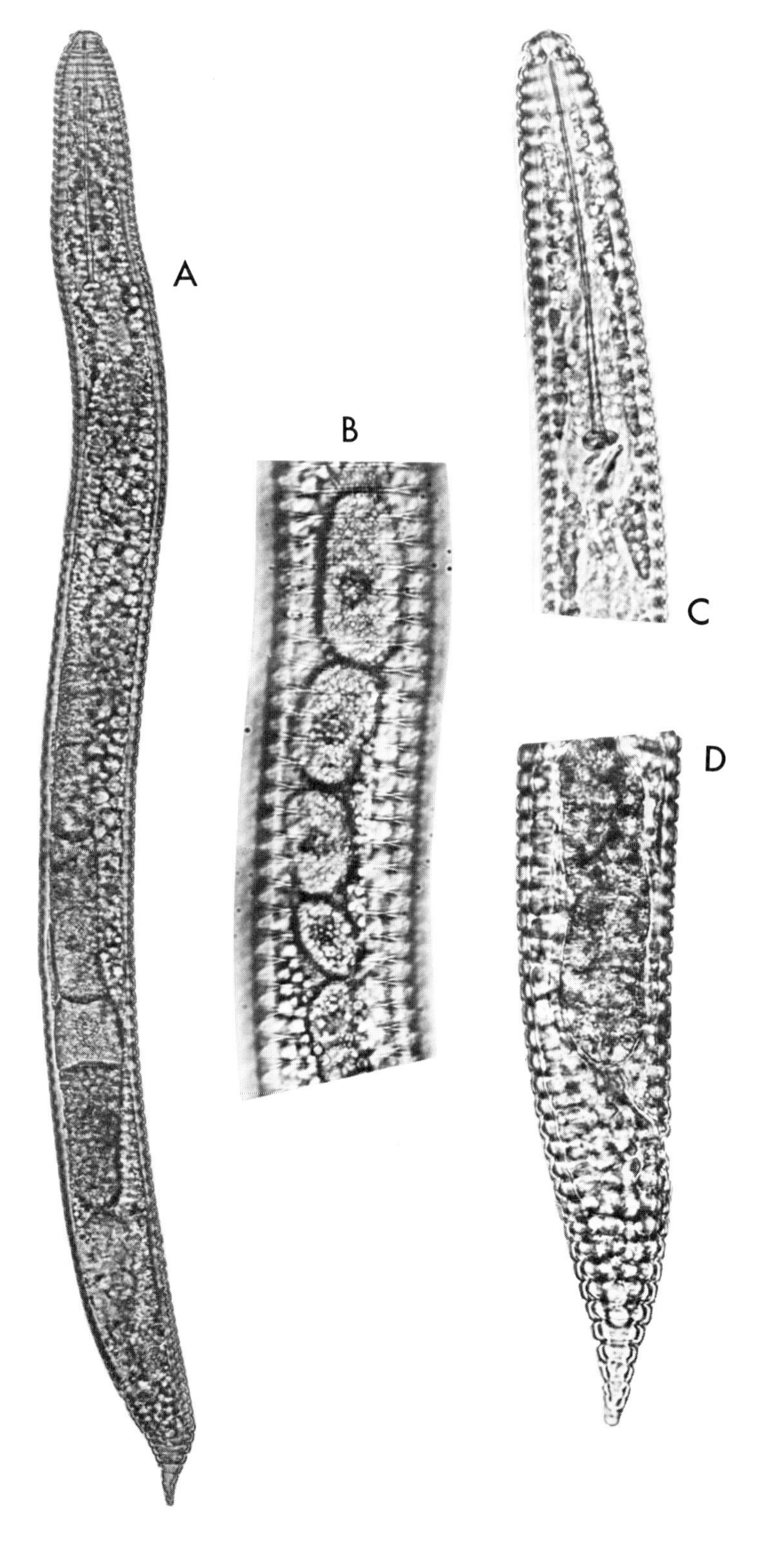
A
B
C
D

41. Annules of female without membranous structures on posterior margins .... (Example: Plate 50) 42
Annules of female with membranous structures on posterior margins ..... (Plate 49) *BAKERNEMA*

Bakernema Wu 1964

Type species: *Bakernema bakeri* (Wu, 1964) Wu 1964

Description: Criconematinae. Body stout, fusiform. Cuticle thick with large transverse annules having membranous structures on the posterior margins. Lip region not set off. Labial framework heavy. Spear elongate, with large basal cupped knobs. Esophagus typical of the subfamily. Vulva in posterior region of the body. Ovary single. Male more slender than female. Spicules paired, bursa and gubernaculum present. (From Wu 1964.)

General Characteristics

As stated by Wu (1964), *Bakernema* possesses characters that fall between *Criconema* and *Criconemoides*. The presence of membranous structures on the posterior edge of the transverse cuticular annules separates *Bakernema* from the latter genera.

The specimens were collected from soil under a sugar maple tree (*Acer saccharum* Marsh) in Ontario, Canada. Information on their pathogenicity is not included.

Plate 49. *Bakernema bakeri*. (Female.) 2. Anterior end. 3 and 4. Lip regions of same specimen, showing indentation on margin of first annule, and incomplete annule. 5. Posterior end, alimentary system not shown. 6. Posterior end, ventral view. 7. Annules 20 to 22, showing excretory pore. 8. Membranous outgrowths, lateral view of body. a. annules 5–9. b. annules 14–15. c. annule 17. (Male.) 9. Anterior end. 10. Posterior half. 11. a and b. Showing some variations of tail. 12. Anal region, showing anal sheath. (After Wu 1964. Courtesy of Canadian Journal of Zoology.)

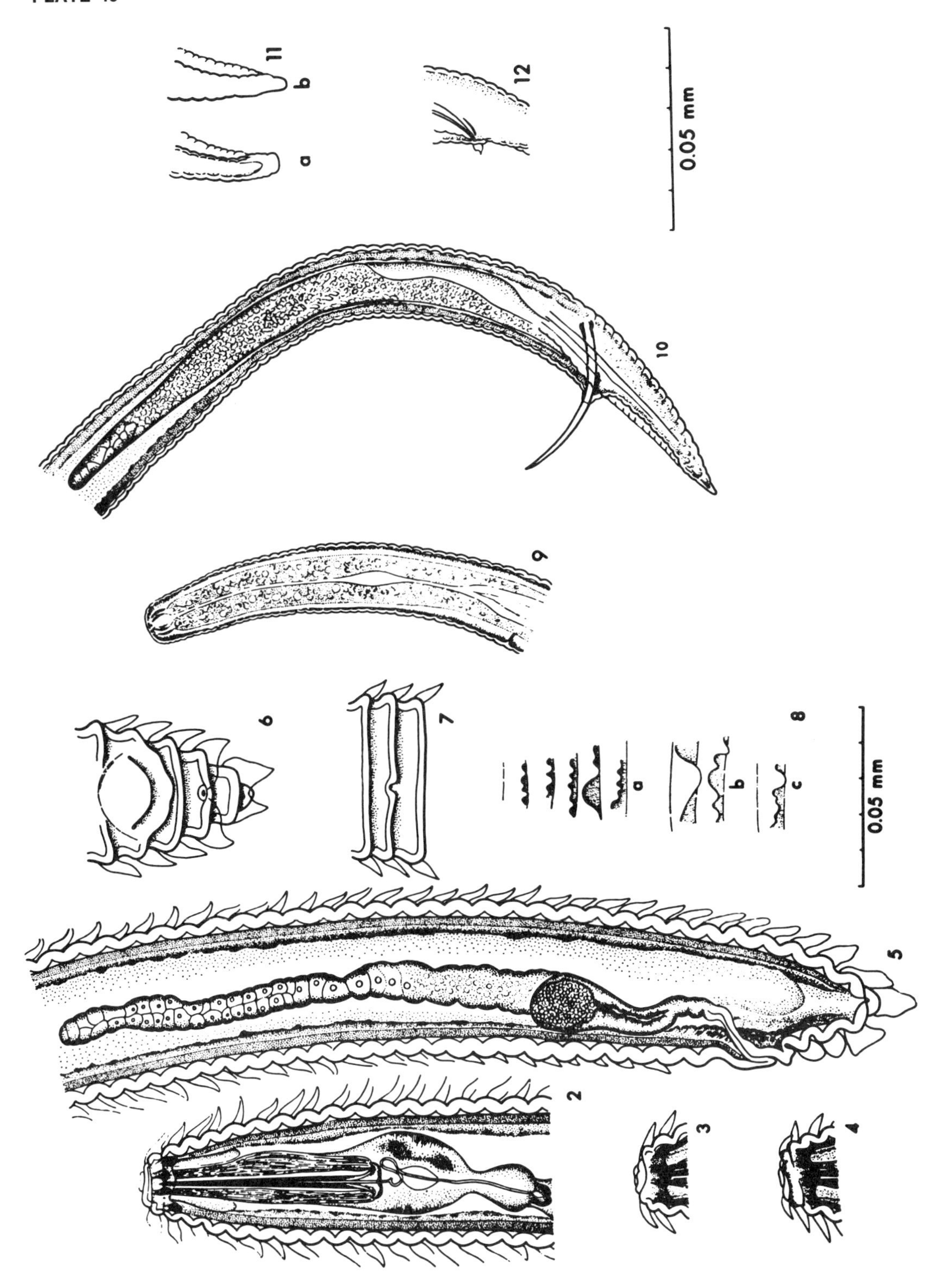
11
a
b
12
0.05 mm
10
9
6
7
8
a
b
c
0.05 mm
5
2
3
4

42. Cuticle of female ornamented with minute tubercules .... (Plate 50) *CACOPAURUS*
    Cuticle of female not ornamented with minute tubercules .... (Example: Plate 51) 43

Cacopaurus Thorne 1943

Type species: *Cacopaurus pestis* Thorne 1943

Description: Paratylenchinae. Female body obese, immobile, generally distorted in form, attached to host by unusually long spear. Female annules ornamented by minute tubercules. Vulva near terminus. Ovary single. Phasmids a scutellumlike area. At senility the female forms a brownish-yellow empty case which remains attached to the root. Excretory pore far forward, opposite median bulb or even to base of spear. Males slender, active, losing spear at last moult. Both males and females about equal in length to fourth-stage larvae, from which they developed. Male cuticle with plain striae. Cloacal region a shallow depression, with outer margins sometimes appearing as a narrow bursa when seen from a subdorsal view. Testis one. (From Thorne 1961.)

General Characteristics

Thorne (1943) erected the genus *Cacopaurus* to include the single species *C. pestis*, and Allen and Jensen (1950) described a second species *C. epacris*. Raski (1962) erected a new genus *Gracilacus*, transferred the species *C. epacris* to *Gracilacus* and designated it as the type species of the new genus. Siddiqi and Goodey (1963) made *Gracilacus* a synonym of *Paratylenchus* and although not mentioned specifically in their paper, apparently transferred *C. epacris* to *Paratylenchus*. They also reinstated *Cacopaurus* as a valid genus, apparently with a single species *C. pestis*. Jenkins and Taylor (1967) agreed that *C. epacris* belongs in the genus *Paratylenchus*.

The chief difference between *Paratylenchus* and *Cacopaurus* is the presence of refractive elements on the cuticle of *Cacopaurus*.

When examining roots, the small attached females of *C. pestis* are easily overlooked unless special techniques are used. Thorne (1943) recommends carefully washing the roots free of soil, then rubbing them briskly between the fingers in a small amount of water, and examining the residues under high power of the steroscopic microscope. If this species is present, the brown shrunken female cases will be found with a few larvae, males, and developing females.

Plate 50. Cacopaurus pestis. A. Anterior portion of young larva, 1475 X. B. Posterior portion of male, slightly subdorsal view with bursa in profile, 1475 X. C. Anterior portion of male, 1475 X. D. Spicula and gubernaculum, ventral view, 1970 X. E. Female head *en face*, 1970 X. F. Cuticular pattern on posterior portion of female, 1475 X. G. Anterior portion of testis; 1970 X. H. Developing female larva, 442 X. I. Female just after last moult, 442 X. J. Adult female showing structure of oviduct and spermatheaca, 442 X. K. Adult gravid female, 442 X. (After Thorne 1943. Courtesy of Helminthological Society of Washington.)

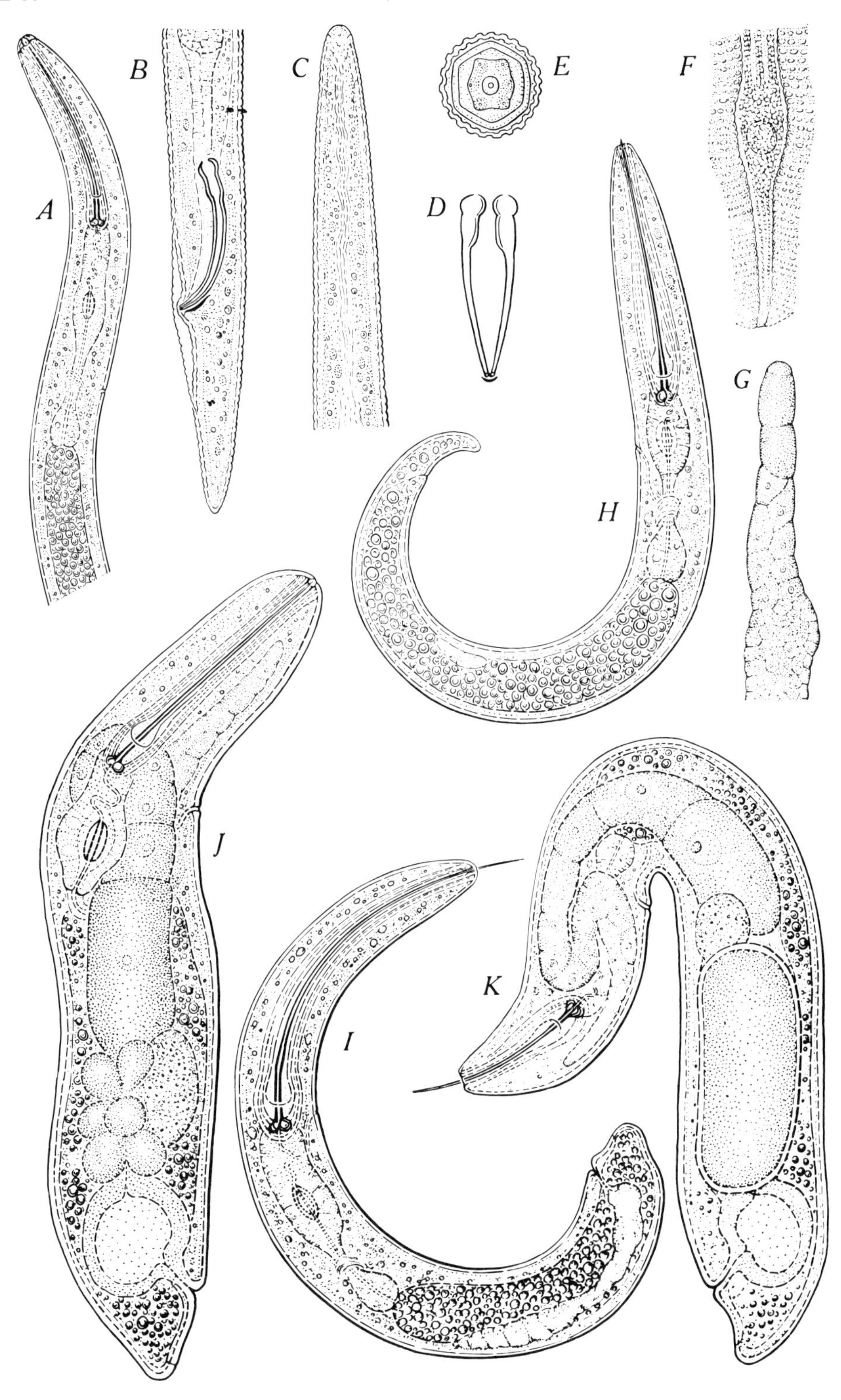
A
B
C
D
E
F
G
H
I
J
K

43. Female stylet 36μ or less . . . . (Plate 51) *PARATYLENCHUS*
    Female stylet 45 to 120μ . . . . (Plate 52) *GRACILACUS*

Paratylenchus (Micoletzky 1922)

Type species: *Paratylenchus bukowinensis* (Micoletzky 1922)

Description: Paratylenchinae. Basal knobs of spear spheroid. Body less than 0.5 mm, elongate-cylindrical, C-shaped at death. Cuticle with fine annulations and lateral fields usually with 4 incisures. Head conical, rounded or truncate, usually without annules. Isthmus slender, leading to a rounded or spathulate bulb. Vulva usually with lateral cuticular membranes. Male with reduced or no spear; esophagus ill defined or nonexistent. Tail usually with a short protruding sheath. Spicules slender and pointed; gubernaculum short. Development through juveniles resembling the female in general body shape. (Mostly from Siddiqi and Goodey 1963.)

General Characteristics

Raski (1962) erected the family Paratylenchidae to include the genera *Paratylenchus, Cacopaurus*, and the new genus *Gracilacus. C. epacris* was transferred from *Cacopaurus* to *Gracilacus* and designated as the type species for *Gracilacus*. Siddiqi and Goodey (1963) made *Gracilacus* a synonym of *Paratylenchus* and, although not mentioned specifically in their paper, they apparently transferred *C. epacris* to *Paratylenchus*. Allen and Sher (1967) did not recognize either the family Paratylenchidae or the genus *Gracilacus*. Jenkins and Taylor (1967) agreed that *Gracilacus* is a synonym of *Paratylenchus* and that *C. epacris* belongs in the genus *Paratylenchus*. Thorne and Malek (1968) re-erected *Gracilacus*.

Large populations of members of *Paratylenchus* are found around plant roots in various soil types widely distributed geographically.

Several species have been shown to be ectoparasites of plant roots. Generally, infection results in an overall reduction in top and root growth without obvious symptoms on the roots.

The younger larval stages of most species possess poorly developed stylets (Raski 1962). According to Rhoades and Linford (1961), individuals of *P. projectus* in the first and preadult stages possess very short delicate stylets and poorly defined esophagi, whereas individuals in the second and third larval stages have well-developed stylets and esophagi. The same authors found that the preadults of *P. projectus* are more tolerant of dessication and sudden exposures to low temperatures than individuals of other stages.

Plate 51. Paratylenchus sp. A and C. Mature female, 710 X. B. Male tail, 1500 X. D. Mature male, 620 X.

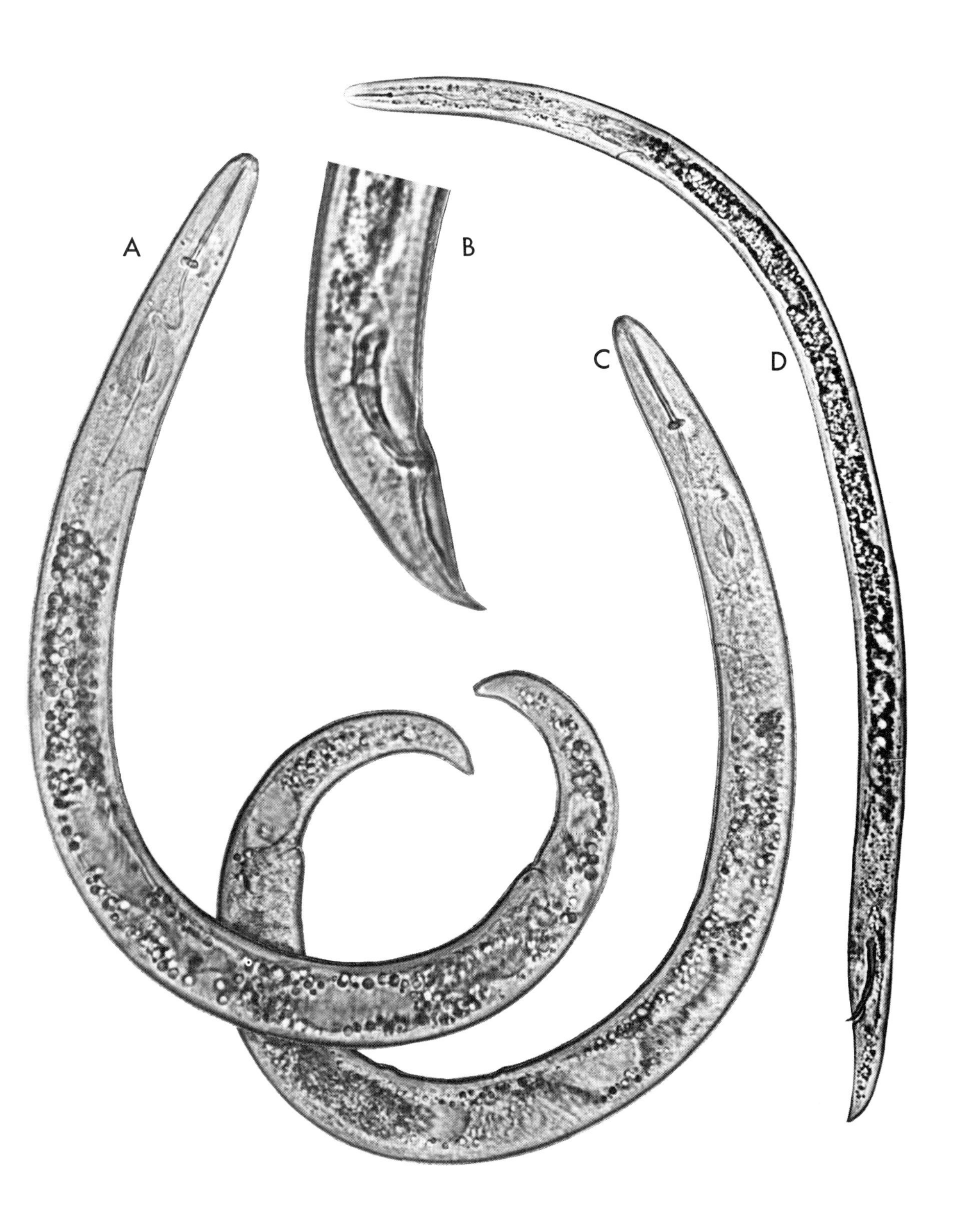
A
B
C
D

Gracilacus Raski 1962

Type species: *Gracilacus epacris* (Allen and Jensen 1950) Raski 1962

Description: Paratylenchidae. Small species less than 0.50 mm. Most larvae with elongate stylet. Female slender to obese with a stylet 48–119μ in length. Body posterior to vulva elongate. Cuticle finely annulated, without ornamentation. Excretory pore generally in region of metacorpus near the valve or further anterior but may be near nerve ring. Male slender, active, stylet absent or much reduced. Caudal alae represented by thickened, cuticular evaginations. Ovary single. Testis one. (From Raski 1962.)

General Characteristics

Raski (1962) erected the genus *Gracilacus* to include several new species, 9 species transferred from the genus *Paratylenchus*, and 1 species transferred from the genus *Cacopaurus*. In addition, he erected the family Paratylenchidae to include the 3 genera *Paratylenchus*, *Gracilacus*, and *Cacopaurus*. Formerly the genera *Paratylenchus* and *Cacopaurus* were classified in the Paratylenchinae, a subfamily of the Criconematidae.

In addition to a description of the new family Paratylenchidae and an emended description of the family Criconematidae, this paper includes keys to the genera of the Paratylenchidae and to species of the genus *Gracilacus*.

Siddiqi and Goodey (1963) did not recognize Paratylenchidae as a family, and *Gracilacus* was reduced to synonymy with *Paratylenchus*. Thorne and Malek (1968) restored *Gracilacus* to generic rank. Golden (1971) considered it a valid genus and indicated that two major differences between *Gracilacus* and *Paratylenchus* are stylet length and position of the excretory pore. In *Gracilacus* the female stylet length is 45 to 120μ, and the excretory pore is typically anterior to the nerve ring, often near the valves of the median bulb; in *Paratylenchus* the female stylet is 36μ or less, and the excretory pore is always near the nerve ring or posterior to it.

Plate 52. Gracilacus audriellus. A. Female, 925 X. B and C. Male tails, 925 X. D. Anterior portion of male, 925 X.

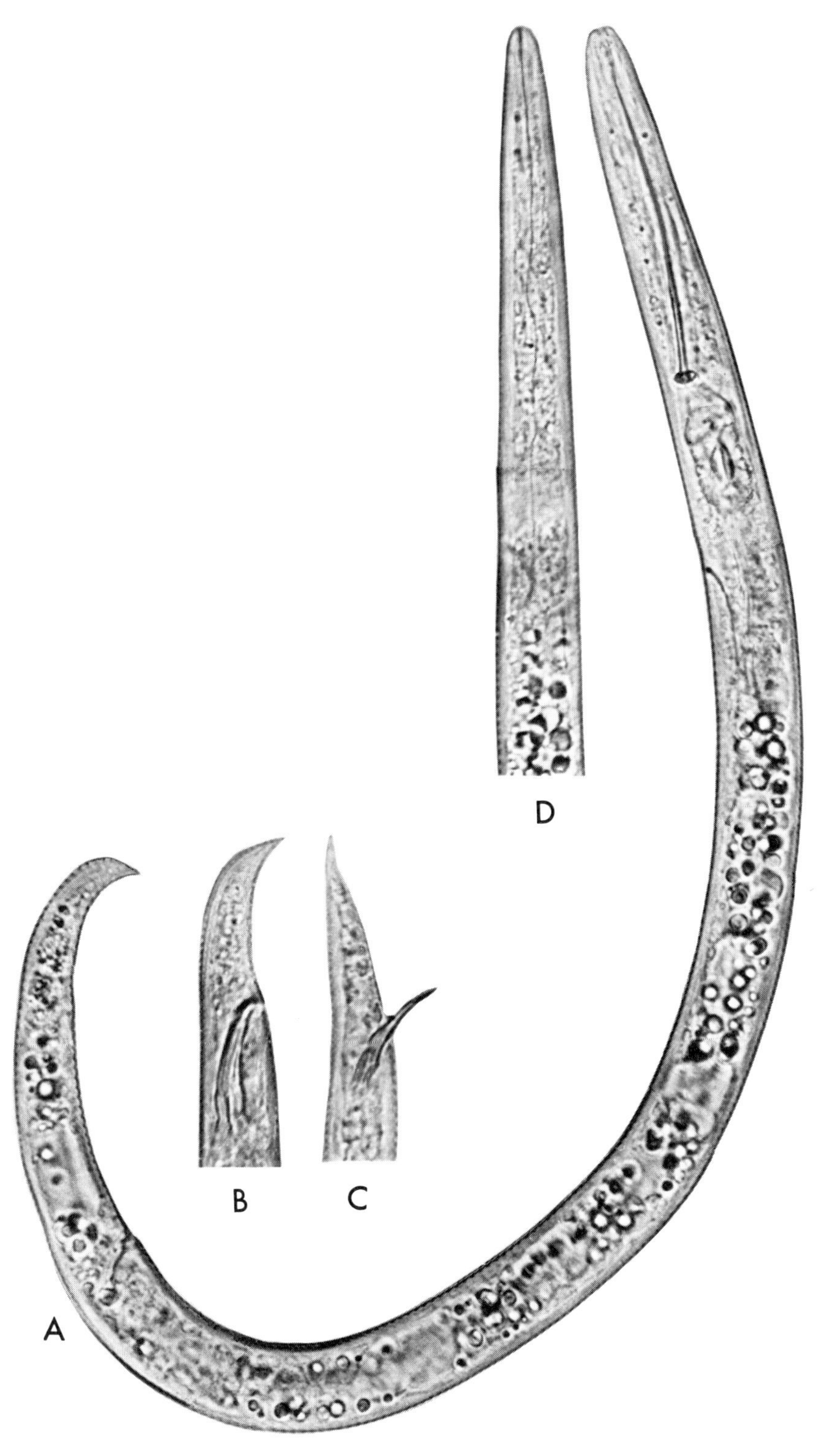
D
B
C
A

44. *s* (stylet length/body diameter at base of stylet) = 2.5 or more . . . . (Example: Plate 53C) 45
s = generally less than 2.5 . . . . (Example: Plate 53A, B, and D) 49
45. Esophageal glands usually unequal in length, overlapping intestine . . . . (Example: Plate 53A, B, and C) 46
Esophageal glands enclosed within a bulb, usually not overlapping intestine . . . . (Example: Plate 53D) 47

Plate 53. Stomatostylets less than 2.5 times width of body at stylet base: A. *Hoplolaimus* sp., 905 X. B. *Meloidodera* sp., 826 X. D. *Tylenchorhynchus* sp., 950 X. Stomatostylet more than 2.5 times width of body at stylet base: C. *Belonolaimus* sp., 650 X.

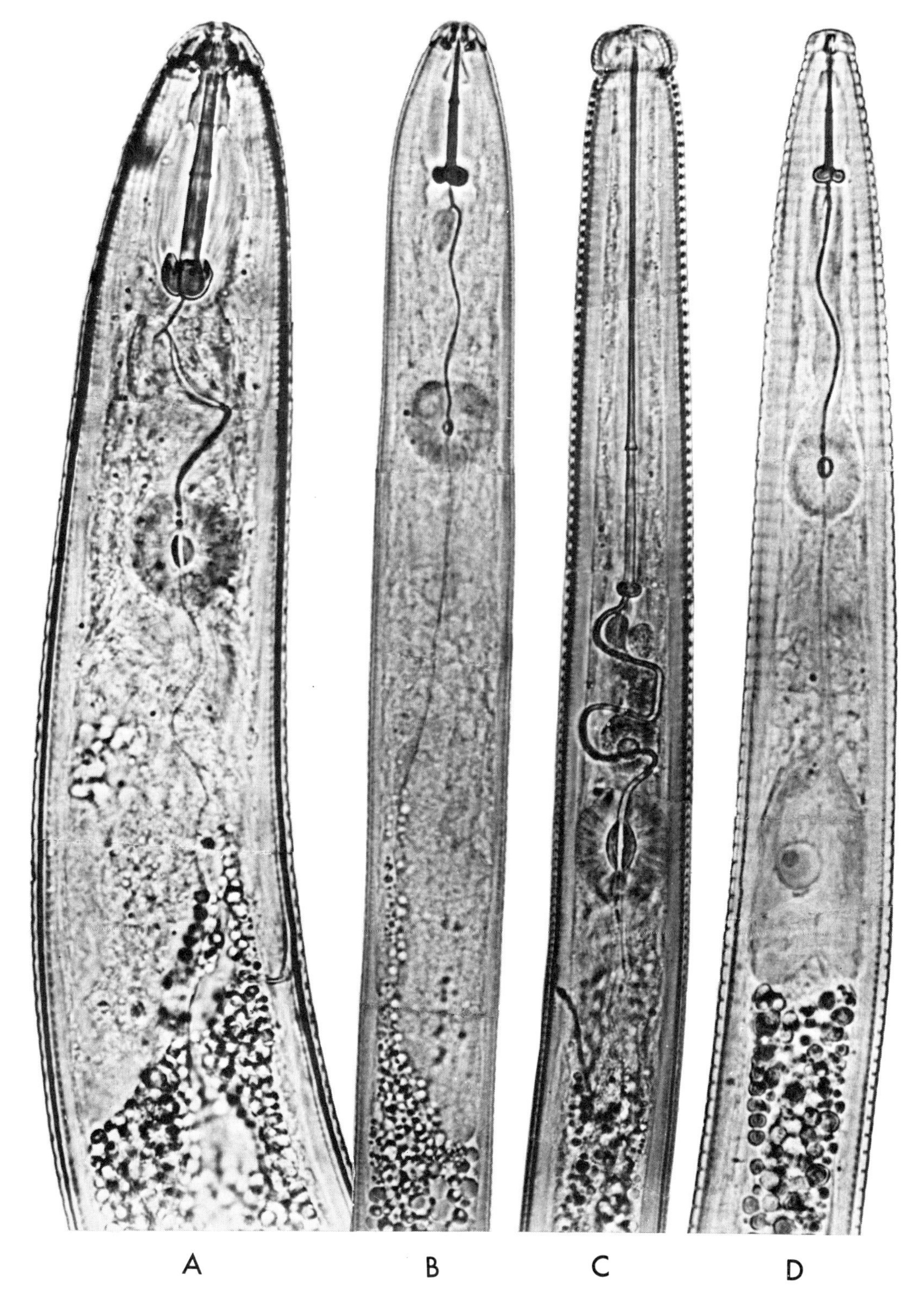
A
B
C
D

46. Average body length usually 1.75 mm or more . . . . (Plate 54) *BELONOLAIMUS*
    Average body length usually less than 1.75 mm 48

Belonolaimus Steiner 1949

Type species: *Belonolaimus gracilis* Steiner 1949

Description: Belonolaiminae. Members of this genus have slender cylindrical bodies with lengths a little more than 2 mm for the females and a little less than 2 mm for the males. The distinctly rounded lip region is divided by lateral, dorsal, and ventral grooves into 4 distinct lobes, each of which bears 6 or more striations. Viewed *en face*, 6 distinct lips can be distinguished, with amphidial openings near the outer edges of the lateral ones. The body is strongly annulated, the annules extending around the terminus of the female tail. The lateral fields are marked in both sexes by a single line. The stylets of both males and females are long and slim with distinct knobs. The esophageal glands form a lobe overlapping the anterior end of the intestine. There are 2 outstretched ovaries with distinct spermathecae. The testis also is outstretched. The spicules are curved; there is a well-developed gubernaculum, and the bursa envelops the tail end. (Mostly from Rau 1958.)

General Characteristics

Belonolaimus spp., the sting nematodes, were reported causing injury to the roots of slash and longleaf pine seedlings in nurseries in Florida and Georgia (Steiner 1949). *Belonolaimus gracilis* is pathogenic to sweet corn, celery, and strawberry (Christie 1959). Among the areas where the sting nematodes are associated with poor plant growth are Virginia (Owens 1951), South Carolina (Graham and Holdeman 1953), and New Jersey (Hutchinson and Reed 1956). A host list is given by Christie (1959).

The field symptoms are growth retardation and generalized root decay. These nematodes do not cause galls, but on some plants, like corn, the main roots may end in enlargements caused by the repeated forming and killing of new branches (Christie 1959).

Two species, differing from all others by having 4 incisures in the lateral field, have been described by Colbran (1960) and Roman (1964).

Plate 54. Belonolaimus sp. A. Mature female, 115 X. B. Male tail, lateral view, 414 X. C. Male tail, ventral view, 414 X. D. Portion of esophagus, 428 X. E. Anterior portion of female, 215 X. a. lumen of esophagus.

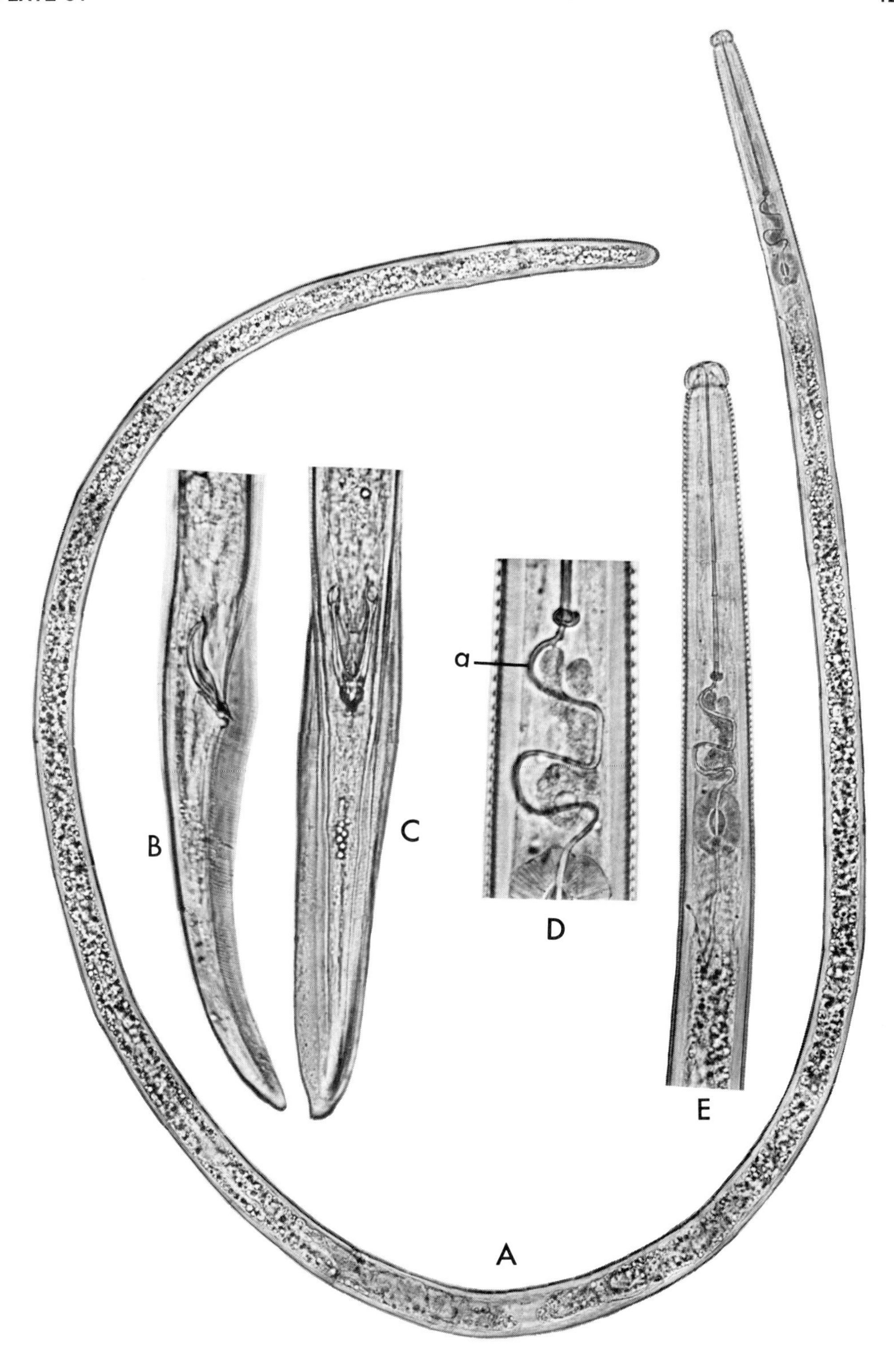
a
B
C
D
E
A

47. Lip region continuous . . . . (Plate 55) *MACROTROPHURUS*
Lip region set off by distinct constriction . . . . (Plate 56) *DOLICHODORUS*

Macrotrophurus Loof 1958

Type species: *Macrotrophurus arbusticola* Loof 1958

Description: Tylenchidae. Nematodes of moderate size (1–2 mm); body in both sexes rather slender, attenuated in front, parallel behind. Lip region smooth, rounded, continuous with body contour. Stylet very long. Opening of dorsal esophageal gland a short distance behind spear base. Middle bulb large, well set off from enlarged precorpus, with conspicuous valves. Isthmus rather long and narrow. Terminal bulb elongate. Tail rounded with very thick cuticle. Ovaries paired, opposed and outstretched. Vulva near middle of body. Male with well-developed bursa. (From Loof 1958.)

General Characteristics

Loof (1958) comments concerning the systematic position of the genus: "The structure of the esophagus shows that it belongs to the subfamily Tylenchinae. From the genera of this group, *Trophurus* and *Tylenchorhynchus* seem to be its nearest relatives. From the former genus, *Macrotrophurus* differs by the elongate stylet and the paired ovaries. Now, *Trophurus,* although monodelphic, has the vulva placed equatorially and therefore stands closely to didelphic forms. As moreover the peculiar shape of head and tail ends does not give the impression of being of strongly adaptive character, *Trophurus* may be regarded as the closest relative of *Macrotrophurus*. This view is expressed in the generic name."

Members of this genus are found around roots of pear and poplar in The Netherlands and from tree nurseries near Zurich, Switzerland.

Plate 55. Macrotrophurus arbusticola. A. Head and neck of female. B. Anterior end of female, medial view, with amphid apertures. C. Female tail, lateral view. D. Male tail, sublateral view. (After Loof 1958. Courtesy of Nematologica.)

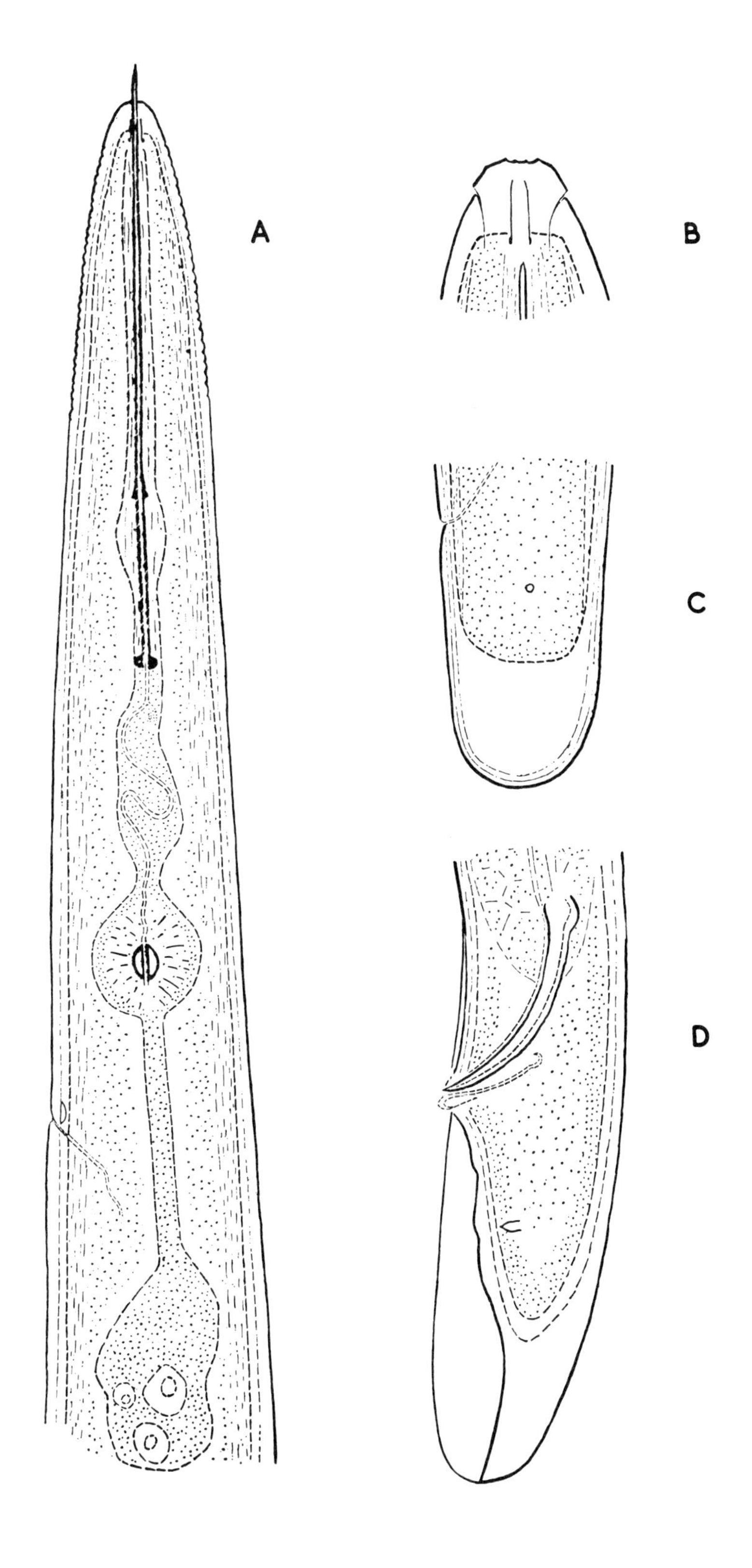
A
B
C
D

Dolichodorus Cobb 1914

Type species: *Dolichodorus heterocephalus* Cobb 1914

Description: Dolichodorinae, with a conspicuous striated lip region set off from the body by constriction or depression. Lip cap present. Supporting sclerotized framework of the lip region massive. Amphid openings at the lateral margins of the lip cap. Cuticle coarsely striated. Lateral incisures 3 or 4 at the middle of the body. Females with 2 ovaries. Caudal alae terminal, striated, and conspicuously lobed. Spicules 2, gubernaculum linear. Stylet in both sexes elongate and bearing conspicuous basal knobs. Corpus of esophagus thick, postcorpus elongate-oval with a strongly developed valvular apparatus. Isthmus of esophagus a slender tube terminating in an elongate posterior basal bulb containing 3 esophageal gland nuclei. (Mostly from Allen 1957.)

General Characteristics

Dolichodorus spp., awl nematodes, are long slender forms very much like the sting nematodes in size, shape, general appearance, and feeding habits (Christie 1959). Hirschmann (1960) describes the differences between the two genera as follows: "*Dolichodorus* has 3 or 4 incisures in the lateral field while *Belonolaimus* possesses only 1 incisure. The esophagus of *Dolichodorus* terminates in an elongate end bulb containing the 3 esophageal glands, whereas in *Belonolaimus*, the end portion of the esophagus consists of 3 much-enlarged gland cells overlapping the anterior end of the intestine. The male of *Dolichodorus* possesses caudal alae which are terminal, striated, and characteristically lobed, while the caudal alae of *Belonolaimus* envelop rather smoothly the tail terminus."

D. heterocephalus causes severe root injury to and correspondingly severe growth reduction of celery, corn, beans, and tomatoes, and moderate injury to pepper plants (Perry 1953). This nematode is primarily a root tip parasite, but does feed along the sides of roots and at the base of the hypocotyl. Injury symptoms, which resemble those of the stem nematodes, are stubby or coarse root, or both, often accompanied by considerable tissue discoloration and destruction (Christie 1959).

Members of this genus have been reported from the southeastern United States, Michigan, California, Akodessewa (Togo) in Africa, South Australia, New Zealand, India, and France.

Plate 56. Dolichodorus sp. A. Anterior portions, 445 X. B. Male tail, 315 X. C. Female, 173 X.

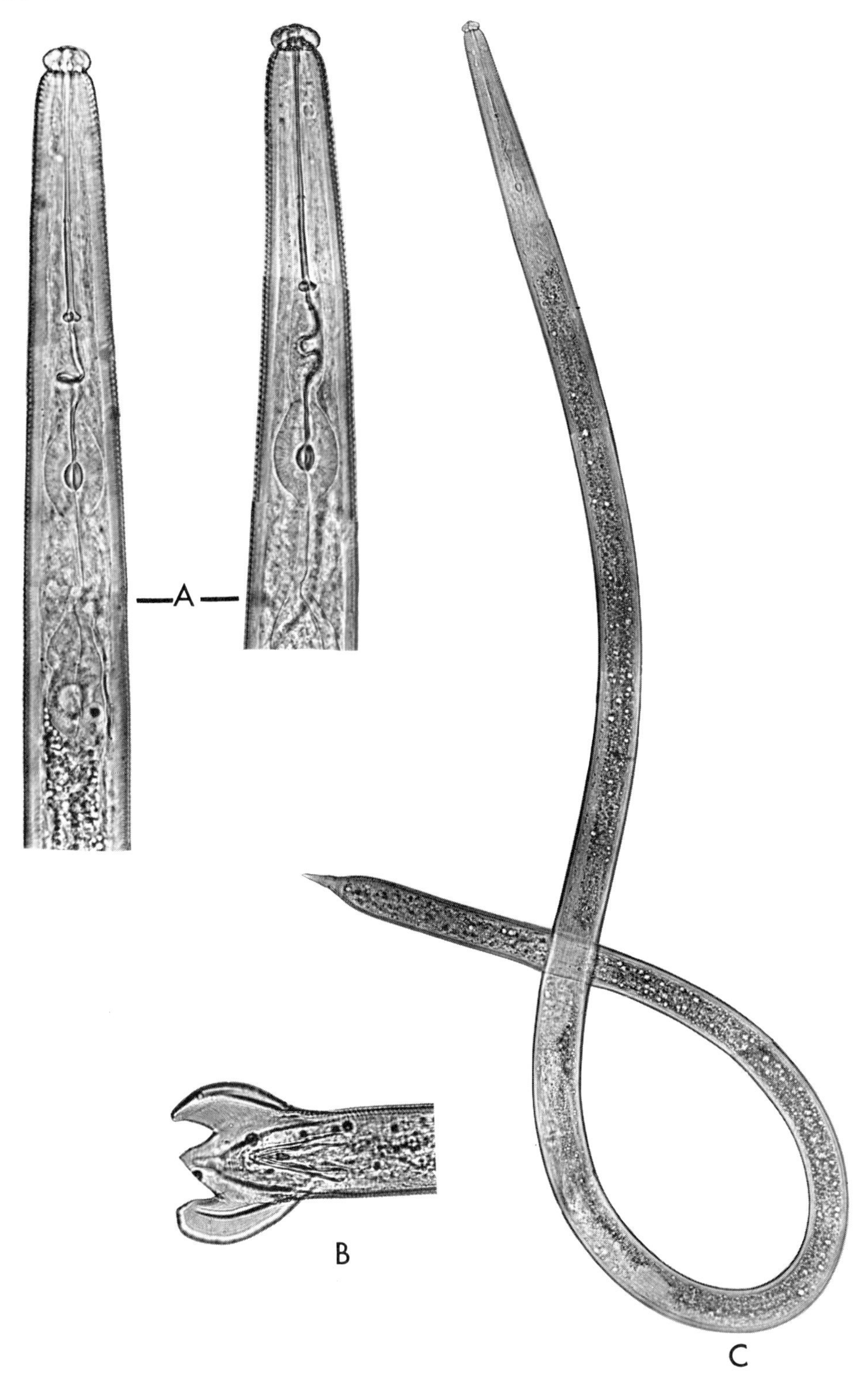
A
B
C

48. Lateral field with 4 incisures . . . . (Plate 57) *MORULAIMUS*
Lateral field with 2 incisures . . . . (Plate 58) *CARPHODORUS*

Morulaimus Sauer 1965

Type species: *Morulaimus arenicolus* Sauer 1965

Description: Belonolaiminae. Cylindroid nematodes of moderate size. Body slightly arcuate when relaxed by heat. Cuticle marked by distinct annulations which are wider in the neck and tail regions than in midbody. Lateral field with 4 lines, transverse striae crossing the outer lines at least in neck and tail regions. Lip region set off with 4 large submedian lobes bearing 6 or 7 annules, surmounted usually by a thin labial disc, which is more or less lemon-shaped due to lateral extensions. Small, apparently unstriated lateral lobes interrupt the first 3 or 4 head annules. Below the level of the lateral lobes the head annules may be continuous, or interrupted by striation or by an irregular series of platelets. Small amphid apertures are located on lips near edges of the labial disc. Labial framework usually lightly sclerotized. Stylet long to very long, slender, with well-developed basal knobs. Stylet guiding apparatus obscure. Dorsal gland orifice close to spear knobs. Median bulb of esophagus subspherical with strong valve apparatus. Isthmus of variable length, surrounded by nerve ring. Terminal lobes of esophagus well developed, overlapping the intestine dorsally and laterally, usually with only one gland nucleus conspicuous. Position of esophago-intestinal junction varies from close to base of isthmus to about halfway along the esophageal lobes. Excretory pore located in the region of esophageal lobes with a conspicuous hemizonid usually several body annules anterior to the pore. Tubules of the excretory system conspicuous throughout the pseudocoele.

Female: Vulva equatorial, transverse, $\frac{1}{3}$ to $\frac{1}{2}$ body width, usually with a well-developed double epitygma forming a distinct vulval flap. Ovaries paired, opposed, outstretched, each with a distinct subspherical spermatheca. Tail conoid or cylindrical, with blunt or nearly hemispherical terminus which is distinctly annulated. Tail length $1\frac{1}{2}$ to 3 times anal body diameter. Phasmids situated just anterior to midpoint of tail.

Male: Slightly smaller than female, but without marked sexual dimorphism. Testis single, outstretched. Spicules strong, arcuate with prominent overlapping ventral and dorsal flanges. Gubernaculum well developed, trough-shaped, cupping the spicules, and heavily sclerotized along the edges of the distal part, with a large distal knob and a smaller proximal knob which may be directed downwards to give a hooked appearance in lateral view. Caudal alae well developed, extending to tip of tail. (From Sauer 1965.)

General Characteristics

A genus, closely related to *Belonolaimus,* was erected by Sauer in 1965. Three new species were described: *M. arenicolus, M. geniculatus*, and *M. sclerus*. He proposed the new combinations *M. hastulatus* (Colbran 1960) Sauer 1965, and *M. whitei* (Fisher 1964) Sauer 1965, and included a key to the 5 species included in this genus. According to Sauer (1965) nematodes of this genus appear to occur widely in eastern Australia, and some specimens he collected appear to represent additional species, although new species cannot be defined at present due to the small number of specimens available.

Plate 57. Morulaimus arenicolus. A. Anterior end of female. B. Female head, lateral. C. Face view, female. D. Male tail, ventral. E. Region of vulva. F. Female tail. G. Male tail. H. End-on view of spicules showing overlapping flanges. (After Sauer 1966. Courtesy of Nematologica.)

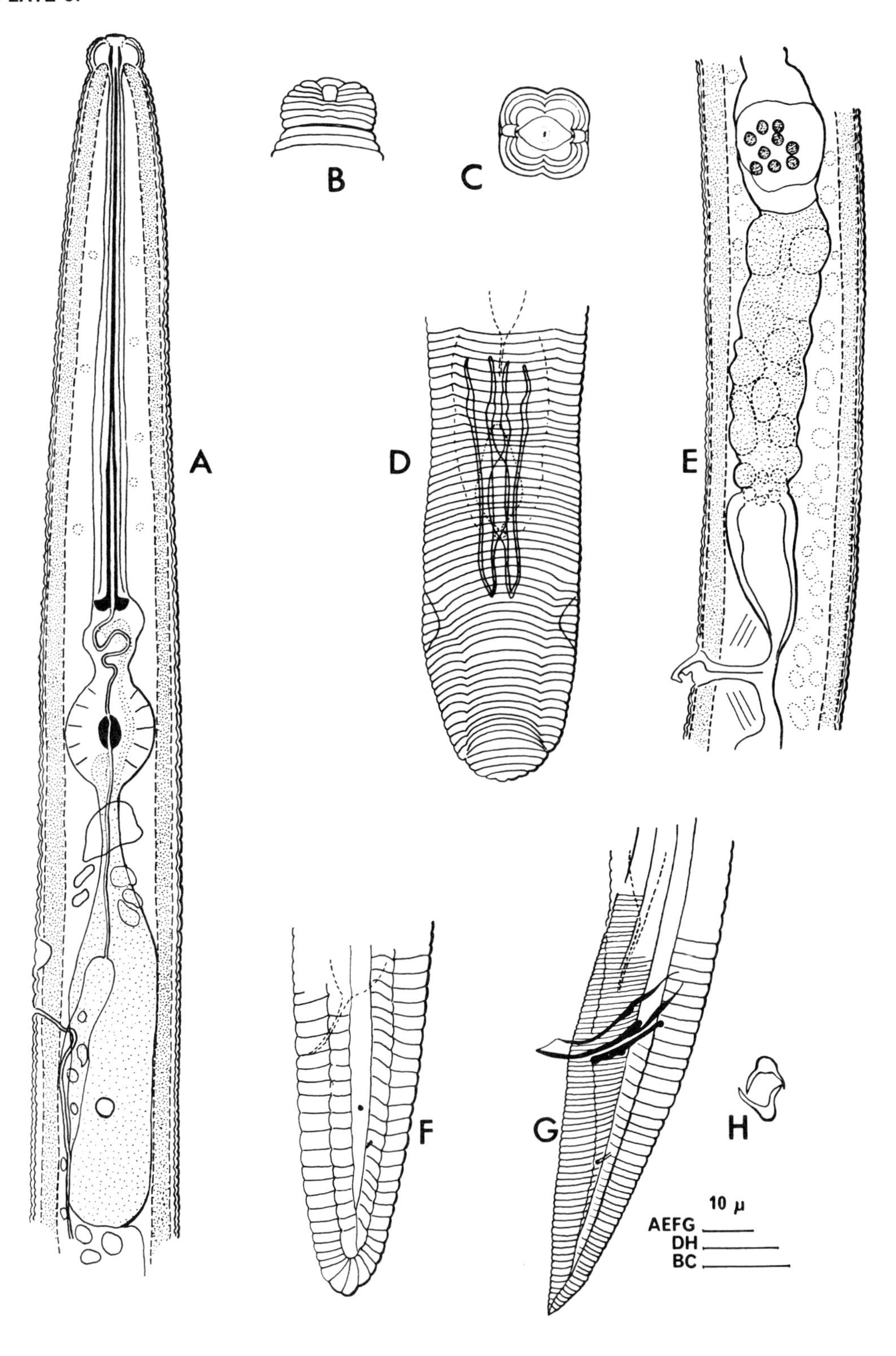
A
B
C
D
E
F
G
H
10 μ
AEFG
DH
BC

Carphodorus Colbran 1965

Type species: *Carphodorus bilineatus* Colbran 1965

Description: Dolichodorinae. Lateral fields with 2 incisures. Internal sclerotization well developed in basal portion of lip region. Esophagus with a short lobe overlapping intestine. Tail subcylindrical, 1.5–2.0 anal body widths in length; terminus broadly rounded. Caudal alae not lobed. Gubernaculum with titillae.

General Characteristics

Colbran found *C. bilineatus*, the type species, and the only species of this genus, in well-drained, sandy soils from eucalypt forests at Pozieres and Landsborough in southern Queensland. Populations were relatively low, and males were more common than females. He also found this nematode in soil around the roots of *Casuarina littoralis* (black sheoak) in a eucalypt forest at Jowarra Park, Landsborough, in Queensland. According to Colbran this species has many of the characters of *Dolichodorus* but possesses an esophagus with a short terminal lobe overlapping the intestine. It resembles *Dolichodorus* in the shape of the lip region, cephalic sclerotization, and position of the hemizonid behind the excretory pore; and the gubernaculum resembles that of *D. adelaidensis*.

Plate 58. Carphodorus bilineatus. A. Female. B. Female head. C. Cross section of lateral field. D. Female tail. E. Male tail. F. Female lateral field. G. Spicules and gubernaculum. H. *En face* view of female. (After Colbran 1965. Courtesy of Queensland Journal of Agricultural and Animal Sciences.)

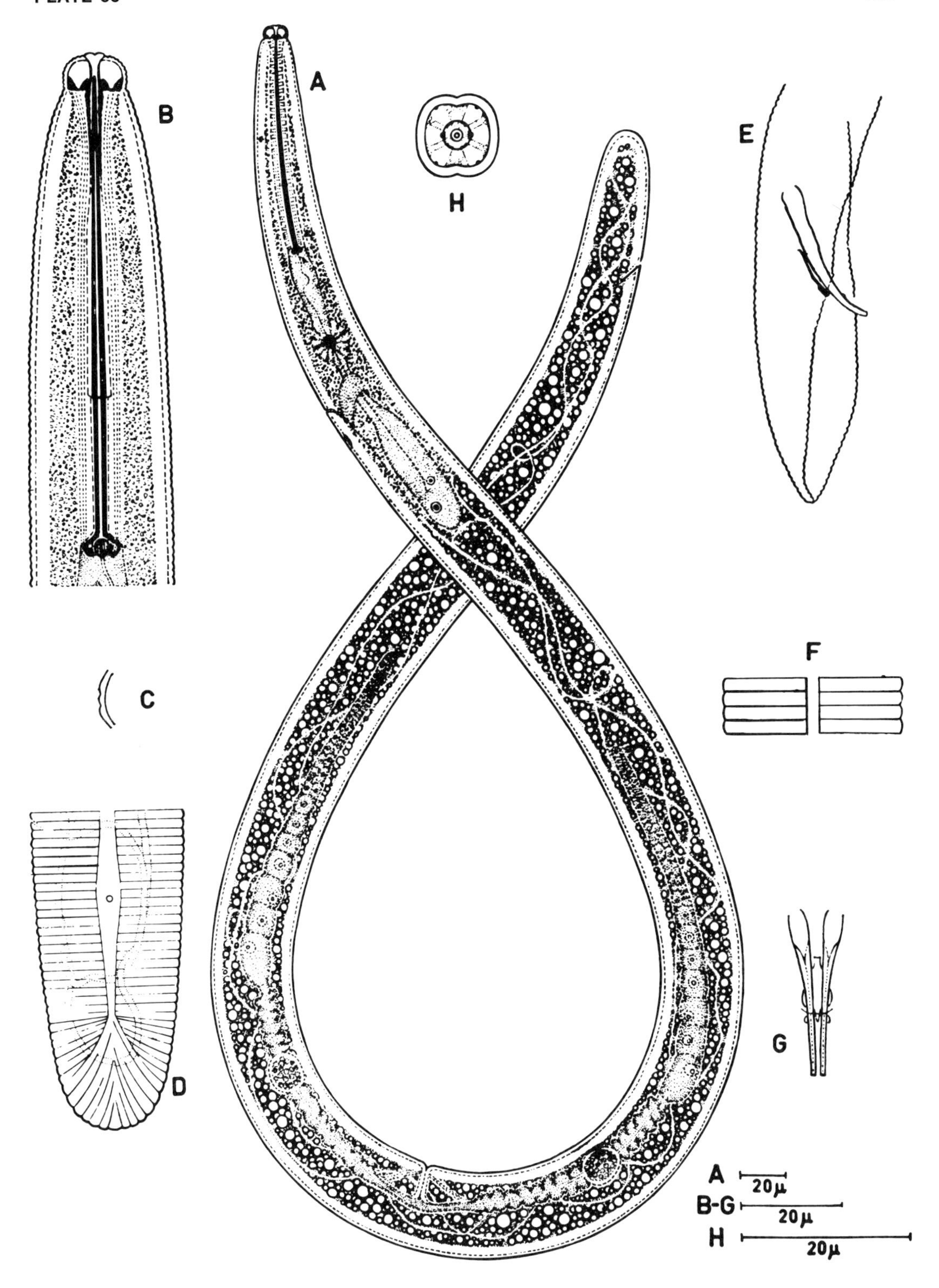
A
B
C
D
E
F
G
H
A 20μ
B-G 20μ
H 20μ

49. Phasmids absent .... (Plate 59) *APHASMATYLENCHUS*
Phasmids present .... (Examples: Plate 5 and Plate 73) 50

Aphasmatylenchus Sher 1965

Type species: *Aphasmatylenchus nigeriensis* Sher 1965

Description: Aphasmatylenchinae. Lip region annulated, set off from body, cephalic framework well developed. Face view exhibits 6 lips, elongated amphid apertures, and large labial disc. Spear well developed, slender. Esophageal gland narrow, overlaps intestine ventrally. Male spear and esophagus not as well developed as female. Four lateral incisures, incompletely areolated. Phasmids and deirids absent. Ovaries paired, outstretched. Female tail length about one and a half times body diameter at anus, terminus hemispherical. Male tail length more than twice body diameter at cloaca. Caudal alae envelops male tail. (Mostly from Sher 1965.)

General Characteristics

In erecting the genus *Aphasmatylenchus*, Sher (1965) placed it in the new subfamily Aphasmatylenchinae. Closely related to the Hoplolaiminae, it differs by the lack of phasmids, presence of elongated amphid apertures, and a narrow ventral esophageal gland.

In Nigeria, the specimens were first reported from cocoa (*Theobroma cacao*); additional specimens were later found associated with rubber, *Hevea braziliensis*. A new species, *A. straturatus*, associated with the roots of peanut, *Arachis hypogaea*, growing in West Africa was described by Germani (1970).

No information is available on pathogenic potential of the nematode.

Plate 59. Aphasmatylenchus nigeriensis. Left. (Female.) A. Entire body, lateral view. B. Tail. C. Face view. D. Anterior end. E. Surface view at vulva. F. Cross section through basal annule of lip region. *Right.* (Male.) A. Entire body, lateral view. B. Face view. C. Tail, lateral view. D. Anterior end. E. Cross section through basal annule of lip region. F. Tail, ventral view. (After Sher 1965. Courtesy of Helminthological Society of Washington.)

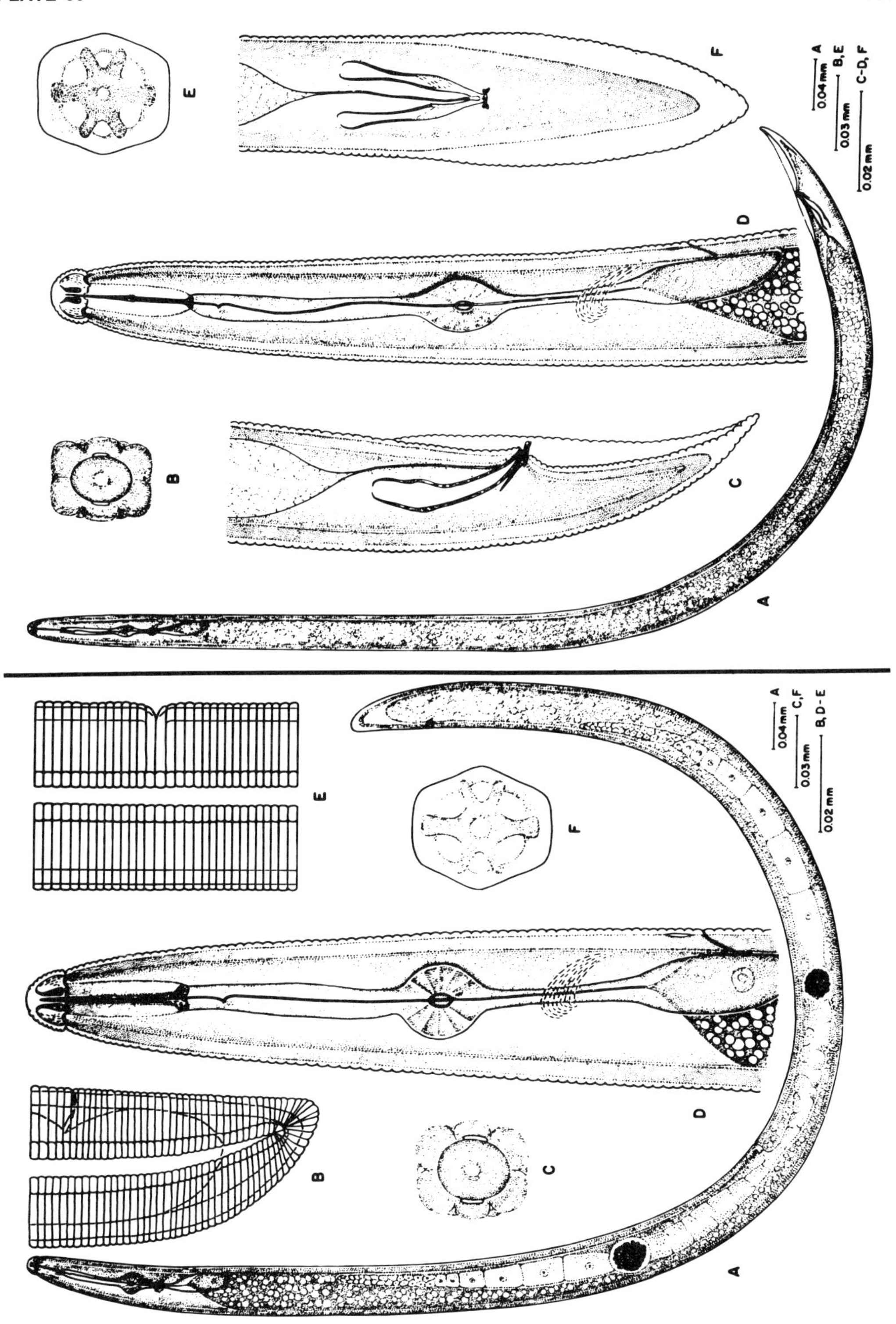
A
B
C
D
E
F
0.04 mm A
0.03 mm B, E
0.02 mm C-D, F
A
B
C
D
E
F
0.04 mm A
0.03 mm C, F
0.02 mm B, D-E

50. Tail generally less than 1.5 times anal body diameter . . . . (Example: Plate 71) 61
    Tail 1.5 or more times as long as anal body diameter . . . . (Example: Plate 60) 51
51. Esophageal glands usually unequal in length, overlapping the intestine dorsally or lateroventrally . . . . (Example: Plate 62) 52
    Esophageal glands usually enclosed within a bulb; if not enclosed, then of about equal length, and therefore considered as not overlapping the intestine . . . . (Example: Plate 68) 59
    [It is possible that certain conditions may cause the lengthening of either the dorsal or the subventral glands which may give the impression of overlapping. Therefore, several specimens should be carefully observed in relation to this character. Some confusion may arise with careful observation of the esophagus because the extent of the variation of this and other morphological characters has not been properly studied in many of the nematode genera described.]
52. No cephalic framework or moderately developed; female head not low or flattened . . . . (Example: Plate 60) 53
    Well-developed cephalic framework, female head low, rounded, or flattened . . . . (Example: Plate 64) 56
53. Well-developed stylet; lateral field with 4 incisures . . . . (Example: Plate 61) 54
    Slender stylet with diverging basal knobs; lateral field with 3 incisures . . . . (Plate 60) *TRICHOTYLENCHUS*

Trichotylenchus Whitehead 1959

Type species: *Trichotylenchus falciformis* Whitehead 1959

Description: Telotylenchinae: Head rounded, annulated, and generally continuous with body. Cephalic framework with little or no sclerotization. Stylet small, slender, sometimes diverging (or somewhat forked) at its base, with basal knobs. Esophageal glands overlapping anterior end of intestine (only to a limited extent in *T. rhopalocercus*). Lateral field with 3 incisures. Ovaries amphidelphic. Female tail about 5 anal body diameters in length, tapering to a rounded terminus. Bursa distinct, extending to or enveloping tail tip. (From Golden 1971.)

General Characteristics

When Whitehead erected this genus in 1959 he indicated that one of the most important diagnostic characters was that the spear base was forked with each arm bearing a rounded, basal knob. He indicated that this genus was closely related to *Belonolaimus*.

Jairajpuri (1971) regarded *Telotylenchus* a junior synonym of *Trichotylenchus*. On the other hand, Seinhorst (1971) regarded the 2 genera as closely related but distinct, because they could be separated on the basis of the appearance of the stylet, the shape of the female tail, the lateral field, and the shape of the gubernaculum in the male. He did not mention that in the original description of this genus Whitehead (1959) indicated that the stylet was forked. Golden (1971) regarded *Trichotylenchus* and *Telotylenchus* as distinct genera and mentioned the forked stylet characteristic in his generic description. According to Tarjan (1973) there is considerable variation among species of *Trichotylenchus* in the overlap of the esophagus. Seinhorst (1971) does not consider this to be an important characteristic.

Plate 60. Trichotylenchus falciformis. A. General structure. B. Anterior end. C. Tail in subventral view. D. Cuticular pattern in the middle of the body. E. Cuticular pattern in the anterior body region. F. Gonads in ventral view. a. muscular part of the uterus; b. quadricolumella; c. spermatheca; d. proximal part of the oviduct. (After Coomans and DeGrisse 1963. Courtesy of Nematologica.)

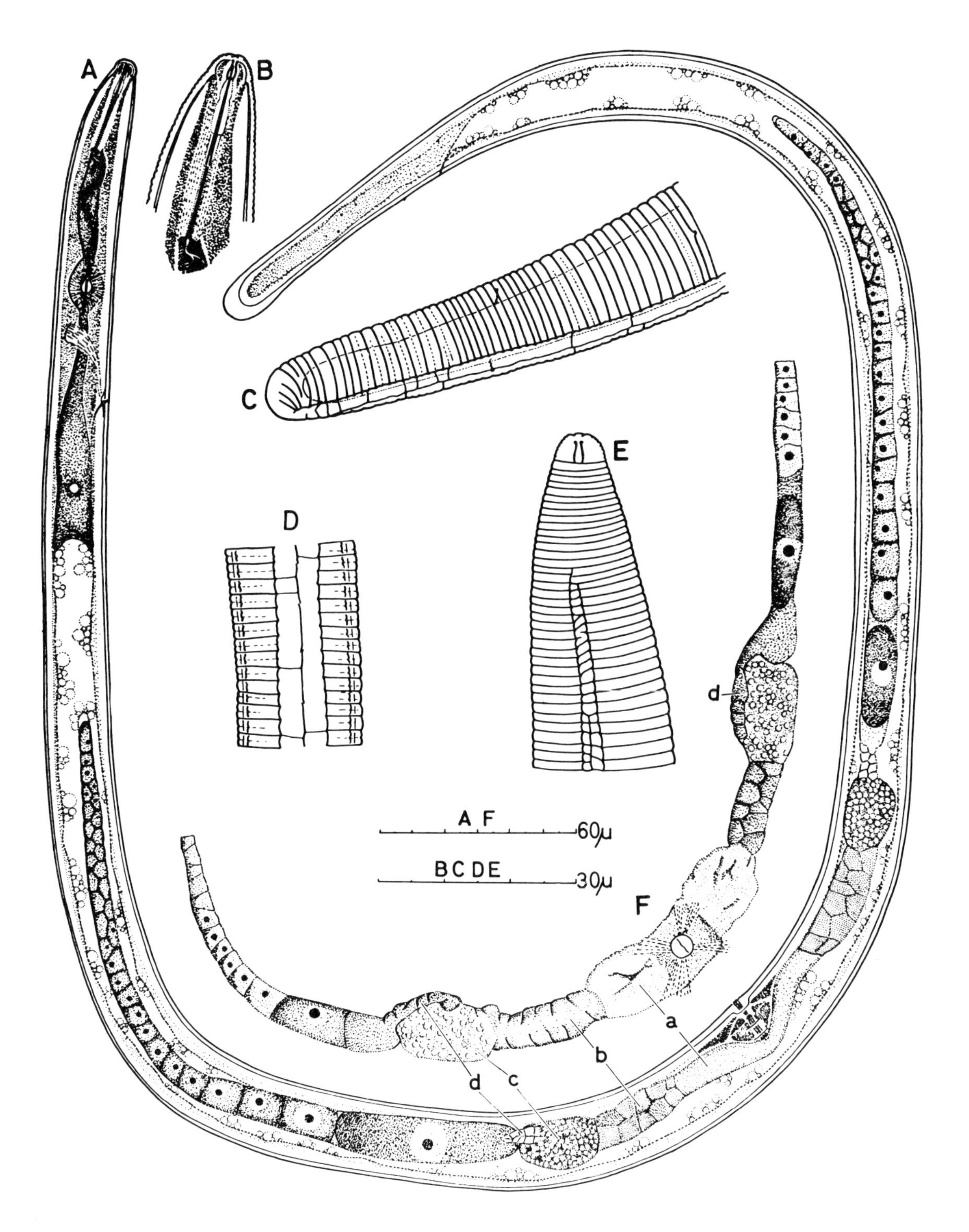
A
B
C
D
E
F
a
b
c
d
A F
60μ
BCDE
30μ

54. Female tail cylindroid with round terminus . . . . (Example: Plate 62) 55
Female tail elongate conoid with blunt terminus . . . . (Plate 61) *TELOTYLENCHUS*

Telotylenchus Siddiqi 1960

Type species: *Telotylenchus indicus* Siddiqi 1960

Description: Telotylenchinae. Sexes similar in length and shape. Body elongate with distinct transverse striae. Lateral field with incisures. Lip region striated with or without lip cap. Stylet well developed and variable in length. Median bulb of esophagus with crescentic valve plates. Esophageal glands contained in a lobe overlapping the intestine. Ovaries paired, outstretched, opposed. Female tail elongate-conoid with blunt terminus. Male tail pointed, completely enveloped by bursa. Spicules and gubernaculum tylenchoid. Phasmids porelike, located on tail. (Mostly from Fisher 1964.)

General Characteristics

Siddiqi (1960) comments concerning the habitat of nematodes assigned to *Telotylenchus indicus*: "Nine females, 4 males and 6 larvae of a species of nematode rather like a *Tylenchorhynchus* were collected early in 1958 from soil around the roots of grass, *Cynodon dactylon* Pers., at Aligarh, North India." Neither feeding habits nor results of plant pathogenicity tests are reported.

Since 1960, 4 additional species were added to this genus: *T. ventralis* (Loof 1963); *T. housei* (Raski et al. 1964); *T. whitei* (Fisher 1964); and in 1964 Fisher classified *Belonolaimus hastulatus* under *Telotylenchus*.

Plate 61. *Telotylenchus indicus*. A. Female. B. Anterior end of female. C. Esophageal glands of female in lateral view. D. Female tail in lateral view. E. Male tail in ventral view. F. Male tail in lateral view. G. Testis. H. Vulva in ventral view. I. Cross section of female showing lateral fields. (After Siddiqi 1960. Courtesy of Nematologica.)

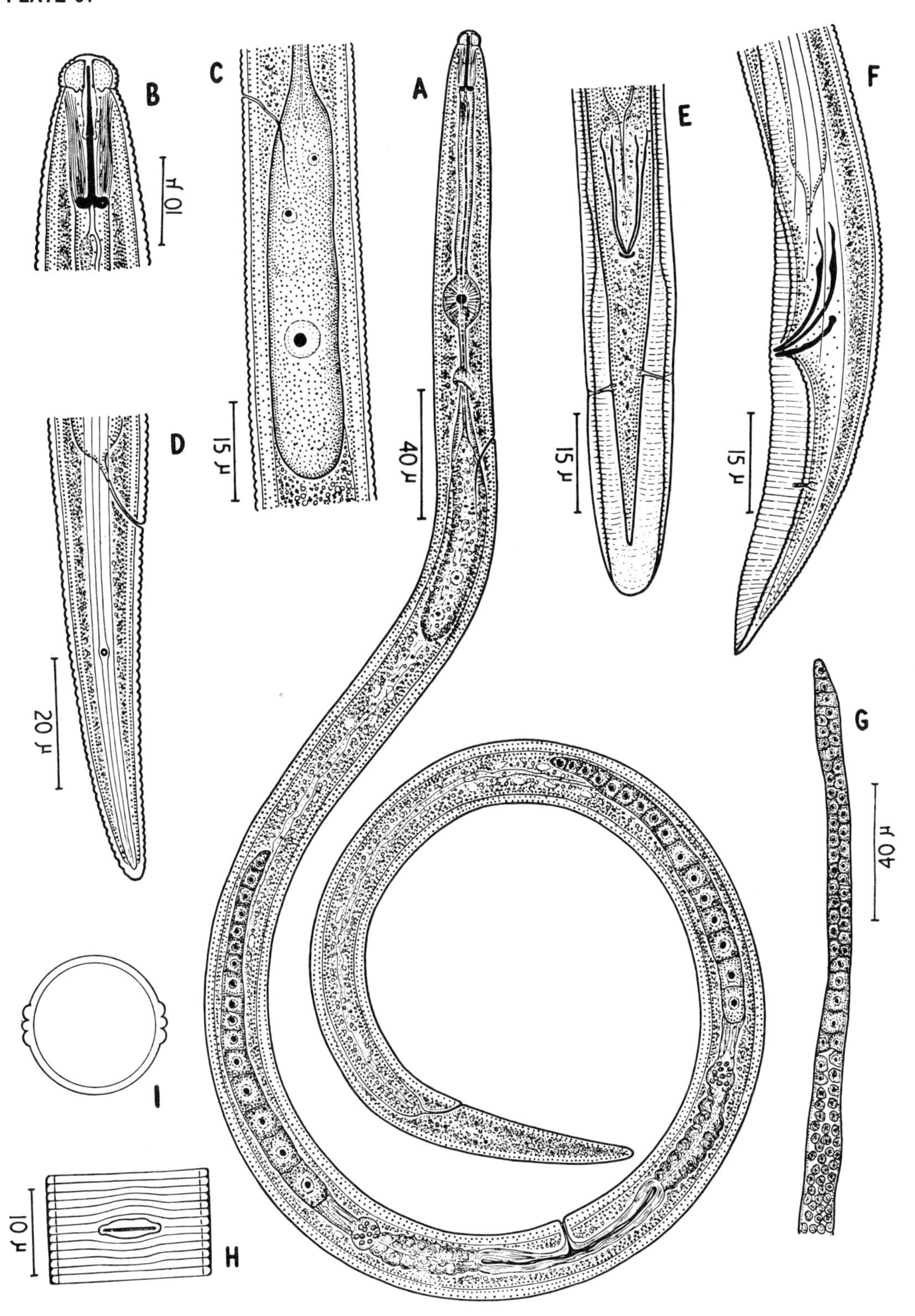
B
10 µ
C
15 µ
A
40 µ
E
15 µ
F
15 µ
D
20 µ
G
40 µ
I
H
10 µ

55. Anterior portion of stylet asymmetrical; tail rather short with broadly rounded terminus .... (Plate 62) *HISTOTYLENCHUS*
Anterior portion of stylet asymmetrical; female tail with broadly rounded to bulbous terminus with strongly thickened cuticle .... (Plate 63) *TELOTYLENCHOIDES*

Histotylenchus Siddiqi 1971

Type species: *Histotylenchus histoides* Siddiqi 1971

Description: Telotylenchinae, Belonolaimidae. Lip region continuous, somewhat rectangular, finely annulated; framework moderately sclerotized; labial disc absent. Lateral fields areolated along the entire length of the body (hence the generic epithet), with 4 incisures. Amphids minute, porelike, close to oral opening; phasmids also porelike, on tail. Spear well developed, with prominent knobs; anterior tapering portion (conus) asymmetrical, with lumen becoming angular near its base as seen in lateral view; its protractor muscles attached to the outer area of the basal plate. Median esophageal bulb large and strongly muscular. Esophageal glands overlapping as in *Telotylenchus*, viz., with large dorsal gland lying on laterodorsal and lateral sides of intestine and small subventrals close to esophago-intestinal junction. Postrectal intestinal sac present. Tail rather short, with cylindroid to a broadly rounded terminus. Ovaries paired, opposed, outstretched; spermatheca round, axial; epitygma present. Spicules well developed, with large ventral flanges distally; gubernaculum also large, protrusible, with proximal portion directed dorsally and distal portion carinate, with titillae. (From Siddiqi 1971.)

General Characteristics

According to Siddiqi (1971) the asymmetrical anterior portion of the spear differentiates *Histotylenchus* from related genera. It further differs from the closely related genus *Telotylenchus* in having a large rectangular head, areolated lateral fields, cylindroid female tail and recurved gubernaculum. *Trichotylenchus*, another closely related genus, has 3 incisures in the lateral fields and a symmetrical, attenuated spear.

In the generic description of *Histotylenchus* it is stated that the esophageal glands overlap the esophagus with the large dorsal gland lying on lateral sides of the intestine and the small subventrals close to the esophago-intestinal junction. However, in several fixed specimens we examined in this laboratory the overlap of the esophagus was difficult to observe.

Plate 62. Histotylenchus histoides. (A–I, female.) A. Head end. B. *En face* view. C. Cross section through basal plate. D. Cross section of body near middle of spear. E. Vulval region and spermatheca. F. Esophageal region. G–I. Tail ends. J and K. Tail ends of male. (After Siddiqi 1971. Courtesy of Nematologica.)

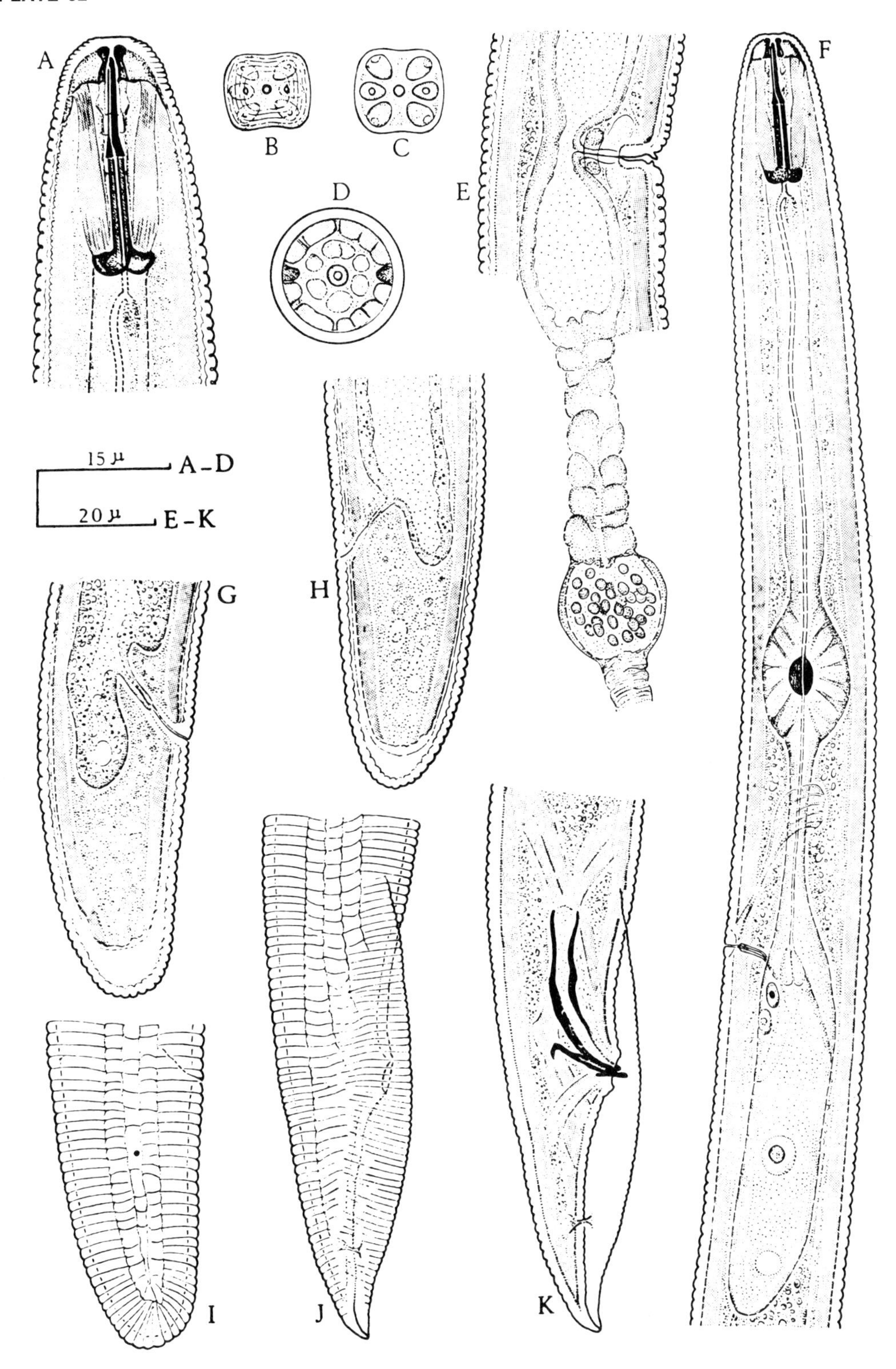
A
B
C
D
E
F
15 µ A-D
20 µ E-K
G
H
I
J
K

Telotylenchoides Siddiqi 1971

Type species: *Telotylenchoides housei* (Raski et al. 1964) Siddiqi 1971

Description: Telotylenchinae, Belonolaimidae. Lip region conically rounded, continuous with body, without lobes or oral disc; framework prominently sclerotized, with lateral sectors wider than submedians. Lateral fields with 4 incisures, not areolated. Phasmids distinct, pore-like. Spear well developed, with large, compact basal knobs. Esophageal gland overlap formed by the dorsal gland extending dorsally and dorsolaterally over the intestine with its large nucleus located closely behind the esophago-intestinal junction; nuclei of the subventral glands indistinct, anterior to esophago-intestinal junction. Ovaries paired outstretched; spermatheca round, axial. Female and larval tail cylindrical, with broadly rounded to bulbous terminus comprised of abnormally thickened cuticle; male tail also with a conspicuous cuticular terminal portion. Bursa well developed, enveloping tail. Spicules cephalated, ventrally arcuate, with large ventral flanges on distal end; gubernaculum large, protrusible, appearing rodlike in lateral view. (From Siddiqi 1971.)

General Characteristics

This genus was proposed by Siddiqi (1971) to include two species, *T. housei*, formerly included in *Telotylenchus*, and *T. lobatus*, formerly classified in *Paratrophurus*.

Telotylenchoides is closely related to *Telotylenchus*. With respect to the differences between *Telotylenchus* and the genera *Telotylenchoides* and *Histotylenchus*, Siddiqi (1971) states: "*Telotylenchoides* differs from *Telotylenchus* in having a continuous conically rounded lip region and the cylindroid to bulbous female and larval tail with abnormally thickened cuticle at the tip. It differs from *Trichotylenchus* in having nonareolated lateral fields with 4 incisures as compared to 3 in the latter, well-developed and compact basal knobs of the spear, broadly rounded female tail with abnormally thickened cuticle at the terminus, and the proximal end of the gubernaculum not directed dorsally. The overlap of the esophageal glands in *Telotylenchoides* appears to be different from that of *Telotylenchus*, *Trichotylenchus*, and *Histotylenchus* in that the nucleus of the dorsal gland lies close behind the esophago-intestinal junction, and the indistinct nuclei of the subventral glands lie anterior to it."

Plate 63. Telotylenchoides lobatus. A. Head end. B. *En face* view. C. Cross section through basal plate. D. Adult. E and F. Tail ends. G–I. *T. housei*. G. Head end of female. H. Tail end of male. I. Tail end of female. (After Siddiqi 1971. Courtesy of Nematologica.)

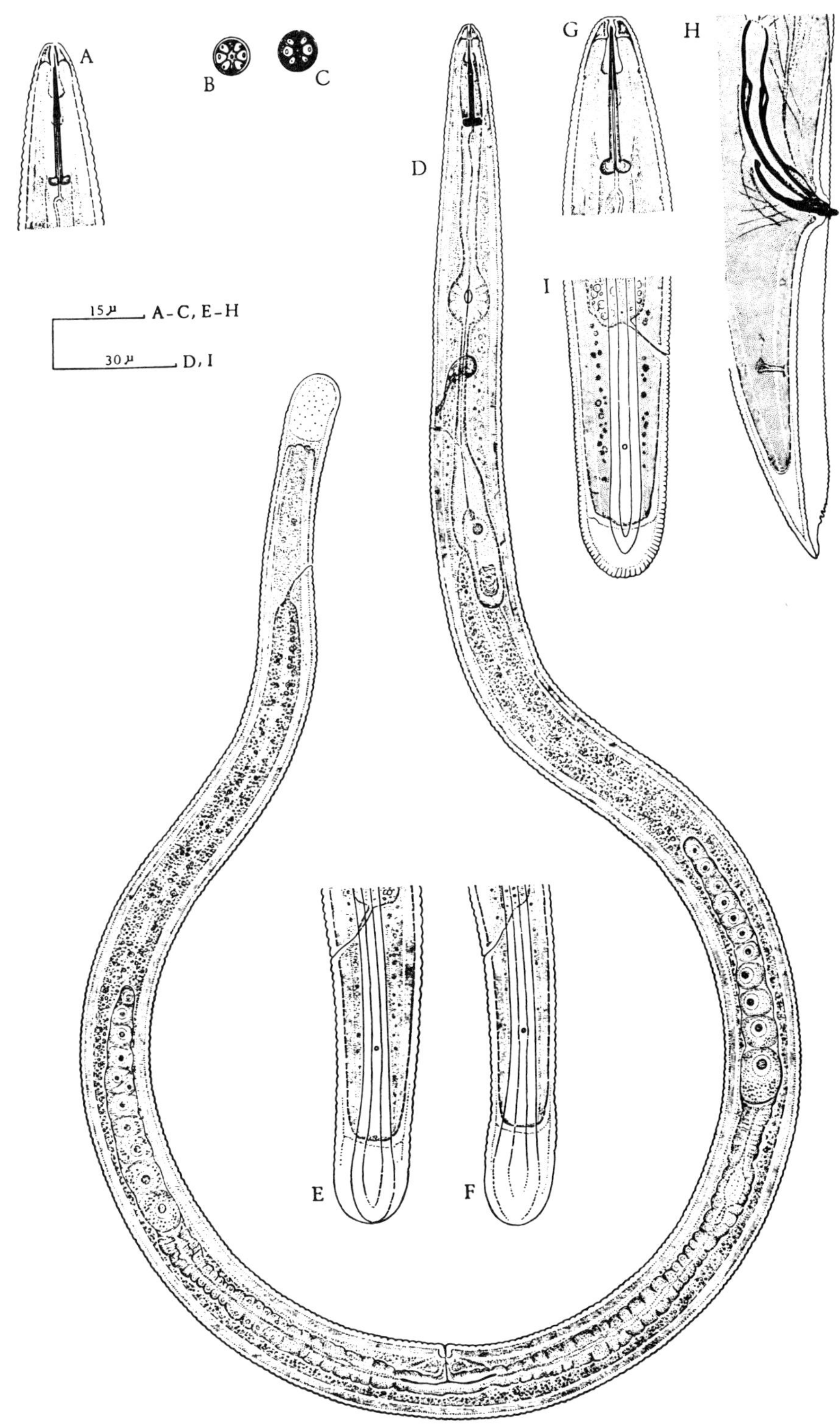
A
B
C
D
G
H
I
15 μ A-C, E-H
30 μ D, I
E
F

56. Esophagus overlapping intestine dorsally . . . . (Example: Plate 65) 57
    Esophagus overlapping intestine ventrally . . . . (Example: Plate 67) 58
57. Short overlap. No marked sexual dimorphism . . . . (Plate 64) *PRATYLENCHOIDES*
    Long overlap. Marked sexual dimorphism . . . . (Plate 65) *RADOPHOLUS*

Pratylenchoides Winslow 1958

Type species: *Pratylenchoides crenicauda* Winslow 1958.

Description: Radopholinae. Labial framework and stylet well developed. Sexual dimorphism present in anterior part of body, male with slightly higher rounded lip region; stylet and esophagus not as developed as female. Deirids present. Esophageal glands overlapping intestine ventrally, laterally, and dorsally, greatest development dorsally with at least 1 esophageal gland nucleus above the esophageal intestinal valve. Ovaries paired. Phasmids in posterior portion of tail. Caudal alae enveloping tail, gubernaculum not projecting from cloaca. (From Sher 1970.)

General Characteristics

Nematodes placed in *P. crenicauda*, the type species of this genus, were found by Winslow (1958) in aqueous extracts from turf of the Rothamsted Agricultural Experiment Station. This species lives endoparasitically, producing root lesions similar to those caused by *Pratylenchus* species. Nematodes and developing eggs were found in cortical tissues of young lateral roots. In pot tests, populations of this nematode increased in association with roots of several plant species.

Sher (1970) gives an emended description of this genus and presents a key to 7 species. With respect to this genus Sher states:

> *Pratylenchoides* differs from the genus *Radopholus* in having deirids; less reduction of the stylet and esophagus in the male; usually broadly rounded stylet knobs, sloping distally; a more variable type of overlapping esophageal glands with at least 1 esophageal nucleus above the esophageal intestinal valve; gubernaculum not protruding from the cloaca; and usually a lower position of the phasmids on the tail.
>
> *Pratylenchoides* can be easily distinguished from *Zygotylenchus* by the dorsal esophageal glands, which overlap the intestine, and by the presence of deirids.
>
> The genus *Pratylenchoides* appears to be almost world-wide in distribution with no known area of large speciation, whereas the genus *Radopholus* is apparently native to Australia (Sher 1968) where it has many species. In addition, the one widely distributed species of *Radopholus*, the very most important plant pest, *R. similis* (Cobb 1893), is usually found in the warmer areas of the world. The genus *Pratylenchoides* is in the cooler, more temperate areas and is not known to have any important, widely distributed plant pathogenic species, although the type species, *P. crenicauda,* has been identified from widely separated areas of the world.

Plate 64. Pratylenchoides crenicauda. A. Fore part of body. B. Male tail. C. Female tail. (After T. Goodey 1932, 1940. Courtesy of Journal of Helminthology.)

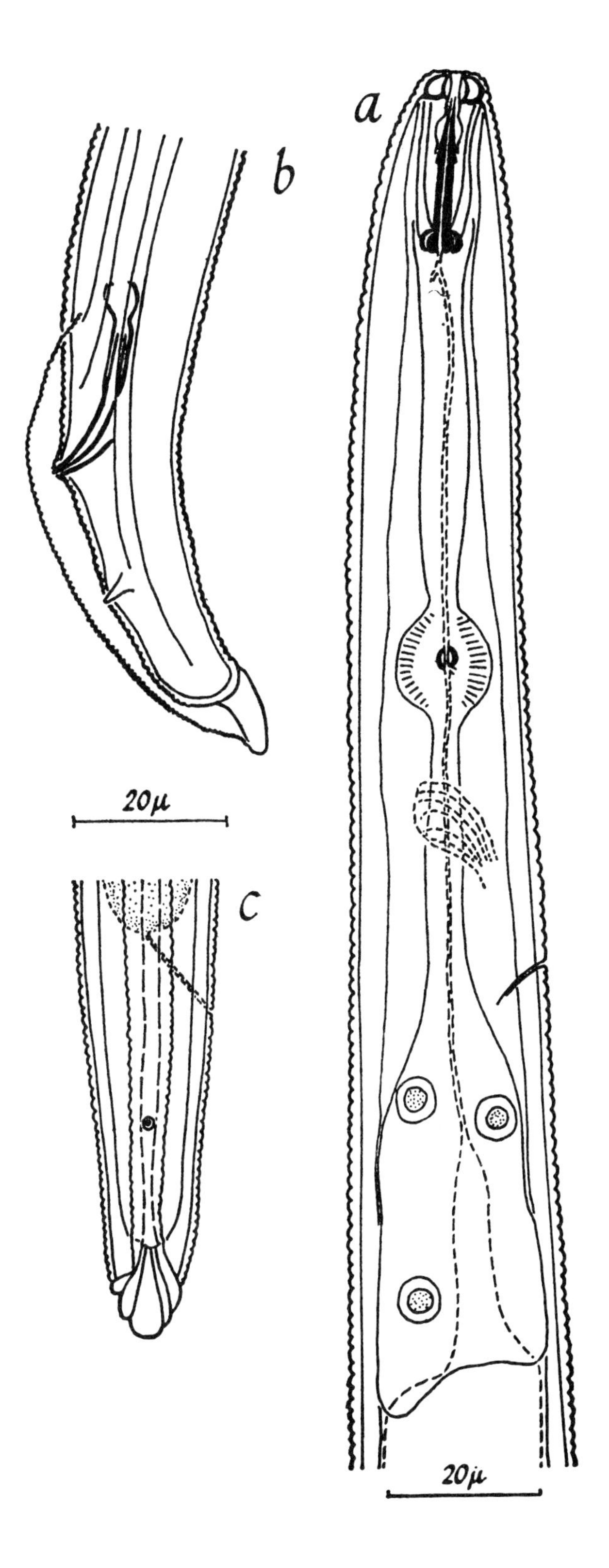
a
b
c
20μ
20μ

Radopholus (Thorne 1949) Luc and Goodey 1962

Type species: *Radopholus similis* (Cobb 1893) Thorne 1949

Description: Radopholinae Allen and Sher, 1967. Labial framework and stylet well developed in female and juveniles. Sexual dimorphism marked; the male has higher rounder lip region and lateral lips, labial framework, stylet, and esophagus reduced. Anterior portion of stylet is long or longer than posterior portion. Dierids absent. Esophageal glands overlapping intestine dorsally. Gonads amphidelphic. Phasmids usually in anterior part of tail. Tail tapering to a rounded or almost pointed terminus, usually 2 to 4 times as long as the body width at the anus or cloaca. Gubernaculum protruding from cloaca. (From Luc and Goodey 1962.)

General Characteristics

Members of this genus closely resemble those of the genus *Pratylenchus*. Individuals of both genera are migratory endoparasitic plant pathogens and are somewhat similar morphologically. One of the chief differences is that females of *Radopholus* have 2 ovaries whereas those of *Pratylenchus* have a single ovary. The type species is *R. similis* (Cobb 1893) Thorne 1949. Sher (1968) revised this genus and presents a key to 8 species. Since 1968 a number of new species have been described.

R. similis, the burrowing nematode, is an important pathogen of banana, sugar cane, pepper, coffee, and edible canna. Feeding results in lesions on roots and other fleshy underground parts. A host list has been compiled by Christie (1959). The nematode is widely distributed in tropical and subtropical countries.

Plate 65. Radopholus similis. A. Anterior portion of female, 615 X. B. Male tail, 5740 X. C. Anterior portion of male, 615 X. D. Female head, 1230 X. E. Female tail, 615 X. (Originals.) F. Young female, 450 X. a. lip region. b. spear guide. c. 3-bulbed spear. d. ampula, salivary gland. e. esophageal lumen. f. esophagus. g. median bulb. h. nerve cells. i. nerve ring. j. excretory pore. k. initial intestinal cells. l. anterior salivary gland. m. end of ovary. n. ovum. o. renette duct. p. posterior salivary gland. q. fat granule, intestine. r. renette cell?. s. terminus. t. phasmid. u. vulva. v. anus. w. crenate cuticle. x. spermatozoa. After Cobb 1915. G. Tylenchoidea. Arrangement of esophageal gland outlets. H. Aphelenchoidea. Arrangement of esophageal gland outlets. After Goodey 1929. (After Thorne 1949. Courtesy of Helminthological Society of Washington.)

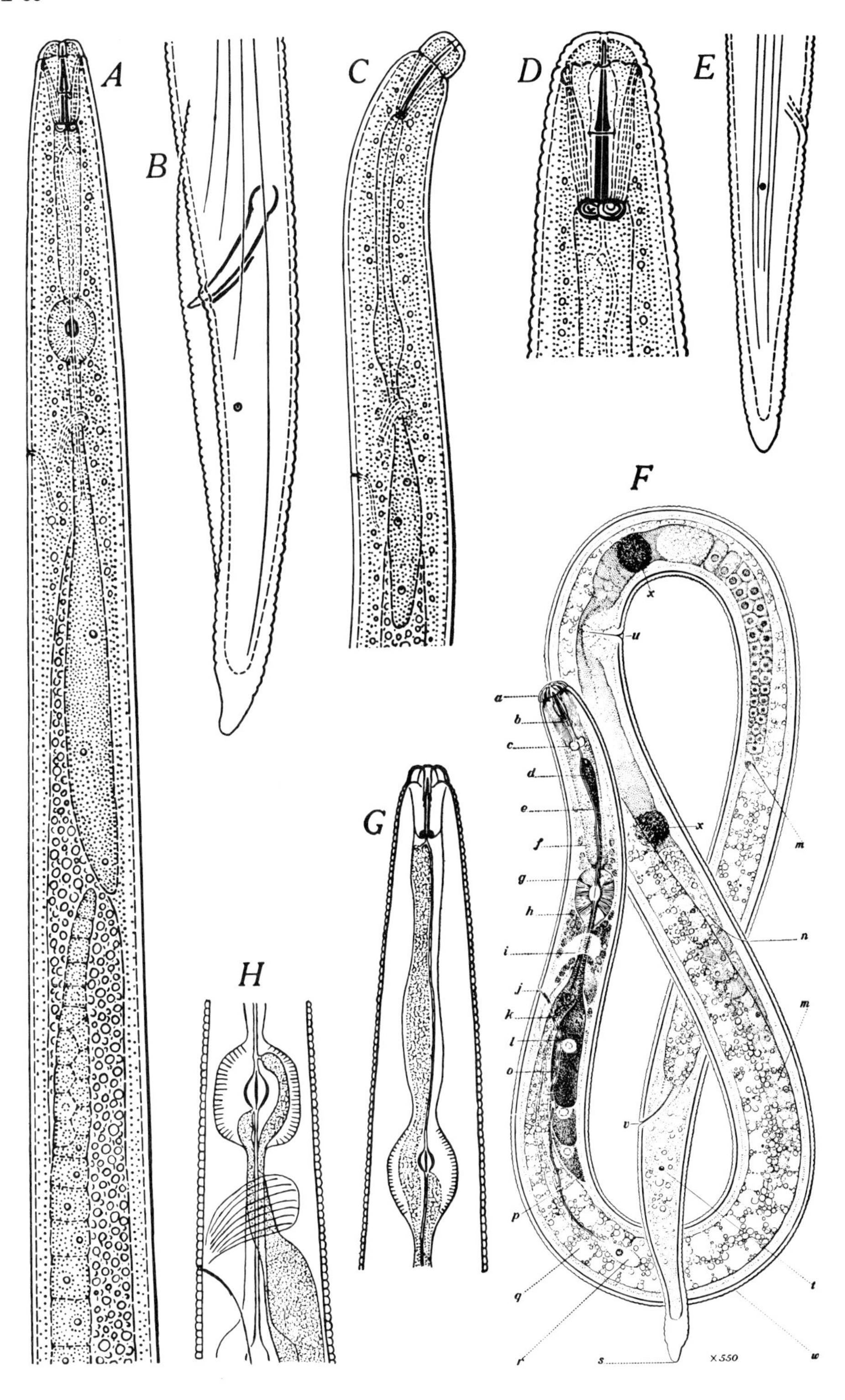
A
B
C
D
E
F
G
H
a
b
c
d
e
f
g
h
i
j
k
l
o
p
q
r
s
t
u
v
w
x
m
n
×550

58. Tail tip mucronate . . . . (Plate 66) *HIRSCHMANNIELLA*
Tail tip not mucronate . . . . (Plate 67) *ZYGOTYLENCHUS*

Hirschmanniella Luc and Goodey 1962

Type species: *Hirschmanniella spinicaudata* (Schuurmans Stekhoven 1944) Luc and Goodey 1962

Description: Pratylenchinae. From 0.9 to 4.2 mm in length as adults. Lip region with well-developed internal framework, not set off from body, flattened anteriorly to hemispherical. Spear 3 to 5 times body diameter at basal plate. Esophageal glands with elongated ventral overlap of the intestine. Lateral field with 4 incisures. Ovaries paired. Phasmids in posterior third of tail. Tail 3 or more times as long as the body width at the anus, tapering, usually terminating in a point or mucro. Caudal alae not enveloping tail. No sexual dimorphism in anterior part of body. (From Sher 1968.)

General Characteristics

Thorne (1949) established the genus *Radopholus* "to receive those didelphic species which most clearly resemble the genus *Pratylenchus*." Luc and Goodey (1962) indicated 2 separate groups of species. The first group, *R. similis*, *R. inaequalis*, and *R. neosimilis*, appear more closely related to the species now assigned to the genus *Pratylenchus*. The second group, *R. lavabri*, *R. gracilis*, *R. oryzae*, and *R. mucronatus*, was transferred to the new genus *Hirschmannia* Luc and Goodey 1962. Siddiqi (1966) described 2 additional species, *H. nana* and *H. magna*.

Due to the previous use of "*Hirschmannia*" as a generic name, the name of the new genus was changed to *Hirschmanniella* Luc and Goodey 1962 (Luc and Goodey 1963). Sher (1968) revised this genus.

Plate 66. Hirschmanniella gracilis. A. Male tail, lateral view, 572 X. B. Male tail, ventral view, 572 X. C. Anterior portion of mature female, 554 X. D. Mature female, 203 X. a. terminal process.

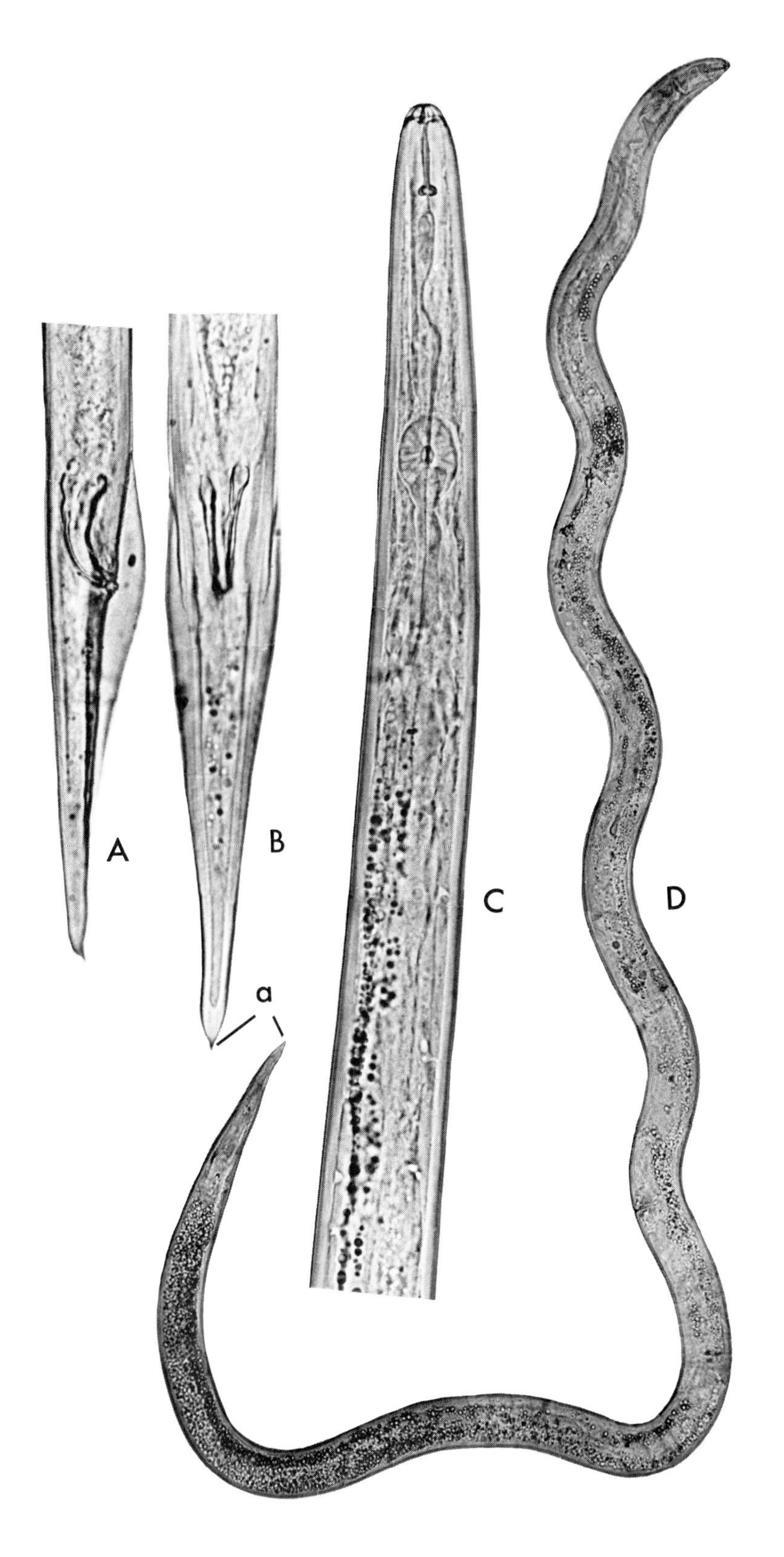
A
B
C
D
a

Zygotylenchus Siddiqi 1963

Type species: *Zygotylenchus guevarai* (Tobar 1963) Braun and Loof 1966

Description: Pratylenchinae. Female with ovaries paired, outstretched, opposite and vulva in submedian position. Labial region flattened anteriorly, more or less conical. Labial sclerotization conspicuous. Spear strong, with rounded basal knobs. Esophageal glands surrounding the beginning of the intestine anteriorly and extending posteriorly as a long lobe on its ventral side. Intestine connected to the esophageal lumen by a small refringent valvula. Female tail cylindrical, ending in a smooth tip, more or less rounded. Male tail tapering to a pointed terminus, completely enveloped by a bursa. Phasmids located near middle of tail in both sexes. No marked sexual dimorphism. (From De Guiran and Siddiqi 1967.)

General Characteristics

Other than the type species, *Z. guevarai*, *Z. toomasinae* is the only other species of this genus (De Guiran and Siddiqi 1967).

Tarjan and Weischer (1965) proposed the synonymy of the genera *Zygotylenchus* and *Mesotylus* with *Pratylenchoides*, in which they include the species described under the former genera. Although Braun and Loof (1966) and De Guiran and Siddiqi (1967) agree that *Mesotylus* is not a valid genus, they maintain that *Zygotylenchus* and *Pratylenchoides* should not be synonymized. With respect to the differences between these 2 genera, De Guiran and Siddiqi state: "In *Pratylenchoides*, the subventral glands and their nuclei are anterior to the dorsal gland and its nucleus respectively. The subventral glands are also anterior to the esophago-intestinal junction and it is mostly the dorsal gland that forms an overlap over the front end of the intestine. In *Zygotylenchus*, on the other hand, this condition is just the reverse. All the glands lie posterior to the esophago-intestinal junction. The subventral glands are posterior to the dorsal gland and form a long, ventral overlap over the intestine."

Plate 67. Zygotylenchus browni. 1. Cephalic end of female, lateral. 2. Female (Holotype). 3. Male (Allotype). 4. Cephalic end of male, lateral. 5. Esophageal region of female, lateral. 6. Transverse striations and lateral field near midbody of female, lateral. 7. a–c. variations in female tails. 8. Caudal end of male, lateral. (After Siddiqi 1963. Courtesy of Zeitschrift fur Parasitenkunde.)

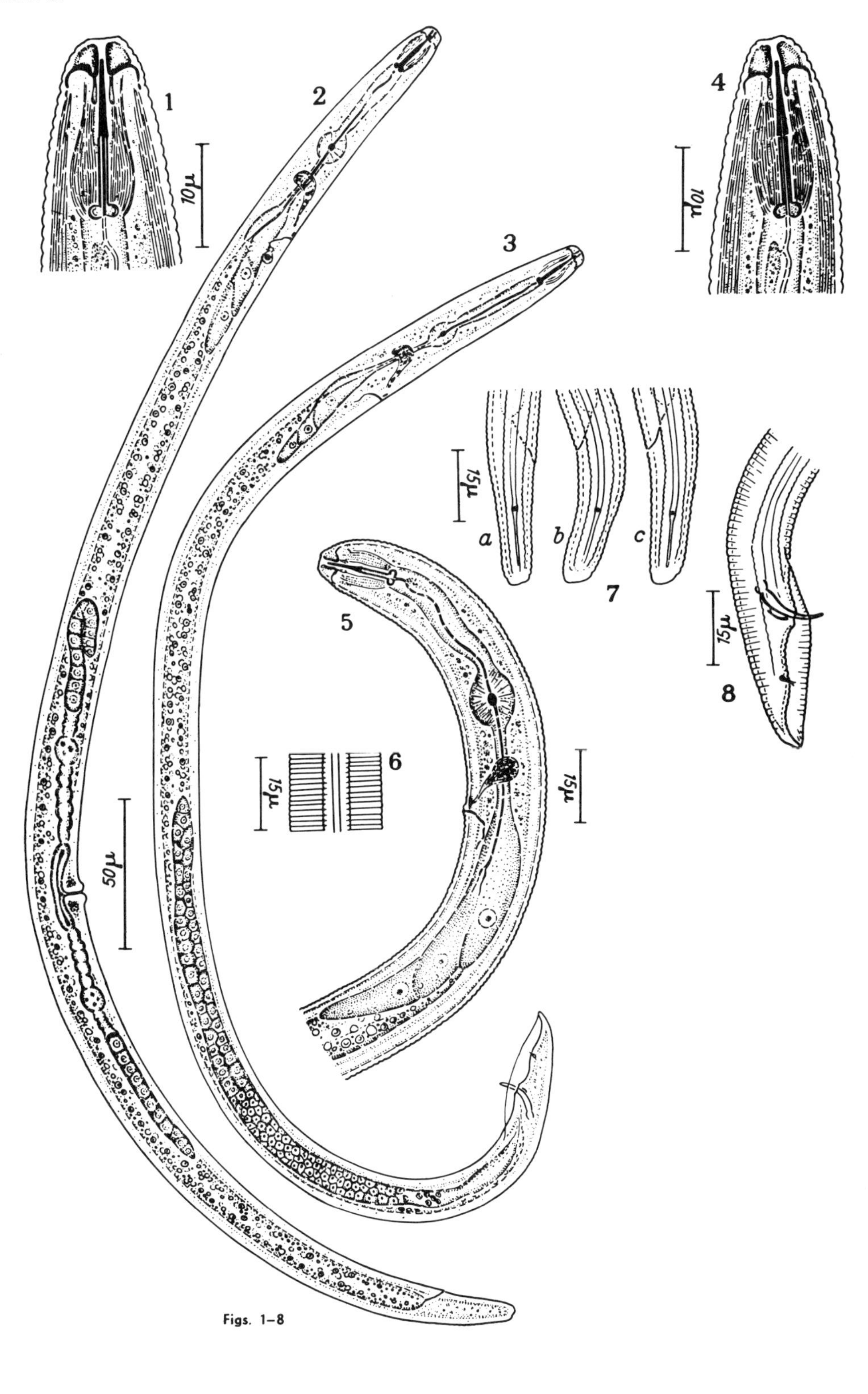

Figs. 1–8

59. Female tail not acute . . . . (Example: Plate 69) 60
Female tail acute or subacute . . . . (Plate 68) *TETYLENCHUS*

Tetylenchus Filipjev 1936

Type species: *Tetylenchus tenuis* (Micoletzky 1922) Filipjev 1936

Description: Tylenchidae without sclerotized cephalic framework. Cuticle finely striated. Ovaries 2, outstretched. Spear of moderate size, with or without basal knobs. Tails of both sexes tapering to an acute or subacute terminus. Deirids and phasmids present, generally easily visible. Bursa subcaudal, extending almost to the terminus. Distance from anterior end to valve of median bulb shorter than that from valve to base of esophagus. Esophagus with elongate-ovate median bulb; unusually long slender isthmus, and elongate-pyriform basal bulb containing the usual 3-gland nuclei. Cardia usually discoid. Spicula tylenchoid; gubernaculum a simple troughlike plate. (From Thorne 1949.)

General Characteristics

In general, the genus *Tetylenchus* resembles *Tylenchorhynchus*. Concerning the differences between these genera, Hirschmann (1960) states: "In contrast to *Tylenchorhynchus*, however, the tail in both sexes [of *Tetylenchus*] tapers to an acute or subacute terminus. The spear is of moderate size and may or may not have knobs. The lip region is without framework and, in face view, has an entirely different pattern of arrangement of lip sectors, amphid apertures, and papillae. The caudal alae of the male are subcaudal, extending almost to the tail terminus."

Tarjan (1973) presents a key to seven species of this genus. Zuckerman (1960) observed individuals of this species feeding on the roots of cranberry (*Vaccinium macrocarpon*). In the presence of cranberry roots, the nematodes increased substantially.

Plate 68. Tetylenchus joctus. A. Mature female, 380 X. B. Male tail, lateral view, 850 X. C. Male tail, ventral view, 850 X. D. Female tail, 850 X. E. Head region, 1520 X. F. Anterior portion of female, 850 X. a. phasmids.

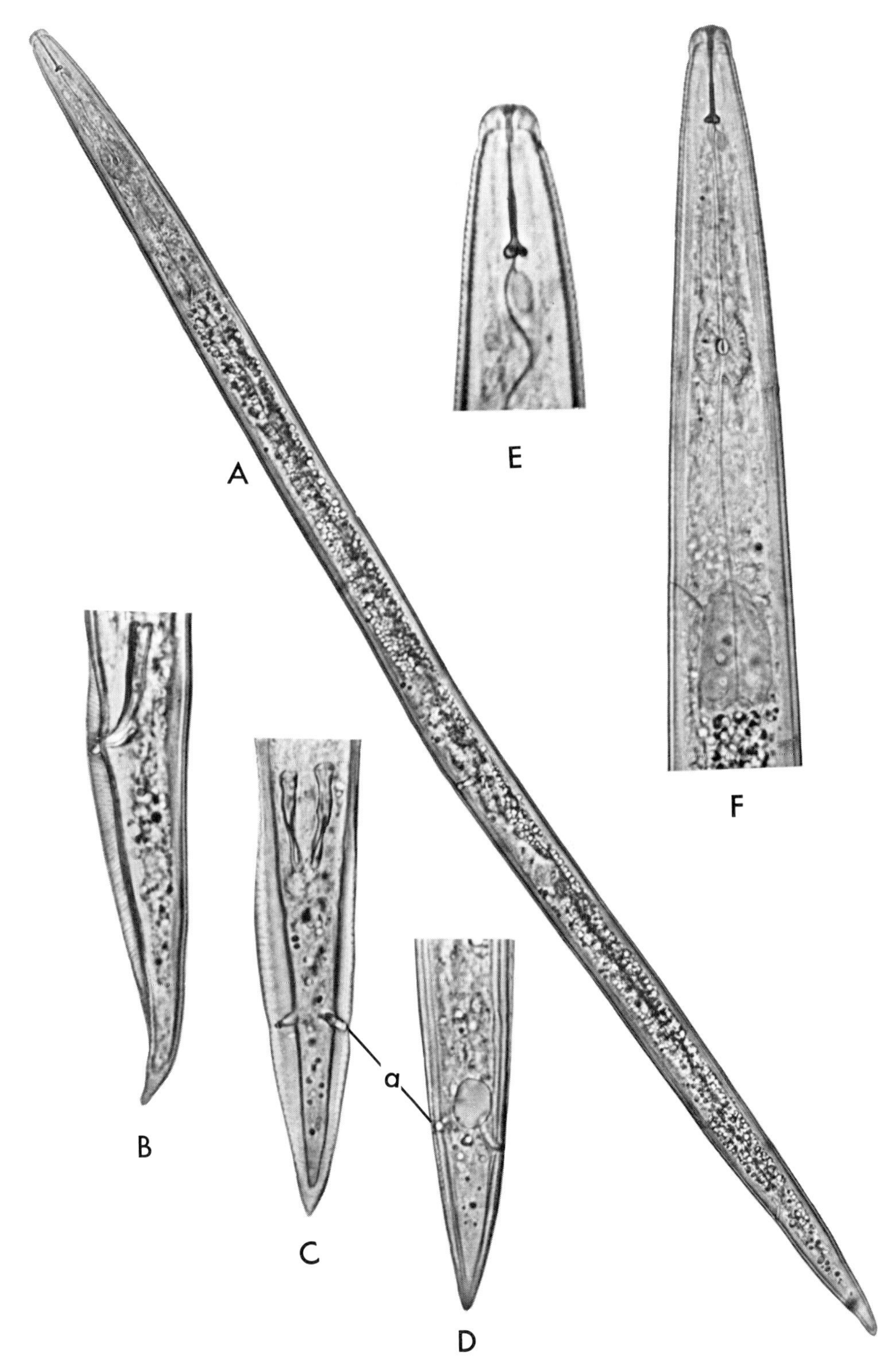
A
B
C
D
E
F
a

60. Female tail conoid, with terminus usually bluntly rounded . . . . . . (Plate 69) *TYLENCHORHYNCHUS*
Female tail cylindroid, with terminus broadly rounded and strongly thickened cuticle . . . . (Plate 70) *PARATROPHURUS*
[*Paratrophurus lobatus* Loof 1970 has overlapping glands, and for this reason it was placed in *Telotylenchoides*, by Siddiqi (1971). A more accurate decision about the correct placement of this species in either of the 2 genera can only be made when the extent of the variation of the esophageal glands is properly studied and its validity as a taxonomic character in this case is established.]

Tylenchorhynchus Cobb 1913

Type species: *Tylenchorhynchus cylindricus* Cobb 1913

Description: Tylenchinae. No sexual dimorphism. Esophageal glands contained in basal bulb, sometimes a lobe of the basal bulb slightly overlapping intestine. Lateral fields marked by 4, 5, or 6 incisures. Stylet well developed, with conspicuous basal knobs. Deirids usually inconspicuous. Phasmids conspicuous, located near middle of tail. Vulva near middle of body. Ovaries 2, outstretched. Female tail cylindrical, conoid, with terminus usually bluntly rounded, not acute. Male tail slightly arcuate, enveloped by bursa; phasmids about middle of tail. Spicula and gubernaculum tylenchoid. Cephalic framework slightly to heavily sclerotized. (From Allen 1955.)

General Characteristics

Members of the genus *Tylenchorhynchus*, the stunt nematodes, are ectoparasites; however, under some conditions, they enter the roots, becoming endoparasites (Steiner 1937). *Tylenchorhynchus* spp. are associated with the roots of many crops, including tobacco, cotton, oats, and corn. *T. claytoni* causes economically important damage to tobacco.

Under a dissecting microscope, the nonoverlapping esophagus, the conical tail, and the strong stylet with distinct basal knobs aid in identifying members of this genus. When relaxed they assume a wide C-shaped.

Geocenamus, *Nagelus*, *Merlinius*, *Quinisulcius*, *Scutylenchus*, and *Uliginotylenchus* are genera closely related to *Tylenchorhynchus*. Individuals of these genera have 2 ovaries, esophageal glands enclosed in a basal bulb, and bluntly rounded female tails.

According to Thorne and Malek (1968), *Nagelus* closely resembles *Tylenchorhynchus* from which it is distinguished by the asymmetrical fine striation of the lip region and lack of labial framework, form of spear knobs, attachment of spear muscles, and presence of epitygma. *N. aberrans* is the only species.

Thorne and Malek (1968) state: "*Geocenamus* is distinctive because of the perioral disc from which the slender spear guide extends back almost $\frac{1}{3}$ the length of the exceedingly slender spear."

The genus *Merlinius* was proposed by Siddiqi (1970) for 32 species of *Tylenchorhynchus*, which have 6 incisures in the lateral fields, rather cylindroid spicules with prominently notched distal end, a nonprotruding gubernaculum, and a moderately developed bursa.

Uliginotylenchus is a genus proposed by Siddiqi (1971) to include 5 species formerly in *Tylenchorhynchus*, with the following characteristics: (1) areolated lateral fields with 3 incisures, (2) clavate to cylindroid female tail with more than 25 annules, and (3) proximal end of gubernaculum bent dorsally.

Another genus, *Quinisulcius*, was proposed by Siddiqi (1971) to include 8 species formerly in *Tylenchorhynchus* in which the lateral fields are not areolated and have 5 incisures, the distal flanges of the spicules are small sized, and the proximal end of the gubernaculum is directed dorsally.

In a discussion of the genus *Scutylenchus*, Tarjan (1973) states: "*Scutylenchus* Jairajpuri 1971 was based on characters of *S. mamillatus* (Tobar-Jimenez 1966) Jairajpuri 1971 [formerly *Tylenchorhynchus mamillatus* Tobar-Jiminez 1966] which included: (1) a subdigitate tail shape, (2) enlarged scutellalike phasmids, (3) areolated lateral field, and (4) sloping stylet knobs."

Plate 69. *Tylenchorhynchus* sp. A. Mature female, 246 X. B. Anterior portion of female, 760 X. C. Lateral field highly magnified. D. Male tail, 644 X.

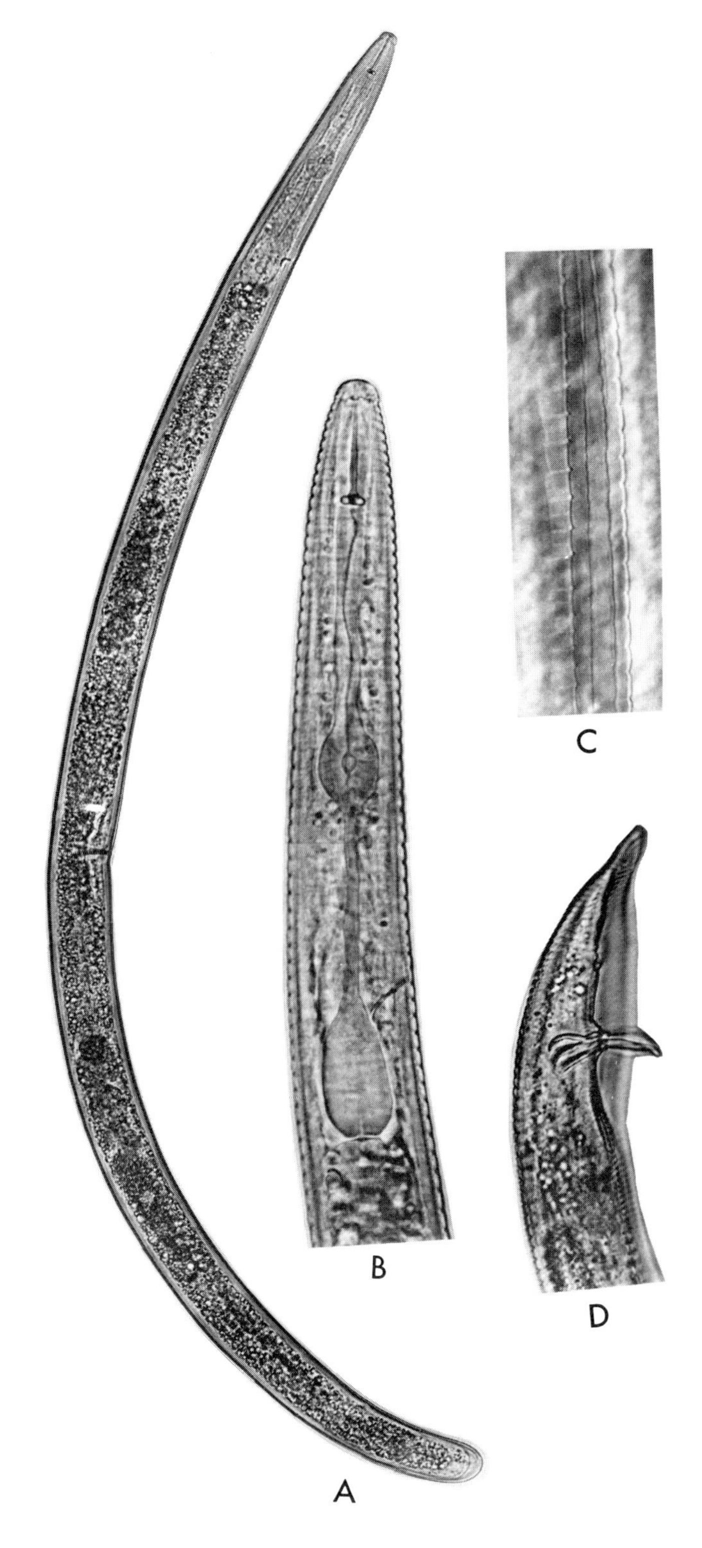
A
B
C
D

Paratrophurus Arias 1970

Type species: *Paratrophurus loofi* Arias 1970

Description: Tylenchidae. Lip region apparently smooth, but showing fine transverse striations in phase contrast at high magnification (1000 X); rounded, continuous with body contour; with moderately developed internal sclerotization. Spear slender, with small but distinct basal knobs. Procorpus cylindrical; median bulb oval with well-developed valves; isthmus cylindrical, shorter than procorpus. Nerve ring about halfway between median and terminal bulb. Deirids invisible. Vulva slightly postequatorial. Female gonads paired, opposed and outstretched. Female tail with rounded tip; cuticle strongly swollen on terminus. Male with tylenchoid spicules and gubernaculum; bursa with crenate edge, completely enveloping tail. (From Arias 1970.)

Siddiqi emends the description of this genus as follows: *Paratrophurus* Arias 1970. Diagnosis, as given by Arias (1970), with these additions and modifications: Trophurinae; Dolichodoridae. Lip region narrow, conoid-rounded. Protractor muscles of the spear rather spindle-shaped, attached to the spear guide which is well developed and to the inner margins of the labial framework. Lateral fields with 4 incisures, not areolated. Spermatheca round, axial. Spicules cephalated, with distal end pointed and prominently flanged ventrally. Gubernaculum large, protrusible. Bursa well developed, with crenate edges, completely enveloping tail which carries a conspicuous nonprotoplasmic terminal portion. (From Siddiqi 1971.)

General Characteristics

P. loofi, obtained from wheat fields in Spain, was described by Arias (1970), and *P. lobatus*, from a cotton field in Sudan, was described by Loof and Yassin (1970). *P. bursifer* and *P. dissitus* were transferred from *Tylenchorhynchus* to *Paratrophurus*.

With respect to relationships with other genera, Siddiqi states: "Arias (1970) placed *Paratrophurus* in the family *Trophurus* Loof 1957. Following Siddiqi's 1970 classification, this genus comes under Trophurinae Paramonov 1967 because of the abnormally thickened cuticle on the tail. The narrow and conoid lip region and the abnormally thickened cuticle on the tail suggest its direct affinities with *Trophurus* and *Macrotrophurus* Loof 1958 and hence it is placed under *Trophurinae*. However, there is a close relationship between *Paratrophurus* and *Tylenchorhynchus* especially in the structure of the esophagus, gonads, bursa, spicula, and gubernaculum."

Plate 70. Paratrophurus lobatus. A. Entire specimen showing body posture in death. B. Neck region. C. Part of genital apparatus. D and E. Tails. (After Loof and Yassin 1970. Courtesy of Nematologica.)

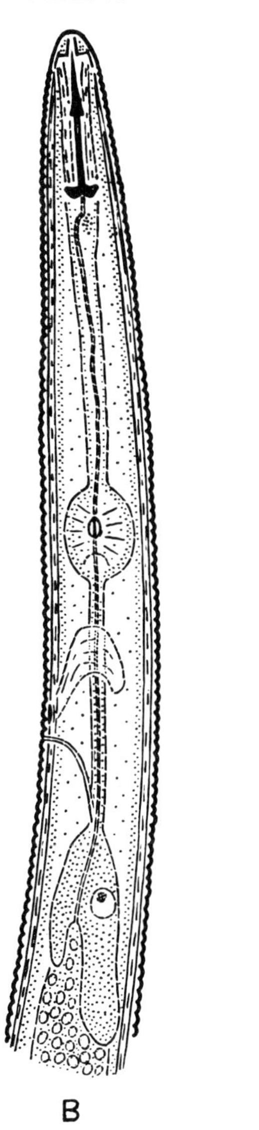
B

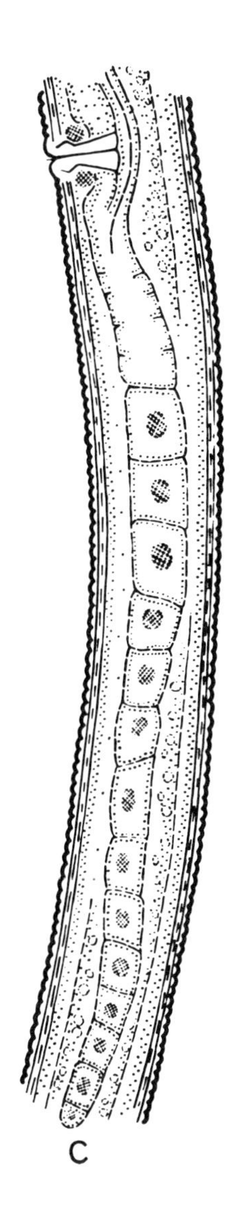
C

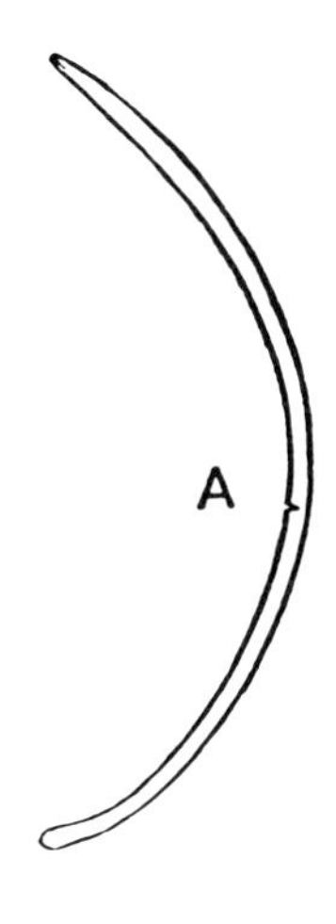
A

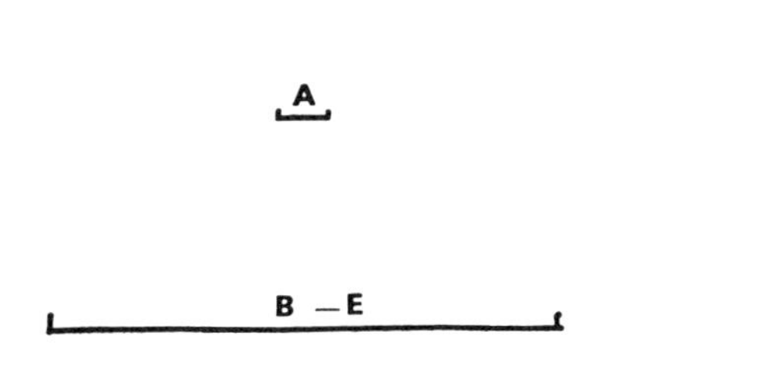
A
B —E

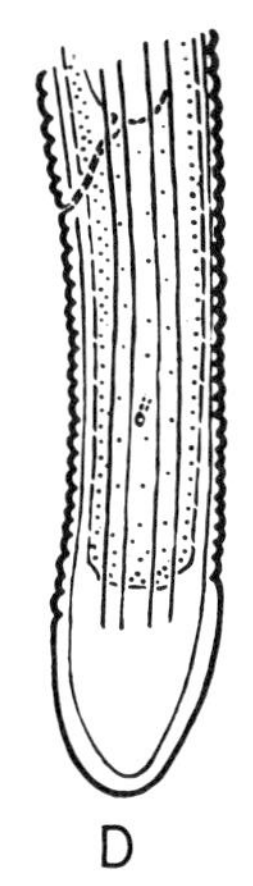
D

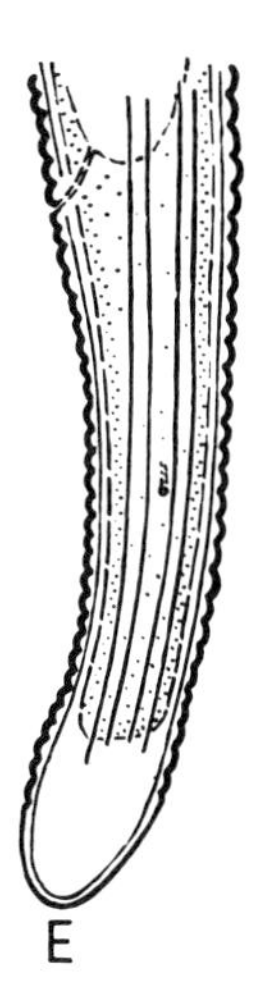
E

61. Phasmids enlarged . . . . (Example: Plate 73) 63
    Phasmids small, porelike . . . . (Example: Plate 71) 62
62. Esophagus overlapping intestine typically dorsally and laterally; lip region with or without annulation or striation; dorsal esophageal gland openings usually less than $\frac{1}{4}$ of the stylet length behind stylet knobs . . . . (Plate 71) *ROTYLENCHUS*
    Esophagus overlapping intestine typically ventrally; lip region without longitudinal striation; dorsal esophageal gland opening usually $\frac{1}{4}$ or more of the stylet length behind stylet knobs . . . . (Plate 72) *HELICOTYLENCHUS*

Rotylenchus Filipjev 1936

Type species: *Rotylenchus robustus* (de Man 1876) Filipjev 1936

Description: Hoplolaiminae. Lip region with or without annulation or striation. Esophageal glands overlap intestine primarily dorsally and laterally. Phasmids small in posterior part of body. Four incisures in lateral field. Female terminus hemispherical, usually slightly more convex dorsally, sometimes with small rounded ventral projection. (From Sher 1965.)

General Characteristics

In a monograph of this genus, Sher (1965) includes a key to 14 species. Concerning this diverse genus, Sher (1965) comments as follows:

> The species proposed in the genus *Rotylenchus* show a wide range of morphological characters. The type species, *R. robustus,* is large, robust; with a long, well-developed spear; well-developed annulation and longitudinal striation of the distinctly set-off lip region; and a male tail longer than the body width at the cloaca. *R. calvus* is half the size of the type species; short spear; lip region without well-developed annulation or longitudinal striations, not set off from the body and a male tail shorter than the body width at the cloaca (unique for the genus).
>
> *R. quartus* and *R. caudaphasmidius* differ from the type species in that they have the phasmids on the tail and a high, not set-off lip region without longitudinal striations.
>
> One species, *R. breviglans*, has only a slight overlap of the esophageal glands over the intestine and a lip region without longitudinal striations not set off from the body. This species resembles *Tylenchorhynchus brevicaudatus* Hopper 1959. An examination of paratypes of *T. brevicaudatus* shows a slight overlap of the glands over the intestine, but the lumen of the esophageal glands appears to be in the center with an equal overlap dorsally and ventrally. *R. breviglans* shows a more dorsal overlap and therefore is considered a *Rotylenchus*. Additional undescribed species closely related to this group have been seen from mountainous areas of western United States.
>
> Although some or all of the above groups may be the basis for future genera, they are not considered as such at present because many of the characters are not unique for the group and there has been limited collecting in this genus.
>
> *Rotylenchus* species are world-wide in distribution although they appear to be primarily a genus from the temperate and colder areas of the world. The genus has wide distribution in North America and Europe and is found in the colder, higher altitudes of more southerly regions.

Golden (1971) includes a description of this genus.

Plate 71. Rotylenchus buxophilus. A. Female, anterior end. B. Female, face view. C. Female, cross section through first annule of lip region. D and E. Female, surface view of posterior end. F. Female. (After Sher 1961. Courtesy of Nematologica.)

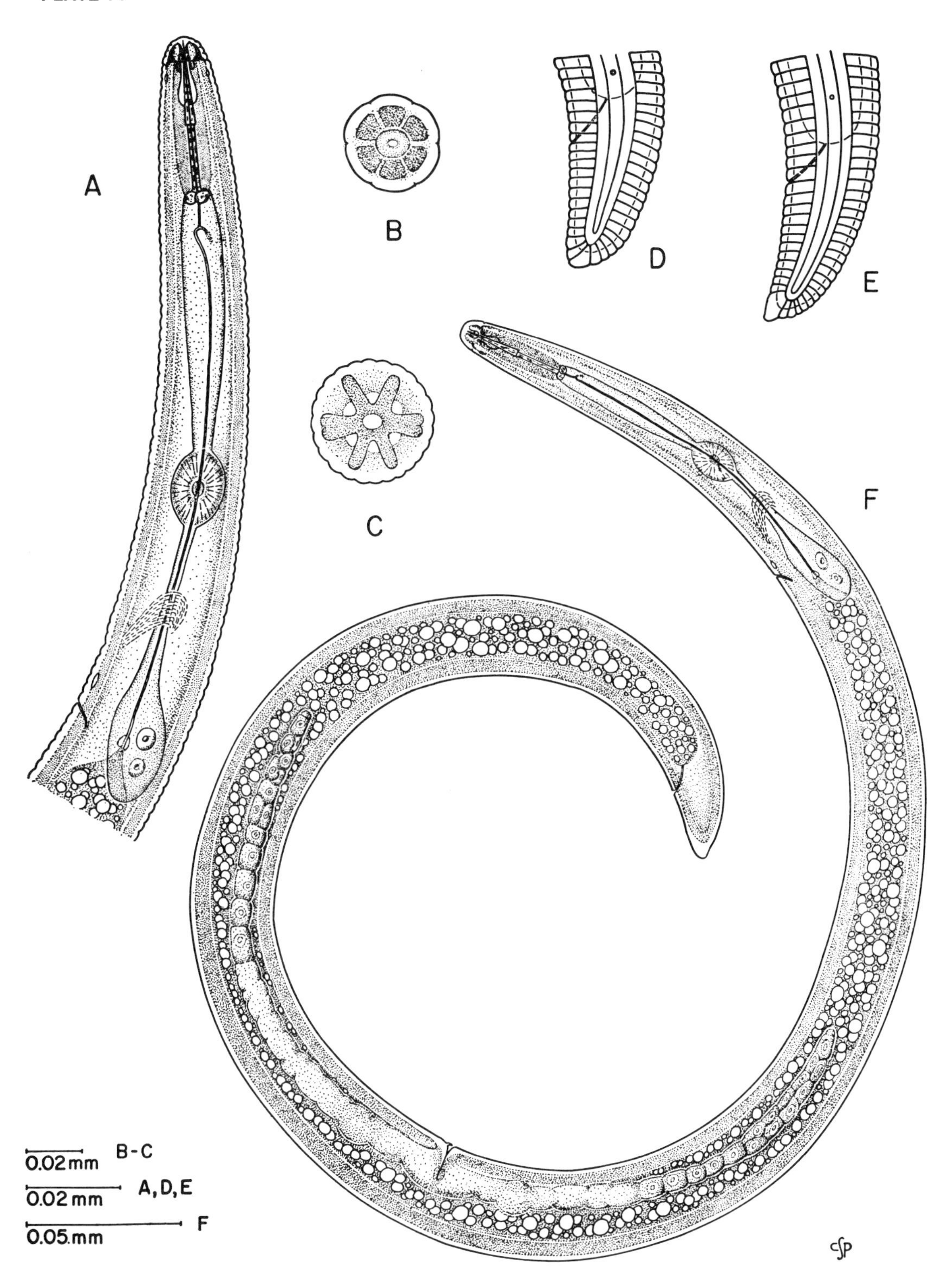
A
B
C
D
E
F
0.02 mm B-C
0.02 mm A, D, E
0.05 mm F

Helicotylenchus Steiner 1945

Type species: *Helicotylenchus dihystera* (Cobb 1893) Sher 1961

Description: Hoplolaiminae. Lip region without longitudinal striations, not set off from body. Dorsal esophageal gland opening usually $\frac{1}{4}$ or more of the spear length behind spear knobs. Esophageal glands often distinct, overlap intestine dorsally, laterally, and ventrally, the longest overlap usually ventral. Two amphidelphic functional ovaries. Female tail usually more curved dorsally, terminus hemispherical to elongated ventrally. Phasmids small, near anus. Lateral field with 4 incisures, usually areolated only anteriorly. (From Sher 1966.)

General Characteristics

Sher (1966) discussed the morphology of nematodes of this genus and includes a key to 38 species. He also states that he has examined many species of *Helicotylenchus* that probably represent additional species of this genus but, because of insufficient and/or poorly preserved material, they are not described in this paper. Siddiqi (1972) discusses this genus and includes a key to more than 70 species.

Nematodes of this genus are found frequently in soil samples collected in the vicinity of plant roots. Usually they are ectoparasites, feeding with not only the stylet inserted into the root but also the anterior part of the body (Thorne 1961). Individuals of some species, however, are migratory endoparasites. Apparently all species are parasitic on roots and other underground parts of plants. When relaxed with gentle heat, these nematodes assume a somewhat spiral form.

Antarctylus, a new genus closely related to *Helicotylenchus* was described by Sher in 1973. *A. humus*, the only species of this genus, was found in peat soil in the subantarctic.

Plate 72. Helicotylenchus dihystera. A. Female. B. Female, face view. C. Female, cross section through first annule of lip region. D. Female, surface view of posterior end. (After Sher 1961. Courtesy of Nematologica.)

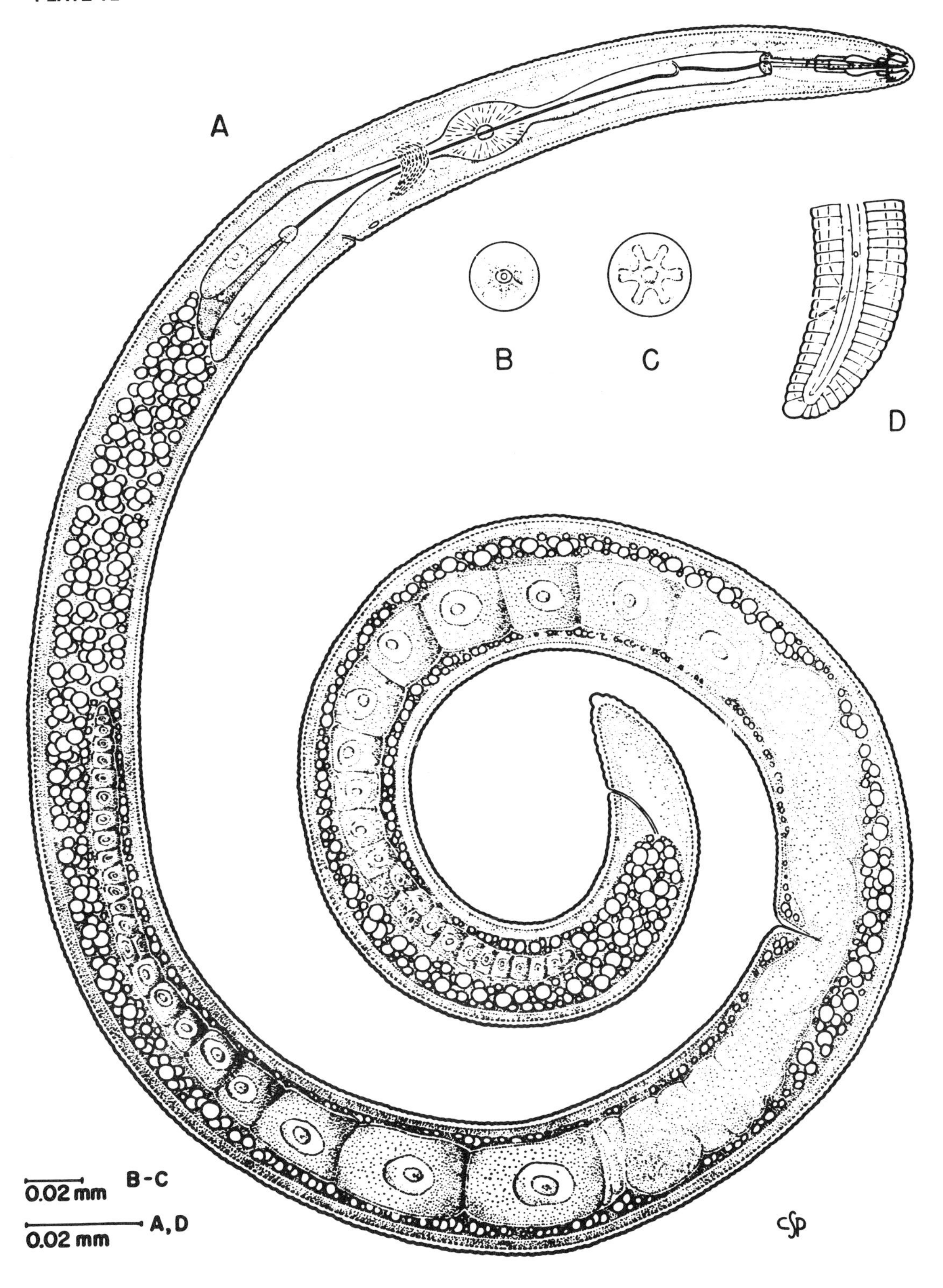
A
B
C
D
0.02 mm B-C
0.02 mm A, D
CSP

63. Both phasmids located posterior to vulva . . . . (Example: Plate 74) 64
One phasmid located anterior to vulva and 1 posterior to vulva . . . . (Example: Plate 76) 65
64. Phasmids opposite or nearly opposite each other in region of anus; lip region with transverse striae . . . . (Plate 73) *SCUTELLONEMA*
Phasmids not opposite each other, anterior to anus; lip region without striae . . . . (Plate 74) *PELTAMIGRATUS*

Scutellonema Andrassy 1958

Type species: *Scutellonema bradys* (Steiner and Le Hew 1933) Andrassy 1958

Description: Hoplolaiminae. Lip region with horizontal, and with or without longitudinal, striations. Esophageal glands overlap intestine dorsally and laterally. Phasmids enlarged, located opposite or nearly opposite one another near anal region. Four incisures in lateral field, areolated primarily in region of phasmids and anteriorly. Female tail rounded. (From Sher 1963.)

General Characteristics

Sher (1963) in a monograph on this genus presents a key to 11 species. Because of errors in this key, Sher (1964) published a revised key to the same 11 species. Timm (1965) described *S. siamense*, and *S. mangiferae* was described by Khan and Basir (1965).

With respect to areolation of lateral fields, Yeates (1967) described *S. magna* which is areolated along the whole length of the lateral fields, and Edward and Rai (1970) described *S. sheri* in which the lateral fields are not areolated.

Concerning this genus, Sher (1963) states the following:

> The genus *Scutellonema* is most closely related to the genus *Aorolaimus* Sher 1963, being distinguished from this genus mainly by the position of the enlarged phasmids.
>
> *Scutellonema* is widely distributed, primarily in the warmer areas of the world. The greatest distribution and speciation appears to be in Africa. Distribution, at least in part, has probably been accomplished by association, often as endoparasites, with roots and underground parts of plants.

All species of this genus are considered to be plant parasites, their feeding habit being either ectoparasitic or endoparasitic. Large numbers of *S. bradys* cause extensive damage to Jamaican yams (Steiner and Le Hew 1933).

Plate 73. Scutellonema brachyurum. A. Mature female, 475 X. B. Female tails, 1900 X. C. Anterior portion, 856 X. a. phasmid. b. dorsal gland outlet.

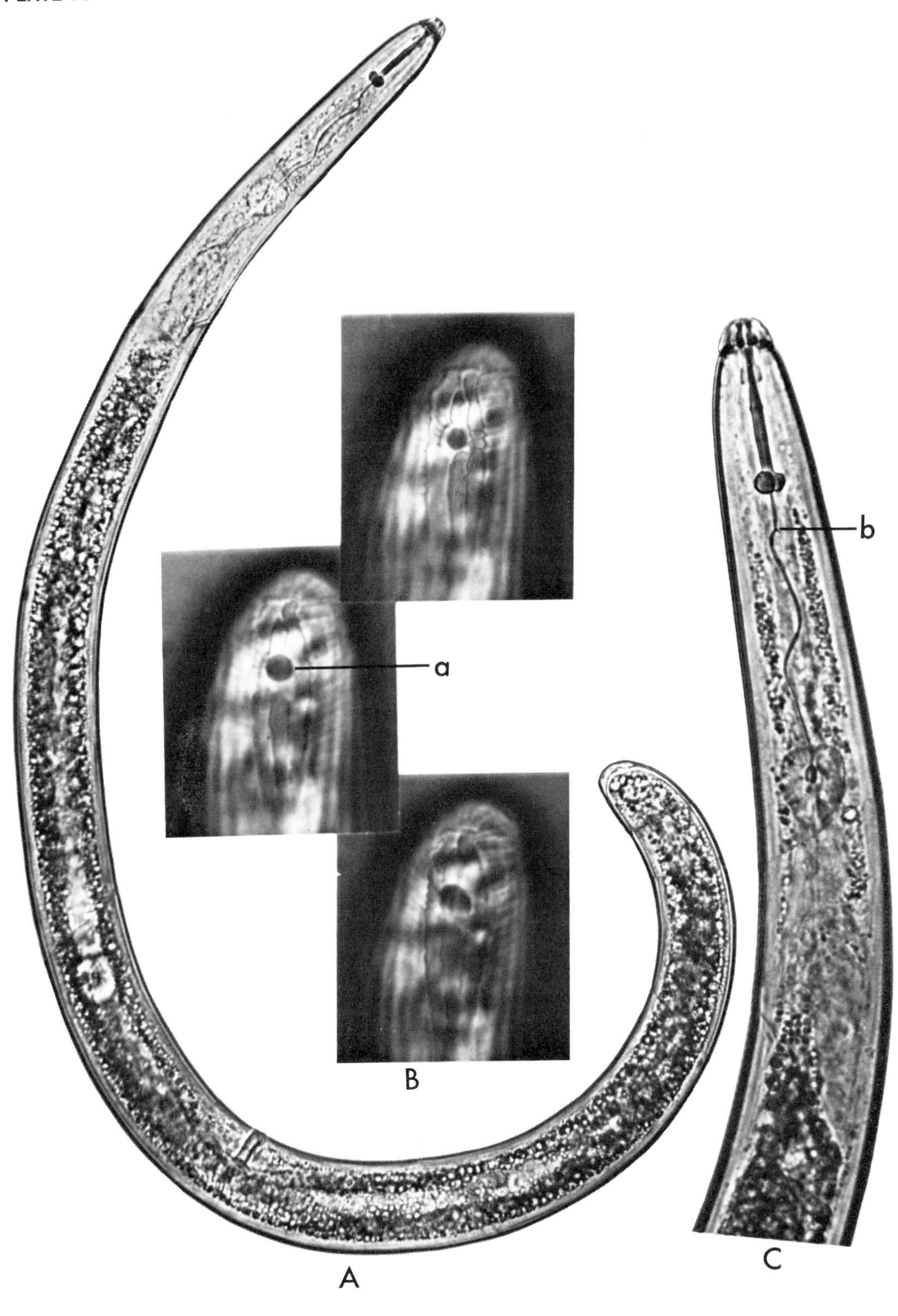
a
b
A
B
C

Peltamigratus Sher 1963

Type species: *Peltamigratus christiei* (Golden and Taylor 1956) Sher 1963

Description: Hoplolaiminae. Lip region without striation. Esophageal glands overlap intestine dorsally and laterally. Phasmids enlarged, located in posterior part of the body, not opposite one another, above the tail. Lateral field with 4 or less incisures, not areolated except most anteriorly. Female tail round. Caudal alae indented. (From Sher 1963.)

General Characteristics

Sher separates *Peltamigratus* from the closely related genera *Scutellonema* Andrassy 1958 and *Aorolaimus* Sher 1963 by the following characters: ". . . the location of the enlarged phasmids in the posterior part of the body above the anus and not opposite one another, the absence of striation on the lip region and the indented caudal alae."

Sher synonymizes *Scutellonema christiei* (Golden and Taylor 1956) Andrassy 1958, with the new genus *Peltamigratus* and uses it as the type species.

Originally described as *Rotylenchus christiei*, the type species was found in soil around roots of Bermuda grass (*Cynodon dactylon* [L.] Pers.) and "spotted spurge weed" (*Euphorbiaceae*). Type locality is the 14th fairway of the Bobby Jones Municipal Golf Course, Sarasota, Florida.

Four additional species proposed by Sher are: *P. luci*—collected from pineapple (*Ananas sativus*) soil in Martinique; *P. holdemani*—from jungle soil in Costa Rica; *P. macbethi*—from soil around potato (*Solanum* sp.) and bean soil (*Phaseolus* sp.) in Venezuela; *P. nigeriensis*—from Savannah grass, unknown tree soil, and tomato (*Lycopersicon esculentum*) soil, all in Nigeria. Sher includes a key to the species of *Peltamigratus*.

Knobloch (1969) describes a new species, *P. thornei*, and presents a key to 8 species of this genus. Smit (1971) describes *P. striatus* which differs from *P. holdemani* on the distinct annulation of the lip region, the nonindented bursa, the position of the hemizonid, the configuration of the lateral field and the presence of constrictions between annules of the female tail. Smit emends the generic diagnosis to accommodate *P. striatus* and other species. He also presents a key to 8 species of this genus.

Plate 74. Peltamigratus christiei. A. Female, posterior region. B. Cross section at vulva. C. Male, posterior end, ventral view. D. Female, posterior end. E. Female, face view. F. Female, cross section through base of lip region. G. Male, anterior end. H. Male, posterior end, lateral view. I. Female, anterior region. (After Sher 1963. Courtesy of Nematologica.)

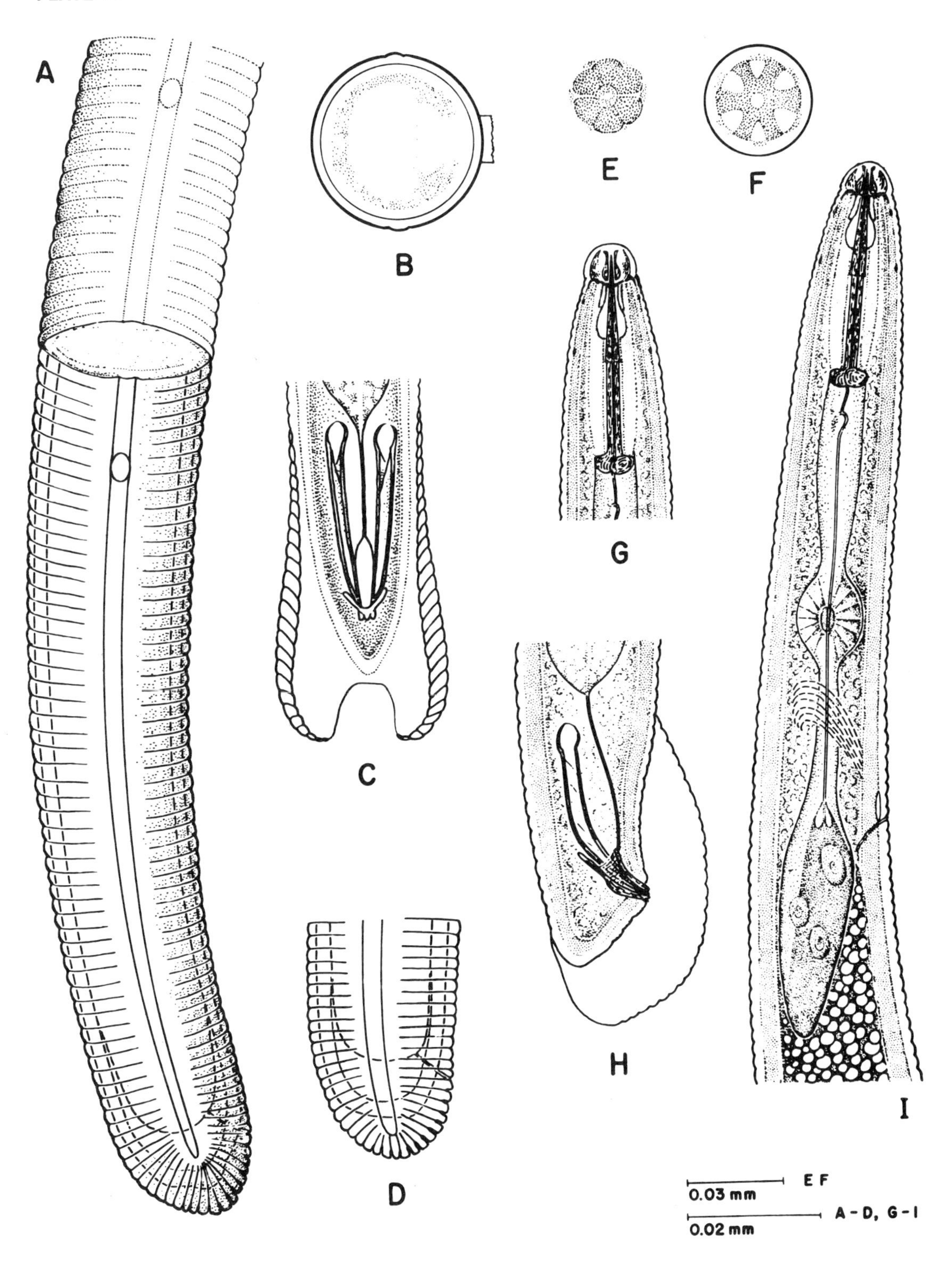
A
B
C
D
E
F
G
H
I
0.03 mm
E F
0.02 mm
A - D, G - I

65. Spear knobs with distinct anterior projections; with 4 or fewer incisures areolated throughout length of lateral field . . . . (Plate 75) *HOPLOLAIMUS*
    Spear knobs rounded or without distinct anterior projections; with 4 incisures areolated at phasmids and anteriorly . . . . (Plate 76) *AOROLAIMUS*

Hoplolaimus Daday 1905

Type species: *Hoplolaimus tylenchiformis* Daday 1905

Description: Hoplolaiminae. Lip region set off, with longitudinal striations; cephalic framework massive. Spear knobs massive, with anterior projections. Dorsal esophageal gland opening near base of spear knobs ($\frac{1}{4}$ or less the spear length). Esophageal glands overlapping intestine dorsally and laterally, with 3 to 6 nuclei. Excretory pore above or below hemizonid. Female tail round, shorter than width of body at anus. Phasmids (scutella) enlarged, not opposite one another on each side of the body. Lateral field with 4 or fewer incisures, areolated. (From Sher 1963.)

General Characteristics

Members of the genus *Hoplolaimus*, the lance nematodes, usually feed some distance back of the root tip and may or may not enter the root. This genus is associated with many kinds of plant roots and is rated as the most important nematode pest of turf in Florida (Christie 1959).

Two distinguishing characteristics of the genus are an offset, caplike lip region, which is divided into minute blocks by longitudinal and lateral striations; and large scutellumlike phasmids, 1 located above the vulva and 1 located below the vulva. The lance nematodes are relatively large (approximately 0.95 to 1.80 mm). The massive stylets possess well-developed forward-pointing knobs. When relaxed with gentle heat, individuals assume a straight or slightly arcuate position.

Plate 75. Hoplolaimus galeatus. A. Adult female, 295 X. B. Anterior portion, 440 X. C. Posterior portion of adult male, 478 X. D. Phasmid in lateral field, 430 X. E. Lateral field highly magnified. a. lip region. b. gubernaculum. c. bursa. d. spicula. e. phasmid. f. incisure.

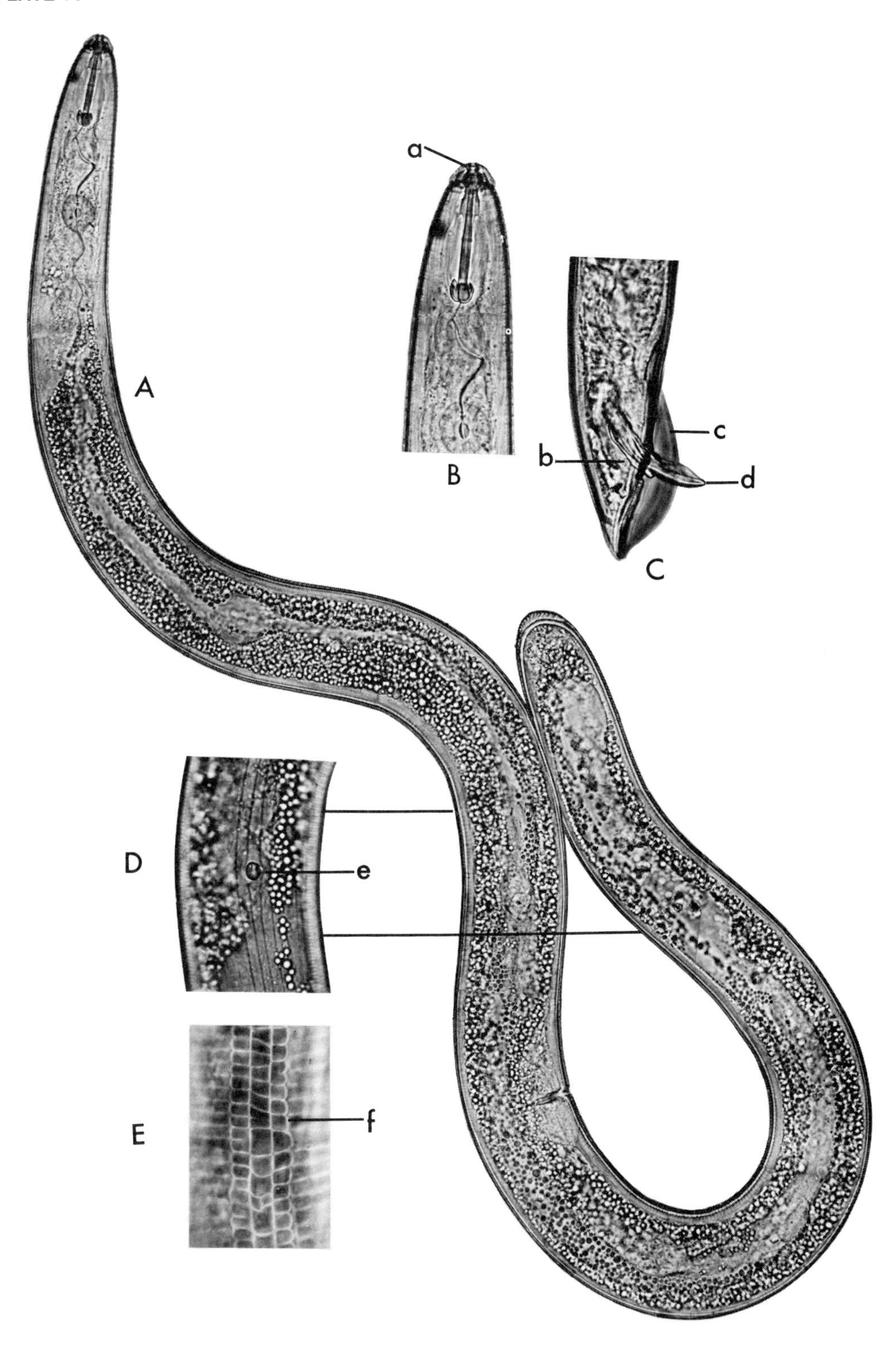
a
A
B
b
c
d
C
D
e
E
f

Aorolaimus Sher 1963

Type species: *Aorolaimus helicus* Sher 1963

Description: Hoplolaiminae. Lip region continuous or set off from body, with or without longitudinal striations; cephalic framework not as massive as *Hoplolaimus*. Spear knobs without distinct anterior projections. Dorsal esophageal gland opening near base of spear knobs ($\frac{1}{4}$ or less the spear length). Esophageal glands overlapping intestine dorsally and laterally with 3 nuclei. Excretory pore below hemizonid. Female tail round, shorter than width of body at anus. Phasmids (scutella) enlarged, 1 in anterior part of body and 1 in posterior part of the body. Lateral field with 4 incisures, not areolated, except at phasmids and anterior end. (From Sher 1963.)

General Characteristics

Sher (1963) erected the genus *Aorolaimus* to include 3 new species, *A. israeli*, *A. helicus*, and *A. leipogammus*. Concerning the characteristics used to separate this new genus from the closely related genera, *Hoplolaimus* and *Scutellonema*, Sher states: "The genus *Aorolaimus* can be distinguished from *Hoplolaimus* by the cephalic framework which is less massive; size and shape of the spear knobs which are less massive and without distinct anterior projections; lateral field with 4 incisures areolated only at the phasmids and anteriorly; and smaller body size, spear, spicules, gubernaculum, and capitulum. *Aorolaimus* can be distinguished from other genera of the Hoplolaiminae by the enlarged phasmids (scutella), 1 in the anterior part and 1 in the posterior part of the body."

Aorolaimus capsici was described by Jimenez-Millan, Arias, and Fijo (1964), *A. baldus* and *A. torpidus* by Thorne and Malek (1968), and *A. intermedius* by Suryawanski (1971). According to Suryawanski (1971) "*A. intermedius* n. sp. while showing affinities to *Aorolaimus* in having the spear knobs without anteriorly directed projections and the lip region continuous with the body contour, differs from all the 5 species described under *Aorolaimus* and resembles *Hoplolaimus* in having no incisures in the lateral field and the excretory pore anterior to the hemizonid."

The enlarged phasmids of *Scutellonema* are both located posterior to the vulva. When relaxed, these nematodes assume a C-shaped or spiral position.

None of these species have been shown experimentally to be plant parasites. However, because they occur around plant roots and are closely related to other plant parasitic species, it is probable that they are root parasites.

Plate 76. Aorolaimus helicus. A. Female, anterior end. B. Female, posterior end. C. Male, anterior end. D. Male, posterior end. E. Female, cross section through basal annule of lip region. F. Female, face view. G. Female. (After Sher 1963. Courtesy of Nematologica.)

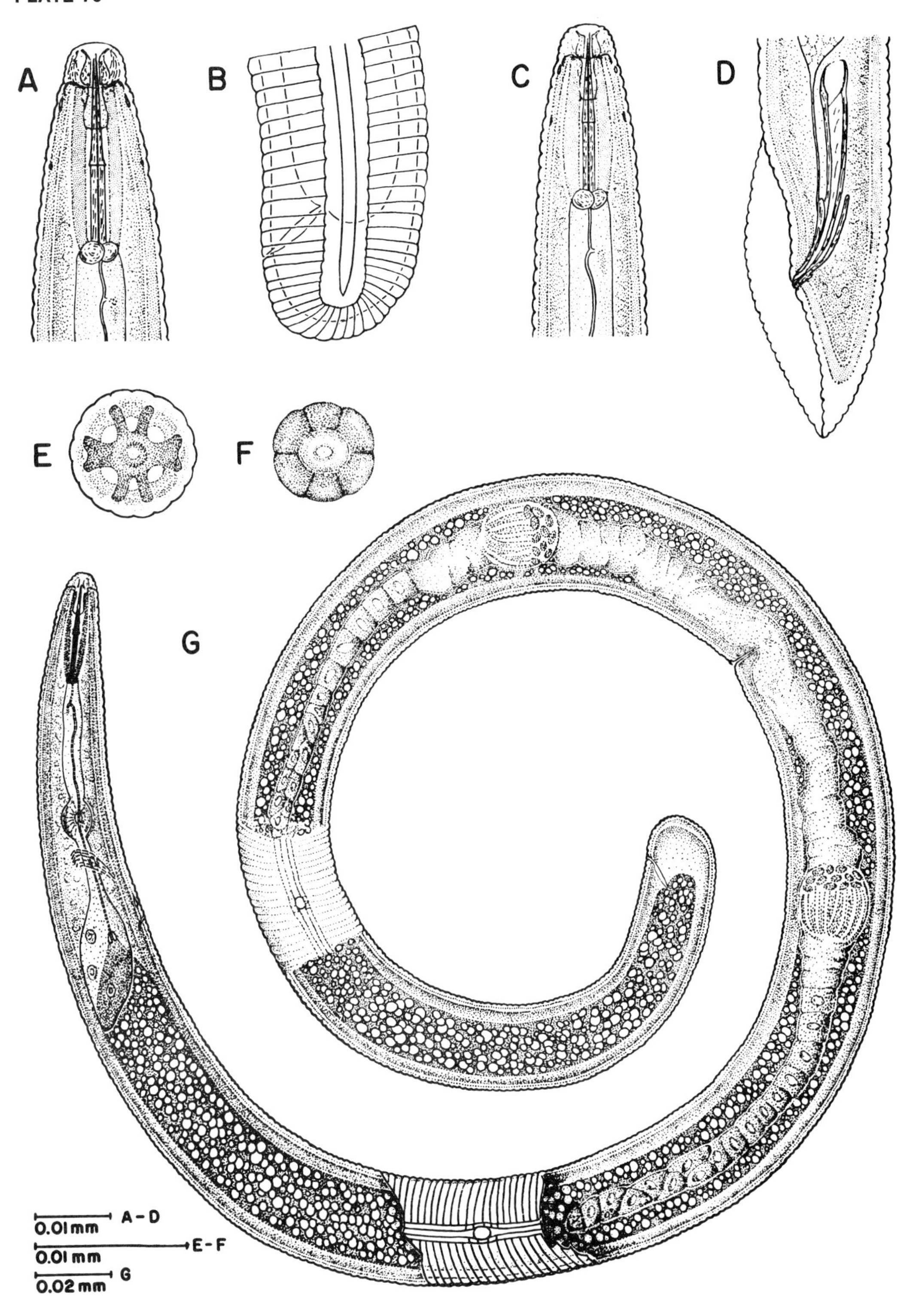
A
B
C
D
E
F
G
0.01mm A-D
0.01mm E-F
0.02mm G

SELECTED REFERENCES

Acontylus

Guiran, G. de. 1967. Description de deux espèces nouvelles du genre *Tylenchorhynchus* Cobb 1913 (Nematoda: Tylenchinae) accompagnée d'une clé des femelles, et précisions sur *T. mamillatus* Tobar-Jimenez 1966. Nematologica 13:217–230.
s'Jacob, J. J. 1959. *Hoplotylus femina* n. g., n. sp. (Pratylenchinae: Tylenchida) associated with ornamental trees. Nematologica 4:317–321.
Linford, M. B., and Juliette M. Oliveira. 1940. *Rotylenchulus reniformis*, nov. gen., n. sp., a nematode parasite of roots. Proc. Helminth. Soc. Wash. 7:35–42.
Luc, M. 1960. Trois nouvelles espèces genre *Rotylenchoides* Whitehead 1958 (Nematoda: Tylenchida). Nematologica 5:7–17.
Meagher, J. W. 1968. *Acontylus vipriensis* n. g., n. sp. (Nematoda: Hoplolaimidae) parasitic on Eucalyptus sp. in Australia. Nematologica 14:94–100.
Thorne, G. 1949. On the classification of the Tylenchida, new order (Nematoda: Phasmidia). Proc. Helminth. Soc. Wash. 16:37–73.
Thorne, G., and M. W. Allen. 1914. *Nacobbus dorsalis* nov. gen., nov. spec. (Nematoda: Tylenchidae) producing galls on the roots of alfileria, *Erodium cicutarium* (L.) L'Her. Proc. Helminth. Soc. Wash. 11:27–31.
Whitehead, A. G. 1958. *Rotylenchoides brevis* n. g., n. sp. (Rotylenchoidinae n. subfam: Tylenchida). Nematologica 3:327–331.

Anguina

Chitwood, B. G. 1935. Nomenclatorial notes, I. Proc. Helminth. Soc. Wash. 2:51–54.
Cobb, N. A. 1932. Nematosis of a grass of the genus Cyanodon caused by a new nema of the genus *Tylenchus* Bastian. J. Wash. Acad. Sci. 22:243–245.
Corbett, D. C. M. 1966. Central African nematodes, III. *Anguina hyparrheniae* n. sp. associated with "Witches" Broom of *Hyparrhenia* spp. Nematologica 12:280–286.
Courtney, W. D., and H. B. Howell. 1952. Investigations on the bent grass nematode, *Anguina agrostis* (Steinbuch 1799) Filipjev 1936. Pl. Dis. Reptr. 36:75–83.
Filipjev, I. N. 1936. On the classification of the Tylenchinae. Proc. Helminth. Soc. Wash. 3:80–82.
Filipjev, I. N., and J. H. Schuurmans Stekhoven, Jr. 1941. A manual of agricultural helminthology. Brill, Leiden.
Goffart, H. 1942. *Anguina klebahni* n. sp. (Tylenchidae), ein Nematode in Bluten von Primula florindae Ward. Zool. Anz. Leipzig 138 (7–8):174–179.
Goodey, J. B. 1959. Gall-forming nematodes of grasses in Britain. J. Sports Turf Res. Inst. 10:1–7.
Goodey, T. 1927. On *Tylenchus graminis* (Hardy 1850) Marcinowski 1909. J. Helminth. 5:163–170.
Goodey, T. 1930. On *Tylenchus agrostis* (Steinbuch 1799). J. Helminth. 8:197–210.
Goodey, T. 1932. The genus *Anguillulina* Gerv. and v. Ben. 1895, vel Tylenchus Bastian 1865. J. Helminth. 10:75–180.
Goodey, T. 1951. Soil and freshwater nematodes. Methuen, London.
Goodey, T. Revised by J. B. Goodey. 1963. Soil and freshwater nematodes. Methuen, London.
Hirschmann, Hedwig. 1960. The genera *Tylenchus, Psilenchus, Ditylenchus, Anguina, Tylenchorhynchus, Tetylenchus, Trophurus*, and *Macrotrophurus*. *In*: J. N. Sasser and W. R. Jenkins, eds. Nematology. Univ. North Carolina Press, Chapel Hill, pp. 171–180.
Kirjanova, E. S. 1944. Plant nematodes of Tadzhikistan. Izvest. Fil. Akad. Nauk Tadzhik, SSR 5:83–94.
Kirjanova, E. S. 1952. A plant nematode—*Anguina poophila* Kirjanova sp. nova. Zool. Zh. 31:223–227.
Kirjanova, E. S. 1955. *Paranguina agropyri* Kirjanova gen. et sp. n. (Nematodes). Trudy Zool. Inst., Akad. Nauk, SSSR 18:42–52.

Kirjanova, E. S., and T. S. Ivanova. 1968. New species of *Paranguina* (Nematoda: Tylenchidae) from the Tadzhik SSR. *In*: "Ushchele Kondara," Dushanbe, Izdatelstvo "Donish," pp. 200–217.
Norton, D. C. 1965. *Anguina agropyronifloris* n. sp., infecting florets of *Agropyron smithii*. Proc. Helminth. Soc. Wash. 32:118–122.
Scopoli, G. A. 1777. Introduction ad historiam natural em sistens genera lapidum, plantarum et animalium hactenus detecta, caracteribus essentialibus donata, in tribus divisa subinde ad leges naturae. Pragae.
Steiner, G. 1937. Opuscula miscellanea nematologica, V. Proc. Helminth. Soc. Wash. 4:33–38.
Steiner, G. 1940. Opuscula miscellanea nematologica, VIII. Proc. Helminth. Soc. Wash. 7:54–62.
Thorne, G. 1934. Some plant-parasitic nemas, with descriptions of three new species. J. Agr. Res. 49:755–763.
Thorne, G. 1949. On the classification of the Tylenchida, new order (Nematoda: Phasmidia). Proc. Helminth. Soc. Wash. 16:37–73.
Thorne, G. 1961. Principles of nematology. McGraw-Hill, New York.
Whitehead, A. G. 1959. *Nothanguina cecidoplastes* n. comb. syn. *Anguina cecidoplastes* (Goodey 1934) Filipjev 1936 (Neotylenchinae: Tylenchida). Nematologica 4:70–75.
Wu, Liang-Yu. 1967. *Anguina calamagrostis*, a new species from grass, with an emendation of the generic characters for the genera *Anguina* Scopoli 1777 and *Ditylenchus* Filipjev 1936 (Tylenchidae: Nematoda). Can. J. Zool. 45:1003–1010.
Yokoo, T., and Y. E. Choi. 1968. On a new species of shoot gall nematode (Tylenchidae: *Anguina*) found from the galls on the leaves of moxa (*Artemisia saiatica* Nakai). Agr. Bull., Saga Univ., No. 26, pp. 1–7.

Aorolaimus

Guiran, G. de, and S. A. Sher. 1968. Sur l'identite d; *Hoplolaimus leiomerus* de Guiran 1963 et *Aorolaimus israeli* Sher 1963. Nematologica 14:313.
Jimenez Millan, F., M. Arias, and M. A. Fijo. 1964. *Aorolaimus capsici* n. sp. (Nematoda: Hoplolaiminae). Boln R. Soc. esp. Hist. Nat. (B) 62:283–287.
Sher, S. A. 1963. Revision of the Hoplolaiminae (Nematoda), II. *Hoplolaimus* Daday 1905 and *Aorolaimus* n. gen. Nematologica 9:267–295.
Suryawanshi, M. V. 1971. Studies on Tylenchida (Nematoda) from Marathwada, India, with descriptions of four new species. Nematologica 17:393–406.
Thorne, G., and R. B. Malek. 1968. Nematodes of the northern great plains, Part I. Tylenchida (Nemata: Secernentea). S. Dak. Agr. Exp. Stn. Tech. Bull. 31:111.
Yuen, Pick H. 1964. The female gonad in the subfamily Hoplolaiminae with a note on the spermatheca of *Tylenchorhynchus*. Nematologica 10:570–580.

Aphasmatylenchus

Germani, G. 1970. *Aphasmatylenchus straturatus* sp. n. (Nematoda: Hoplolaimidae) from West Africa. Proc. Helminth. Soc. Wash. 37:48–51.
Sher, S. A. 1965. *Aphasmatylenchus nigeriensis* n. gen., n. sp. (Aphasmatylenchinae n. subfam., Tylenchoidea: Nematoda) from Nigerian soil. Proc. Helminth. Soc. Wash. 32:172–176.

Aphelenchoides

Allen, M. W. 1952. Taxonomic status of the bud and leaf nematodes related to *Aphelenchoides fragariae* (Ritzema Bos 1891). Proc. Helminth. Soc. Wash. 19:108–120.
Allen, M. W. 1960. The superfamily Aphelenchoidea. *In*: J. N. Sasser and W. R. Jenkins, eds. Nematology. Univ. North Carolina Press, Chapel Hill, pp. 220–221.
Andrassy, I. 1952. Freilebende Nematoden aus dem Bukk-Gebirge. Annls. Hist., Nat. Mus. Natn. Hungary (Ser. Nov.) 2:13–65.
Andrassy, I. 1957. *Aphelenchoides citri* n. sp. ein neuer Wurzelparasit der Zitrone. Nematologica 2:237–240.
Andrassy, I. 1958. Erd- und Susswassernematoden aus Bulgarien. Acta Zool., Budapest 4(1/2):1–88.
Andrassy, I. 1959. Neue und wenig bekannte Nematoden aus Jugoslawien. Annls. Hist., Nat. Mus. Natn. Hungary (Ser. Nov.) 51:259–275.
Baranovskaya, I. A. 1963. Two new species of *Aphelenchoides* Fischer 1894 (Nematoda: Aphelenchoididae). *In*: Helminths of man, animals and plants and their control. Papers on helminthology presented to Academician K. I. Skryabin on his 85th birthday. Moscow, Izdatelstvo Akad. Nauk, SSSR, pp. 480–483.
Baranovskaya, I. A., and M. M. Haque. 1968. *Aphelenchoides graminis* n. sp. (Nematoda: Aphelenchoididae). Zool. Zh. 47 (4):631–634.
Bessarabova, L. M. 1966. *Aphelenchoides conimucronatus* n. sp. from leguminous plants in the Moscow region. Zool. Zh. 45 (10):1569–1570.
Brzeski, M. 1962. A rare nematode species, *Aphelenchoides kungradensis* Karimova, and a nomenclatorial note on *A. spinocaudatus* Skarbilovich (Nematoda: Aphelenchoididae). Bull. Acad. Pol. Sci. Cl. II Ser. Sci. Biol. 10(11):479–481.
Carvalho, J. C. 1953. *Aphelenchoides coffeae* em raizes de geranio. Revta Inst. Adolfo Lutz 13:33–35.
Cayrol, J. C. 1967. Etude du cycle évolutif d'*Aphelenchoides composticola*. Nematologica 13:23–32.
Chawla, M. L., B. L. Bhamburkar, E. Khan, and S. K. Prasad. 1968. One new genus and seven new species of nematodes from India. Labdev. J. Sci. Technol., Ser. B. 6(2):86–100.
Christie, J. R. 1932. Recent observations on the strawberry dwarf nematode in Massachusetts. Pl. Dis. Reptr. 16:113–114.

SELECTED REFERENCES

Christie, J. R. 1939. Predaceous nematodes of the genus *Aphelenchoides* from Hawaii. J. Wash. Acad. Sci. 29:161–170.

Christie, J. R. 1942. A description of *Aphelenchoides besseyi*, n. sp., the summer-dwarf nematode of strawberries, with comments on the identity of *Aphelenchoides subtenuis* (Cobb 1926) and *Aphelenchoides hodsoni* Goodey 1935. Proc. Helminth. Soc. Wash. 9:82–84.

Christie, J. R. 1959. Plant nematodes, their bionomics and control. H. and W. B. Drew, Jacksonville, Florida.

Christie, J. R., and Louise Crossman. 1936. Notes on the strawberry strains of the bud and leaf nematode, *Aphelenchoides fragariae*. Proc. Helminth. Soc. Wash. 3:69–72.

Cobb, N. A. 1927. Note on a new nema *Aphelenchus retusus* with a proposed division of *Aphelenchus* into three subgenera. J. Parasit. 14:57–58.

Das, V. M. 1960. Studies on the nematode parasites of plants in Hyderabad (Andhra Pradesh, India). Z. Parasitenk. 19:553–605.

Dmitrenko, M. A. 1966. Two new phytohelminth species. Zool. Zh. 45:764–766.

Edward, J. C., and S. L. Misra. 1969. Occurrence of some new species of *Aphelenchoides* in the rhizosphere of certain field crops of Uttar Pradesh, India, with a note on an intersex. Allahabad Fmr. 43(1):1–6.

Eroshenko, A. S. 1967. The nematode fauna of oats in the Primorsk territory. Mater. nauch. Konf. vses. Obshch. Gel'mint., Year 1966, Part 5, pp. 159–160.

Eroshenko, A. S. 1967. Three new species of *Aphelenchoides* (Nematoda: Aphelenchoididae). Zool. Zh. 46(4):617–620.

Eroshenko, A. S. 1968. Five new species of *Aphelenchoides* Fischer 1894 (Nematoda: Aphelenchoididae). Soobshch. dal'nevost. Fil. V. L. Komarova sib. Otdel. Akad. Nauk, SSSR, No. 26, pp. 58–66.

Eroshenko, A. S. 1968. Three new species of *Aphelenchoides* (Nematoda: Aphelenchoididae). *In*: Parasites of animals and plants. Moscow, "Nauka," No. 4, pp. 224–228.

Fischer, M. 1894. Ueber eine Clematis-Krankheit. Berichte Physiol. Lab., Landwirthsch. Inst., Univ. Halle, 11:1–11.

Franklin, Mary T. 1952. A disease of *Scabiosa caucasia* caused by the nematode *Aphelenchoides blastophthorus* n. sp. Ann. Appl. Biol. 39:54–60.

Franklin, Mary T. 1957. *Aphelenchoides composticola* n. sp. and *A. saprophilus* n. sp. from mushroom compost and rotting plant tissues. Nematologica 2:306–313.

Franklin, Mary T. 1959. Plant-parasitic nematodes of the genus *Aphelenchoides* Fischer 1894. Plant Nematology, Bull. Minist. Agr. Fish. Fd., London 7:71–77.

Franklin, Mary T. and M. R. Siddiqi. 1963. *Aphelenchoides trivialis* n. sp. from South India. Nematologica 9:15–18.

Fortuner, R. 1970. On the morphology of *Aphelenchoides besseyi* Christie 1942 and *A. siddiqii* n. sp. (Nematoda: Aphelenchoides). J. Helminth. 44:141–152.

Goffart, H. 1928. Zur Systematik und Biologie von *Aphelenchus ritzemabosi* Schwartz (Nemat.). Zool. Anz. 76:242–250.

Goffart, H. 1930. Die Aphelenchen der kulturpflanzen. Monog. Pflchutz, No. 4, Berlin.

Goodey, J. B. 1960. The classification of the Aphelenchoidea Fuchs 1937. Nematologica 5:111–126.

Goodey, T. 1927. On the nematode genus *Aphelenchus*. J. Helminth. 5:203–220.

Goodey, T. 1928. The species of the *Aphelenchus*. J. Helminth. 6:121–160.

Goodey, T. 1929. On some details of comparative anatomy in *Aphelenchus, Tylenchus*, and *Heterodera*. J. Helminth. 7:223–230.

Goodey, T. 1935. *Aphelenchoides hodsoni* n. sp., a nematode affecting Narcissus bulbs and leaves. J. Helminth. 13:167–172.

Goodey, T. 1951. Soil and freshwater nematodes. Methuen, London.

Goodey, T. Revised by J. B. Goodey. 1963. Soil and freshwater nematodes. Methuen, London.

Gritsenko, V. P. 1971. *Ditylenchus tenuidens* n. sp. and *Aphelenchoides curiolis* n. sp. (Nematoda: Tylenchidae and Aphelenchoididae) from Kirgizia. Zool. Zh. 50(9):1402–1405.

Heyns, J. 1964. *Aphelenchoides helicus* n. sp. and *Ditylenchus equalis* n. sp., two new soil-inhabiting nematodes. S. Afr. J. Agr. Sci. 7(1):147–150.

Hooper, D. J. 1958. *Aphelenchoides dactylocercus* n. sp. and *A. sacchari* n. sp. (Nematoda: Aphelenchoidea). Nematologica 3:229–235.

Hooper, D. J., and R. F. Myers. 1971. *Aphelenchoides rutgersi* n. sp. (Nematoda: Aphelenchoidea), description and morphometrics, with observations on *A. dactylocercus* Hooper 1958 and *A. cibolensis* Riffle 1970. Nematologica 17:295–302.

Husain, S. I., and A. M. Khan. 1967. On the status of the genera of the superfamily Aphelenchoidea (Fuchs 1937) Thorne 1949, with the descriptions of six new species of nematodes from India. Proc. Helminth. Soc. Wash. 34:167–174.

Karimova, I. S. 1957. Eelworms of crops on the left bank of the Amu-Darya basin. *In*: A. I. Zemlyanskaya, L. V. Tikhonova, and I. S. Karimova, eds. Eelworms of agricultural crops in the Uzbek SSR Akad. Nauk Uzbek, SSR, Tashkent, pp. 133–208.

Khak, M. M. 1967. *Paraphelenchoides capsuloplanus* n. g., n. sp. (Nematoda: Aphelenchoididae). Zool. Zh. 46(8):1251–1253.

Khak, M. M. 1968. *Aphelenchoides echinocaudatus* n. sp. (Nematoda: Aphelenchoididae). Zool. Zh. 47(2):287–289.

Krall, E. 1959. New and little known Tylenchus (Nematoda: Tylenchida), including a description of gynandromophism in the genus *Aphelenchoides*. Eesti NSV Tead. Akad. Toim., Seer. Biol. 8(3):190–198.

Lisetskaya, L. F. 1971. *Aphelenchoides menthae* n. sp. (Nematoda: Aphelenchoididae). *In*: Parazity zhivotnykh i rastenii. Kishinev, RIO Akad. Nauk Moldavskoi, SSR, No. 6, pp. 123–126.

Loof, P. A. A. 1959. Ueber das Vorkommen von Endotokia matricida bei Tylenchida. Nematologica 4:238–240.
Micoletzky, H. 1914. Freilbende Susswasser-Nematoden der Ost-Alpen. Zool. Jb. f. syst. 36(4/5):331–546.
Micoletzky, H. 1917. Freilbende Susswasser-Nematoden der Bukowina. Zool. Jb. f. syst. 40(6):443–586.
Paesler, F. 1957. Beschreibung einiger Nematoden aus Champignonbeeten. Nematologica 2:314–328.
Riffle, J. W. 1970. *Aphelenchoides cibolensis* (Nematoda: Aphelenchoididae), a new mycophagous nematode species. Proc. Helminth. Soc. Wash. 37:78–80.
Ritzema Bos, J. 1891. Zwei neue Nematodenkrankheiten der Erdbeerpflanze. Z. Pflkrankh. Pflpath. Pflschutz 1:1–16.
Sanwal, K. C. 1959. A simple method for rearing pure populations of the foliar nematode *Aphelenchoides ritzemabosi* in the laboratory. Can. J. Zool. 37:707–711.
* Sanwal, K. C. 1961. A key to the species of the nematode genus *Aphelenchoides* Fischer 1894. Can. J. Zool. 39:143–148.
Sanwal, K. C. 1961. *Aphelenchoides subparietinus* n. sp. (Nematoda: Aphelenchoididae) from diseased lily bulbs. Can. J. Zool. 39:573–577.
Sanwal, K. C. 1965. Appraisal of taxonomic characters of "Parietinus group" of species of the genus *Aphelenchoides* Fischer 1894 (Nematoda: Aphelenchoididae). Can. J. Zool. 43:987–995.
Sanwal, K. C. 1965. Two new species of the genus *Aphelenchoides* Fischer 1894 (Nematoda: Aphelenchoididae) from the Canadian Arctic. Can. J. Zool. 43:933–940.
Shavrov, G. N. 1967. Three new species of *Aphelenchoides* Fischer 1894 (Nematoda: Aphelenchoididae). Zool. Zh. 46(5):762–764.
Siddiqi, M. R., and Mary T. Franklin. 1967. *Aphelenchoides goodeyi* n. sp. (Nematoda: Aphelenchoidea), a mycophagous nematode from South India. Nematologica 13:125–130.
Siddiqi, M. R., S. I. Husain, and A. M. Khan. 1967. *Seinura propora* n. sp. and *Aphelenchoides aligarhiensis* n. sp. (Nematoda: Aphelenchoididae) from North India. Nematologica 13:287–290.
Siddiqui, I. A., and D. P. Taylor. A redescription of *Aphelenchoides bicaudatus* (Imamura 1931) Filipjev and Schuurmans Stekhoven 1941 (Nematoda: Aphelenchoididae), with a description of the previously undescribed male. Nematologica 13:581–585.
Singh, S. D. 1967. On two new species of the genus *Aphelenchoides* Fischer 1894 (Nematoda: Aphelenchoididae) from North India. J. Helminth. 41:63–70.
Singh, S. P. 1969. A new species of the genus *Aphelenchoides* Fischer 1894 (Nematoda: Aphelenchoididae) from rootlets of radish (*Raphanus sativus*) in Lucknow. J. Helminth. 43:193–196.
Slankis, A. 1967. *Aphelenchoides macromucrons* n. sp. (Tylenchida) from Ips typographus L. Mater. nauch. Konf. vses. Obshch. Gel'mint., Part 5, pp. 279–282.
Steiner, G. 1935. Opuscula miscellanea nematologica, II. Proc. Helminth. Soc. Wash. 2:104–110.
Steiner, G. 1936. Opuscula miscellanea nematologica, IV. Observations of nematodes in bulbs of an *Iris tingitana* hybrid. Proc. Helminth. Soc. Wash. 3:74–80.
Steiner, G. 1937. Opuscula miscellanea nematologica, VI. 1. The status of the nematode *Aphelenchoides coffeae* (Zimmerman 1898), new comb. 2. The occurrence of the bud and leaf nematode, *Aphelenchoides fragariae* (Ritzema Bos 1891) Christie 1932, on the peony and oriental poppy in the U.S.A. Proc. Helminth. Soc. Wash. 4:48–52.
Steiner, G. 1941. Nematodes parasitic on and associated with roots of marigolds (*Tagetes* hybrids). Proc. Biol. Soc. Wash. 54:31–34.
Steiner, G. 1945. Opuscula miscellanea nematologica, VIII. 3. Further notes on *Aphelenchoides limberi* Steiner 1936. Proc. Helminth. Soc. Wash. 7:54–62.
Steiner, G., and Edna M. Buhrer. 1934. *Aphelenchoides xylophilus* n. sp., a nematode associated with blue-stain and other fungi in timber. J. Agr. Res. 48:949–951.
Strumpel, H. 1967. Beobachtungen zur Lebensweise von *Aphelenchoides fragariae* in Lorraine-Begonien. Nematologica 13:67–72.
Suryawanshi, M. V. 1971. Studies on Aphelenchoidea (Nematoda) from Marathwada, India, with descriptions of three new species and a discussion on the validity of *Aphelenchus radicicolus* (Cobb 1913) Steiner 1931. Nematologica 17:417–427.
Tandon, R. S., and S. P. Singh. 1970. On two new nematodes (Aphelenchoidea) from tobacco roots in India. J. Helminth. 44:323–328.
Thorne, G. 1961. Principles of nematology. McGraw-Hill, New York.
Tikyani, M. G., S. Khera, and G. C. Bhatnager. 1970. *Aphelenchoides jodhpurensis* n. sp. from soil of great millet from Rajasthan, India. Zool. Anz. 184(3/4):239–241.
Timm, R. W., and Mary T. Franklin. 1969. Two marine species of *Aphelenchoides*. Nematologica 15: 370–375.
Wasilewska, L. 1969. *Aphelenchoides dubius* sp. n. from Poland (Nematoda: Aphelenchoidea). Bull. Acad. Pol. Sci. Cl. II Ser. Sci. Biol. 17(7): 455–458.
Yokoo, T. 1964. On a new species of *Aphelenchoides* (Aphelenchidae: Nematoda), parasite of bulb of lily, from Japan. Agr. Bull., Saga Univ., No. 20, pp. 67–69.

Aphelenchus

Andrassy, I. 1952. Freilebende Nematoden aus dem Bukk-Gebirge. Annls. Hist., Nat. Mus. Natn. Hungary (Ser. Nov.) 2:13–65.
Bastian, H. C. 1865. Monograph on the Anguillulidae or free nematoids, marine, land, and freshwater; with descriptions of 100 new species. Linn. Soc. London 25(2): 73–184.

Christie, J. R., and C. H. Arndt. 1936. Feeding habits of the nematodes *Aphelenchoides parietinus* and *Aphelenchus avenae*. Phytopathology 26:698–701.
Das, V. M. 1960. Studies on the nematode parasites of plants in Hyderabad (Andhra Pradesh, India). Z. Parasitenk. 19:553–605.
Filipjev, I. N., and J. H. Schuurmans Stekhoven, Jr. 1941. A manual of agricultural helminthology. Brill, Leiden.
Foreman, P. L., and D. P. Taylor. 1966. Comparative morphology of female gonads of a species of *Paraphelenchus* and *Aphelenchus avenae*. Nematologica 12:92. (Abst.)
Goodey, J. B. 1960. The classification of the *Aphelenchoides* Fuchs 1937. Nematologica 5:111–126.
Goodey, J. B., and D. J. Hooper. 1965. A neotype of *Aphelenchus avenae* Bastian 1865 and the rejection of *Metaphelenchus* Steiner 1943. Nematologica 11:55–65.
Goodey, T. 1927. On the nematode genus *Aphelenchus*. J. Helminth. 5:203–220.
Goodey, T. 1928. The species of the genus *Aphelenchus*. J. Helminth. 6:121–160.
Goodey, T. 1929. On some details of comparative anatomy in *Aphelenchus*, *Tylenchus*, and *Heterodera*. J. Helminth. 7:223–230.
Goodey, T. 1951. Soil and freshwater nematodes. Methuen, London.
Goodey, T. Revised by J. B. Goodey. 1963. Soil and freshwater nematodes. Methuen, London.
Pillai, J. K., and D. P. Taylor. 1967. Influence of ten fungus species on morphometrics of five mycophagous nematodes. Nematologica 13:149. (Abst.)
Rhoades, H. L., and M. B. Linford. 1959. Control of pythium root rot by the nematode *Aphelenchus avenae*. Pl. Dis. Reptr. 43:323–328.
Sanwal, K. C., and P. A. A. Loof. 1967. A neotype of *Aphelenchus agricola* de Man 1881 (Nematoda: Aphelenchidae), morphological variation in the species and its taxonomic status. Nematologica 13:73–78.
Steiner, G. 1931. On the status of the nemic genera *Aphelenchus* Bastian, *Pathoaphelenchus* Cobb, *Paraphelenchus* Micoletzky, *Parasitaphelenchus* Fuchs, *Isonchus* Cobb, and *Seinura* Fuchs. J. Wash. Acad. Sci. 21:468–475.
Steiner, G. 1936. The status of the nematode *Aphelenchus avenae* Bastian 1865, as a plant parasite. Phytopathology 26:294–295.
Thorne, G. 1961. Principles of nematology. McGraw-Hill, New York.
Timm, R. W. 1956. Nematode parasites of rice in East Pakistan. Pakist. Rev. Agr. 3:115–118.

Atalodera

Wouts, W. M., and S. A. Sher. 1971. The genera of the subfamily Heteroderinae (Nematoda: Tylenchoidea) with a description of two new genera. J. Nematol. 3:129–144.

Bakernema

Raski, D. J., and A. M. Golden. 1965. Studies on the genus *Criconemoides* Taylor 1936 with descriptions of eleven new species and *Bakernema variabile* n. sp. (Criconematidae: Nematoda). Nematologica 11:501–565.
Wu, Liang-Yu. 1964. *Criconema bakeri* n. sp. (Criconematidae: Nematoda). Can. J. Zool. 42:53–57.
Wu, Liang-Yu. 1964. *Bakernema* n. gen. (Criconematidae: Nematoda). Can. J. Zool. 42:921.

Belonolaimus

Christie, J. R. 1959. Plant nematodes, their bionomics and control. H. and W. B. Drew, Jacksonville, Florida.
Colbran, R. C. 1960. Studies of plant and soil nematodes, 3. *Belonolaimus hastulatus*, *Psilenchus tumidus*, and *Hemicycliophora labiata*, three new species from Queensland. Qd. J. Agr. Anim. Sci. 17:175–181.
Goodey, T. Revised by J. B. Goodey. 1963. Soil and freshwater nematodes. Methuen, London.
Graham, T. W., and Q. L. Holdeman. 1953. The sting nematode *Belonolaimus gracilis* Steiner: a parasite on cotton and other crops in South Carolina. Phytopathology 43:434–439.
Hirschmann, Hedwig. 1960. The genera *Dolichodorus* and *Belonolaimus*. *In*: J. N. Sasser and W. R. Jenkins, eds. Nematology. Univ. North Carolina Press, Chapel Hill, pp. 191–195.
Hutchinson, M. T., and J. P. Reed. 1956. The sting nematode, *Belonolaimus gracilis*, found in New Jersey. Pl. Dis. Reptr. 40:1049.
Loof, P. A. A. 1958. Some remarks on the status of the subfamily Dolichodorinae, with description of *Macrotrophurus arbusticola* n. g., n. sp. (Nematoda: Tylenchidae). Nematologica 3:301–307.
Owens, J. V. 1951. The pathological effects of *Belonolaimus gracilis* on peanuts in Virginia. Phytopathology 41:29. (Abst.)
Rau, G. J. 1958. A new species of sting nematode. Proc. Helminth. Soc. Wash. 25:95–98.
Rau, G. J. 1961. Amended descriptions of *Belonolaimus gracilis* Steiner 1949, and *B. longicaudatus* Rau 1958 (Nematoda: Tylenchida). Proc. Helminth. Soc. Wash. 28:198–200.
Rau, G. J. 1963. Three new species of *Belonolaimus* (Nematoda: Tylenchida) with additional data on *B. longicaudatus* and *B. gracilis*. Proc. Helminth. Soc. Wash. 30:119–128.
Rau, G. J., and G. Fassuliotis. 1966. Methods for demonstrating differences in the relation of bivariate characters of species and populations in the genus *Belonolaimus*. Nematologica 12:96–97. (Abst.)
Rau, G. J., and G. Fassuliotis. 1967. The use of 95%-tolerance ellipses and regression coefficients to show relationships of *Belonolaimus longicaudatus* and *B. maritimus* populations in different environments. Nematologica 13:150. (Abst.)

Roman, J. 1964. *Belonolaimus lineatus* n. sp. (Nematoda: Tylenchida). J. Agr., Univ. P. Rico 48:131–134.
Steiner, G. 1949. Plant nematodes the grower should know. Proc. Soil Sci. Soc. Fla. Bull. 113:3–47.
Thorne, G. 1961. Principles of nematology. McGraw-Hill, New York.
Whitehead, A. G. 1960. *Trichotylenchus falciformis* n. g., n. sp. (Belonolaiminae n. subfam: Tylenchida Thorne 1949) an associate of grass roots (*Hyparrhenia* sp.) in southern Tanganyika. Nematologica 4:279–285.

Brachydorus

Golden, A. M. 1971. Classification of the genera and higher categories of the order Tylenchida (Nematoda). *In*: B. M. Zuckerman, W. F. Mai, and R. A. Rohde, eds. Plant-parasitic nematodes, Vol. I. Morphology, anatomy, taxonomy, and ecology. Academic Press, New York and London, pp. 191–232.
Guiran, G. de, and G. Germani. 1968. *Brachydorus tenuis* n. g., n. sp. (Nematoda: Dolichodorinae), associée à Ravenala madagascariensis sur la Cote Est Malgache. Nematologica 14:447–452.

Cacopaurus

Allen, M. W., and H. J. Jensen. 1950. *Cacopaurus epacris*, a new species (Nematoda: Criconematidae), a nematode parasite of California black walnut roots. Proc. Helminth. Soc. Wash. 26:1–8.
Christie, J. R. 1959. Plant nematodes, their bionomics and control. H. and W. B. Drew, Jacksonville, Florida.
Goodey, T. 1951. Soil and freshwater nematodes. Methuen, London.
Goodey, T. Revised by J. B. Goodey. 1963. Soil and freshwater nematodes. Methuen, London.
Jenkins, W. R., and D. P. Taylor. 1967. Plant nematology. Reinhold, New York.
Oostenbrink, M. 1960. The family Criconematidae. *In*: J. N. Sasser and W. R. Jenkins, eds. Nematology. Univ. North Carolina Press, Chapel Hill, pp. 196–205.
Raski, D. J. 1962. Paratylenchidae n. fam. with descriptions of five new species of *Gracilacus* n. g. and an emendation of *Cacopaurus* Thorne 1943, *Paratylenchus* Micoletzky 1922, and Criconematidae Thorne 1943. Proc. Helminth. Soc. Wash. 29:189–207.
Siddiqi, M. R., and J. B. Goodey. 1963. The status of the genera and subfamilies of the Criconematidae (Nematoda); with a comment on the position of *Fergusobia*. Nematologica 9:363–377.
Thorne, G. 1943. *Cacopaurus pestis* nov. gen., nov. spec. (Nematoda: Criconematinae), a destructive parasite of the walnut, *Juglans regia* Linn. Proc. Helminth. Soc. Wash. 10:78–83.
Thorne, G. 1961. Principles of nematology. McGraw-Hill, New York.

Caloosia

Loos, C. A. 1948. Notes on free-living and plant-parasitic nematodes of Ceylon, 3. Ceylon J. Sci. (B) 23: 119–124.
* Mathur, V. K., E. Khan, S. Nand, and S. K. Prasad. 1969. Two new species of *Caloosia* Siddiqi and Goodey (Nematoda: Hemicycliophoridae) from India. Entomological Soc. of India Bull. 10:27–31.
Siddiqi, M. R. 1961. Studies on species of Criconematinae (Nematoda: Tylenchida) from India. Proc. Helminth. Soc. Wash. 28:19–34.
Siddiqi, M. R., and J. B. Goodey. 1963. The status of the genera and subfamilies of the Criconematidae (Nematoda); with a comment on the position of *Fergusobia*. Nematologica 9:363–377.

Carphodorus

Colbran, R. C. 1965. Studies of plant and soil nematodes, 11. *Carphodorus bilineatus* n. g., n. sp. (Nematoda: Dolichodorinae) from eucalypt forest in Queensland. Qd. J. Agr. Anim. Sci. 22:481–484.

Criconema

Andrassy, I. 1962. Neue nematoden-arten aus Ungarn, I. Zehn neue arten der unterklasse Secernentea (Phasmidia). Acta Zool., Budapest 8:1–23.
Chitwood, B. G. 1957. Two new species of the genus *Criconema* Hofmanner and Menzel, 1914. Proc. Helminth. Soc. Wash. 24:57–61.
Colbran, R. C. 1962. Studies of plant and soil nematodes, 5. Four new species of Tylenchoidea from Queensland pineapple fields. Qd. J. Agr. Anim. Sci. 19:231–239.
Colbran, R. C. 1963. Studies of plant and soil nematodes, 6. Two new species from citrus orchards. Qd. J. Agr. Anim. Sci. 20:469–474.
Colbran, R. C. 1965. Studies of plant and soil nematodes, 8. Two new species of *Criconema* (Nematoda: Criconematidae) from Queensland. Qd. J. Agr. Anim. Sci. 22:83–87.
Coninck, L. A. P. de. 1943. *Criconema schuurmans-stekhoveni* n. sp. (Criconematinae: Nematoda). Bull. Mus. Roy. Hist., Nat. Belg. 19(53):8.
Coninck, L. A. P. de. 1943. Wetenschappelijke resultaten der studiereis van Prof. Dr. P. van Oye op Ijsland, XIV. Sur quelques espèces nouvelles de nematodes libres des eaux et des terres saumatre de l'Islande. Biol. Jaarb. 10:193–220.
Coomans, A. 1966. Some nematodes from Congo. Rev. Zool. Bot. Afr. 74:287–312.
* Diab, K. A., and W. R. Jenkins. 1965. Description of *Neocriconema adamsi* n. gen., n. sp. (Criconematidae: Nematoda) with a key to the species of *Neocriconema*. Proc. Helminth. Soc. Wash. 32:193–197.

Edward, J. C., and S. L. Misra. 1963. *Criconema mangiferum* n. sp. associated with roots of mango in India. Nematologica 9:222–224.

Edward, J. C., and S. L. Misra. 1965. *Criconema vishwanathum* n. sp. and four other hitherto described Criconematinae. Nematologica 11:566–572.

Edward, J. C., S. L. Misra, E. Peter, and B. B. Rai. 1971. A new species of *Criconema* associated with pomegranate (*Punica granatum* L.). Indian J. Nematol. 1:59–62.

Edward, J. C., S. L. Misra, and B. B. Rai. 1970. *Criconema coffeae* n. sp. associated with the roots of coffee in Mysore State, India. Allahabad Fmr. 44(1/2):13–15.

Flies, Michele. 1968. *Criconema aquit anense* n. sp. (Nematoda: Criconematidae) Nematologica 14:47–54.

Goodey, T. 1951. Soil and freshwater nematodes. Methuen, London.

Goodey, T. Revised by J. B. Goodey. 1963. Soil and freshwater nematodes. Methuen, London.

Golden, A. M., and W. Friedman. 1964. Some taxonomic studies on the genus *Criconema* (Nematoda: Criconematidae). Proc. Helminth. Soc. Wash. 31:47–59.

Grisse, A. de. 1964. *Criconema microdorum* n. sp. (Nematoda: Criconematidae). Nematologica 10: 164–167.

Gunhold, P. 1953. Drei neue Nematoden aus den Ostalpen. Zool. Anz. 150:35–38.

Heyns, J. 1970. South African Criconematinae, Part 2. Genera *Criconema*, *Hemicriconemoides*, and some *Macroposthonia* (Nematoda). Phytophylactica 2:129–136.

Hoffman, J. K. 1973. *Criconema proclivis* n. sp. (Nematoda: Criconematinae) from woodlands. J. Nematol. 5:155–157.

Hofmanner, B., and R. Menzel. 1914. Neue Arten freilebender Nematoden aus der Schweiz. Zool. Anz. 44(2):80–91.

Hofmanner, B., and R. Menzel. 1915. Die Freilebenden Nematoden der Schweiz. Rev. Suisse Zool. 23(5): 109–243.

Jairajpuri, M. S. 1963. *Criconema simlaensis* n. sp. (Nematoda: Criconematidae) from India. Z. Parasitenk. 23(3):235–238.

Jairajpuri, M. S. 1964. *Criconema taylori* n. sp. (Nematoda: Criconematidae) from South India. Nematologica 10:108–110.

Khan, E., and M. R. Siddiqi. 1963. *Criconema laterale* n. sp. (Nematoda: Criconematidae) from Srinagar, Kashmir. Nematologica 9:584–586.

Khan, E., and M. R. Siddiqi. 1963. *Criconema serratum* n. sp. (Nematoda: Criconematidae), a parasite of peach trees in Almore, North India. [Correspondence.] Current Science, Bangalore 32(9):414–415.

Kirjanova, E. S. 1948. Ten new species of nematodes from the family Ogmidae Southern, 1914. Publn. Ded. Mem. Acad. Sergei Alexeivich Zernov, Acad. Sci., USSR, pp. 346–358.

Krall, E. 1963. *Criconema kirjanovae* n. sp. from soil in the Estonian SSR Izvest. Akad. Nauk Estonskoi, SSR, Seer. Biol. 12(4):342–344.

Loos, C. A. 1949. Notes on free-living and plant-parasitic nematodes of Ceylon, No. 4. J. Zool. Soc. India 1:17–22.

Luc, M. 1959. Nouveaux Criconematidae de la zone intertropicale (Nematoda: Tylenchida). Nematologica 4:16–22.

* Mehta, U. K., and D. J. Raski. 1971. Revision of the genus *Criconema* Hofmanner and Menzel 1914 and other related genera (Criconematidae: Nematoda). Indian J. Nematol. 1:145–198 [Nematology Section, Agricultural College and Research Inst. Coimatore, Tamil Nadu, India.]

Oostenbrink, M. 1960. The family Criconematidae. *In*: J. N. Sasser and W. R. Jenkins, eds. Nematology. Univ. North Carolina Press, Chapel Hill, pp. 196–205.

Siddiqi, M. R. 1961. Studies on species of Criconematinae (Nematoda: Tylenchida) from India. Proc. Helminth. Soc. Wash. 28:19–34.

Siddiqi, M. R., and J. F. Southey. 1962. *Criconema palmatum* n. sp. (Nematoda: Criconematidae) from North Devon, England. Nematologica 8:221–224.

Siddiqi, M. R., and J. B. Goodey. 1963. The status of the genera and subfamilies of the Criconematidae (Nematoda); with a comment on the position of *Fergusobia*. Nematologica 9:363–377.

Siddiqi, M. R. 1965. *Criconemoides citricola* n. sp. (Nematoda: Criconematidae), with a redescription of *Criconema murrayi* Southern 1914. Nematologica 11:239–243.

Steiner, G. 1949. Plant nematodes the grower should know. Proc. Soil Sci. Soc. Fla. Bull. 113:3–47.

Taylor, A. L. 1936. The genera and species of the Criconematinae, a subfamily of the Anguillulinidae (Nematoda). Trans. Am. Microscop. Soc. 55:391–421.

Thorne, G. 1961. Principles of nematology. McGraw-Hill, New York.

Trave, J. 1954. Criconematidae (Nematodea: Tylenchoidea) nouveaux pour la France. Vie et Milieu 5:250–257.

Wu, Liang-Yu. 1960. *Criconema celetum* n. sp. (Nematoda: Criconematidae) from African violets in Canada. Can. J. Zool. 38:913–916.

Wu, Liang-Yu. 1964. *Criconema bakeri* n. sp. (Criconematidae: Nematoda). Can. J. Zool. 42:53–57.

Wu, Liang-Yu. 1965. *Criconema seymouri* n. sp. (Criconematidae: Nematoda). Can. J. Zool. 43:215–217.

Criconemoides

Adams, R. E., and N. A. Lapp. 1967. *Criconemoides grassator* n. sp. from yellow poplar (*Liriodendron tulipifera*) in West Virginia. Nematologica 13:63–66.

Allen, M. W., and S. A. Sher. 1967. Taxonomic problems concerning the phytoparasitic nematodes. *In*: Annual Review of Phytopathology, Vol. 5. Annual Reviews, Palo Alto, California, pp. 247–264.

Andrassy, I. 1952. Freilebende Nematoden aus dem Bukk-Gebirge. Annls. Hist., Nat. Mus. Natn. Hungary (Ser. Nov.) 2:13–65.
Andrassy, I. 1959. Nematoden aus der Tropfsteinhohle "Baradla" bei Aggtelek (Ungarn), nebst einer Ubersicht der bisher aus Hohlen bekannten freilebenden Nematoden-Arten. (Biospeologica Hungarica I.) Acta Zool., Budapest 4:253–277.
Andrassy, I. 1962. Neue nematoden-arten aus ungarn, I. Zehn neue arten der unterklasse Secernentea (Phasmidia). Acta Zool., Budapest 8:1–23.
Andrassy, I. 1963. The zoological results of Gy. Topal's collectings in South Argentina, 2. Nematoda. Neue und einige seltene Nematoden-Arten aus Argentinien. Annls. Hist., Nat. Mus. Natn. Hungary 55:243–273.
Andrassy, I. 1964. Ergebnisse der zoologischen Forschungen von Dr. Z. Kaszab in der Mongolei, 4. Einige Bodennematoden aus der Mongolei. Annls. Hist., Nat. Mus. Natn. Hungary 56:241–255.
Andrassy, I. 1965. Verzeichnis und Bestimmungsschlussel der Arten der Nematodengattungen *Criconemoides* Taylor 1936 und *Mesocriconema* n. gen. Opusc. Zool., Budapest 5(2):153–171.
Arias Delgado, M., F. Jimenez Millan, and J. M. Lopez Pedregal. 1965. Tres neuvas especies de nematodos posibles fitoparasitos en suelos espanoles. Publnes. Inst. Biol. Apl., Barcelona 38:47–58.
Arias, Delgado, M., J. M. Lopez Pedregal, and F. Jimenez Millan. 1963. Nematodos periradiculares en la vid. Boln. R. Soc. esp. Hist. Nat. (B) 61:35–43.
Chitwood, B. G. 1949. Ring nematodes (Criconematinae). A possible factor in decline and replanting problems of peach orchards. Proc. Helminth. Soc. Wash. 16:6–7.
Coomans, A. 1966. Some nematodes from Congo. Rev. Zool. Bot. Afr. 74:287–312.
Diab, K. A. 1965. Variation in De Man's "*a*" and "*b*" values in *Criconemoides*. Nematologica 11:35–36. (Abst.)
* Diab, K. A., and W. R. Jenkins. 1965. Description of *Neocriconema adamsi* n. gen., n. sp. (Criconematidae: Nematoda) with a key to the species of *Neocriconema*. Proc. Helminth. Soc. Wash. 32:193–197.
Diab, K. A., and W. R. Jenkins. 1966. Three new species of *Criconemoides* (Nematoda: Criconematidae). Proc. Helminth. Soc. Wash. 33:5–7.
Edward, J. C., and S. L. Misra. 1963. *Criconemoides nainitalense* n. sp. (Nematoda: Criconematidae). Nematologica 9:218–221.
Edward, J. C., and S. L. Misra. 1964. *Criconemoides magnoliae* n. sp. and *C. juniperi* n. sp. (Nematoda: Criconematidae) from Kumaon region, Uttar Pradesh, India. Nematologica 10:95–100.
Edward, J. C., and S. L. Misra. 1965. *Criconema vishwanathum* n. sp. and four other hitherto described Criconematinae. Nematologica 11:566–572.
Edward, J. C., S. L. Misra, and G. R. Singh. 1968. *Criconemoides michieli* n. sp. and *C. rihandi* n. sp. (Nematoda: Tylenchida) two new species with a note on *C. nainitalense*. Bull. Ent., Loyola Coll. 9(1): 39–44.
Fassuliotis, G., and C. E. Williamson. 1959. *Criconemoides axeste* n. sp. associated with roses in commercial greenhouses in New York State. Nematologica 4:205–210.
Geraert, E. 1968. The synonymy of the families Tylenchulidae and Criconematidae and establishment of the superfamily Criconematoidea. International Nematology Symposium (8th), Antibes, Sept. 8–14, 1965, Reports, p. 33. (Abst.)
Goodey, T. 1951. Soil and freshwater nematodes. Methuen, London.
Goodey, T. Revised by J. B. Goodey. 1963. Soil and freshwater nematodes. Methuen, London.
Grisse, A. de. 1963. *Criconemoides flandriensis* n. sp. (Nematoda: Criconematidae). Nematologica 9: 547–552.
Grisse, A. de. 1964. Morphological observations on *Criconemoides*, with a description of four new species found in Belgium (Nematoda). Meded. LandbHoogesch. OpzoekStns., Gent 29:734–761.
Grisse, A. de. 1967. Description of fourteen new species Criconematidae with remarks on different species of this family. Biol. Jaarb. 35:66–125.
Grisse, A. de. 1969. Contribution to the morphology and the systematics of the Criconematidae (Taylor 1936) Thorne 1949. Faculty of Agricultural Sciences, Ghent.
Grisse, A. de, and H. Koen. 1964. *Criconemoides pseudohercyniensis* n. sp. (Nematoda: Criconematidae). Nematologica 10:197–200.
Grisse, A. de, and P. A. A. Loof. 1965. Revision of the genus *Criconemoides* (Nematoda). Meded. LandbHoogesch. OpzoekStns., Gent 30:577–603.
Grisse, A. de, and P. A. A. Loof. 1968. Revision of the genus *Criconemoides*. International Nematology Symposium (8th), Antibes, Sept. 8–14, 1965, Reports, p. 31. (Abst.).
Grisse, A. T. de, and P. W. Maas. 1970. *Macroposthonia longistyleta* n. sp. and *Discocriconemella surinamensis* n. sp. from Surinam (Nematoda: Criconematidae). Nematologica 16:123–132.
Guiran, G. de. 1963. Quatre espèces nouvelles du genre *Criconemoides* Taylor (Nematoda: Criconematidae). Revue Path. Veg. Ent. Agr. Fr. 42:1–11.
Heyns, J. 1962. Two new species of Criconematidae from South Africa. Nematologica 8:21–24.
Heyns, J. 1970. South African Criconematinae, Part 1. Genera *Nothocriconema Lobocriconema*, *Criconemella*, *Xenocriconemella*, and *Discocriconemella* (Nematoda). Phytophylactica 2(1):49–56.
Heyns, J. 1970. South African Criconematinae, Part 2. Genera *Criconema*, *Hemicriconemoides* and some *Macroposthonia* (Nematoda). Phytophylactica 2(2):129–136.
Hofmanner, B., and R. Menzel. 1914. Neue Arten freilebender Nematoden aus der Schweiz. Zool. Anz., Leipzig 44:80–91.
Jairajpuri, M. S. 1963. Two new species of the genus *Criconemoides* Taylor 1936 (Nematoda: Criconematidae) from North India. Nematologica 9:381–385.

Jairajpuri, M. S. 1964. *Criconemoides basili* nom. nov. (Syn.: *Criconemoides goodeyi* Jairajpuri, preoccupied). Nematologica 10:183.
Jenkins, W. R., and D. P. Taylor. 1967. Plant nematology. Reinhold, New York.
Khan, E., and C. K. Nanjappa. 1972. Four new species of Criconematoidea (Nematoda) from India. Indian J. Nematol. 2:59–68.
Khan, E., A. R. Seshadri, B. Weischer, and K. Mathen. 1971. Five new nematode species associated with coconut in Kerala, India. Indian J. Nematol. 1:116–127.
Khan, S. H. 1963. *Criconemoides siddiqii* n. sp. (Nematoda: Criconematidae) from North India. Zool. Anz. 173:342–345.
Kirjanova, E. S. 1948. Ten new species of nematodes from the family Ogmidae Southern 1914. Publn. Ded. Mem. Acad. Sergei Alexeivich Zernov, Acad. Sci. USSR, pp. 346–358.
Krnjaic, D. 1967. *Discocriconemella yossifovchi* n. sp. (Nematoda: Criconematidae) na vinovoj lozi u Jugoslaviji. Zast. Bilia, 18(93/95):155–160.
Loof, P. A. A. 1964. Four new species of *Criconemoides* from the Netherlands. Versl. Meded. Plziektenk. Dienst Wageningen 141:160–168.
Loof, P. A. A. 1964. Free-living and plant-parasitic nematodes from Venezuela. Nematologica 10:201–300.
Loof, P. A. A. 1965. Zur taxonomie von *Criconemoides rusticus* (Micoletzky) und *C. informis* (Micoletzky). Mitt. Zool. Mus., Berlin 41:183–192.
Loof, P. A. A., and A. de Grisse. 1965. Observations on *Nothocriconema princeps* (Andrassy 1962). Meded. LandbHoogesch. OpzoekStns., Gent 30:1405–1409.
Loof, P. A. A., and A. de Grisse. 1967. Re-establishment of the genus *Criconemoides* Taylor 1936 (Nematoda: Criconematidae). Meded. Rijksfac. LandbWet., Gent 32(3/4):466–475.
Loos, C. A. 1949. Notes on free-living and plant-parasitic nematodes of Ceylon, No. 4. J. Zool. Soc. India 1:17–22.
Luc, M. 1959. Nouveaux Criconematidae de la zone intertropicale (Nematoda: Tylenchida). Nematologica 4:16–22.
Luc, M. 1970. Contribution à l'étude du genre *Criconemoides* Taylor 1936 (Nematoda: Criconematidae). Cah. O.R.S.T.O.M., Ser. Biol. 11:69–131.
Maas, P. W. T., P. A. A. Loof, and A. deGrisse. 1971. *Nothocriconemoides lineolatus* n. gen., n. sp. (Nematoda: Criconematidae). Meded. Fak. LandbWet., Gent 36(2): 711–715.
Oostenbrink, M. 1960. The family Criconematidae. *In*: J. N. Sasser and W. R. Jenkins, eds. Nematology. Univ. North Carolina Press, Chapel Hill, pp. 196–205.
Prasad, S. K., E. Khan, and V. K. Mathur. 1965. *Criconemoides georgii* n. sp. (Nematoda: Criconematidae) from India. Current Science 34:667–668.
Raski, D. J. 1952. On the morphology of *Criconemoides* Taylor 1936, with descriptions of six new species (Nematoda: Criconematidae). Proc. Helminth. Soc. Wash. 19:85–99.
Raski, D. J. 1958. Nomenclatorial notes on the genus *Criconemoides* (Nematoda: Criconematidae) with a key to the species. Proc. Helminth. Soc. Wash. 25:139–142.
* Raski, D. J., and A. M. Golden. 1965. Studies on the genus *Criconemoides* Taylor 1936 with descriptions of eleven new species and *Bakernema variable* n. sp. (Criconematidae: Nematoda). Nematologica 11:501–565.
Raski, D. J., and J. W. Riffle. 1967. Two new species and further notes on *Criconemoides* Taylor 1936 (Criconematidae: Nematoda). Proc. Helminth. Soc. Wash. 34:212–219.
Seshadri, A. R. 1964. Histological investigations on the ring nematode *Criconemoides xenoplax* Raski 1952 (Nematoda: Criconematidae). Nematologica 10:519–539.
Seshadri, A. R. 1964. Investigations on the biology and life cycle of *Criconemoides xenoplax* Raski 1952 (Nematoda: Criconematidae). Nematologica 10:540–562.
Siddiqi, M. R. 1961. Studies on species of Criconematinae (Nematoda: Tylenchida) from India. Proc. Helminth. Soc. Wash. 28:19–34.
Siddiqi, M. R. 1965. *Criconemoides citricola* n. sp. (Nematoda: Criconematidae) with a redescription of *Criconema murrayi* Southern 1914. Nematologica 11:239–243.
Siddiqi, M. R., and J. B. Goodey. 1963. The status of the genera and subfamilies of the Criconematidae (Nematoda); with a comment on the position of *Fergusobia*. Nematologica 9:363–377.
Smart, G. C., Jr. 1965. The aperture and lumen of the spear in three members of Tylenchoidea (Nemata). Nematologica 11:45–46. (Abst.)
Tarjan, A. C. 1966. A compendium of the genus *Criconemoides* (Criconematidae: Nemata). Proc. Helminth. Soc. Wash. 33:109–125.
Taylor, A. L. 1936. The genera and species of the Criconematinae, a subfamily of the Anguillulinidae (Nematoda). Trans. Am. Microscop. Soc. 55:391–421.
Thomas, H. A. 1959. On *Criconemoides xenoplax* Raski, with special reference to its biology under laboratory conditions. Proc. Helminth. Soc. Wash. 26:55–59.
Thorne, G. 1961. Principles of nematology. McGraw-Hill, New York.
Timm, R. W. 1956. Nematode parasites of rice in East Pakistan. Pakist. Rev. Agr. 2:115–118.
Trave, J. 1954. Criconematidae (Nematodea: Tylenchoidea) nouveaux pour la France. Vie et Milieu 5:251–257.
Williams, J. R. 1960. Studies on the nematode soil fauna of sugar cane fields in Mauritius, 4. Tylenchoidea (partim). Occ. Paper, Maurit. Sug. Ind. Res. Inst., No. 4.
Winslow, R. D. 1960. Some aspects of the ecology of free-living and plant-parasitic nematodes. *In*: J. N. Sasser and W. R. Jenkins, eds. Nematology. Univ. North Carolina Press, Chapel Hill, pp. 341–415.

Wu, Liang-Yu. 1965. Five new species of *Criconemoides* Taylor 1936 (Criconematidae: Nematoda) from Canada. Can. J. Zool. 43:203–214.
Yokoo, T. 1964. On a new species of ring nematode from Japan, II. Agr. Bull., Saga Univ., No. 20, pp. 63–65.
Zyubin, B. N. 1969. Plant nematodes of the opium poppy in Kirgizia. Frunze, Izdatelstvo "ILIM."

Cryphodera

Colbran, R. C. 1966. Studies of plant and soil nematodes, 12. The eucalypt cystoid nematode *Cryphodera eucalypt* n. g., n. sp. (Nematoda: Heteroderidae), a parasite of eucalypts in Queensland. Qd. J. Agr. Anim. Sci. 23:41–47.
Franklin, Mary T. 1971. Taxonomy of Heteroderidae. *In*: B. M. Zuckerman, W. F. Mai, and R. A. Rohde, eds. Plant-parasitic nematodes, Vol. I. Morphology, anatomy, taxonomy, and ecology. Academic Press, New York and London, pp. 139–162.
Wouts, W. M. 1972. A revision of the family Heteroderidae (Nematoda: Tylenchoidea), I. The family Heteroderidae and its subfamilies. Nematologica 18:439–446.
Wouts, W. M. 1973. A revision of the family Heteroderidae (Nematoda: Tylenchoidea), II. The subfamily Meloidoderinae. Nematologica 19:218–235.

Ditylenchus

Anderson, R. V., and H. M. Darling. 1964. Embryology and reproduction of *Ditylenchus destructor* Thorne, with emphasis on gonad development. Proc. Helminth. Soc. Wash. 31:240–256.
Anderson, R. V., and H. M. Darling. 1964. Spear development in *Ditylenchus destructor* Thorne. Nematologica 10:131–135.
Andrassy, I. 1958. Erd- und Susswassernematoden aus Bulgarien. Acta Zool., Budapast 4(1/2):1–88.
Barker, K. R. 1959. Studies on the biology of the stem nematode. Phytopathology 49:315.
Barraclough, Ruth, and R. E. Blackith. 1962. Morphometric relationships in the genus *Ditylenchus*. Nematologica 8:51–58.
Christie, J. R. 1959. Plant nematodes, their bionomics and control. H. and W. B. Drew, Jacksonville, Florida.
Cobb, N. A. 1922. Two tree-infesting nemas of the genus *Tylenchus*. An. Zool. Aplic., Chile 9:27–35.
Faulkner, L. R., and H. M. Darling. 1961. Pathological histology, hosts and culture of the potato-rot nematodes. Phytopathology 51:778–786.
Filipjev, I. N. 1934. The classification of the free-living nematodes and their relations to the parasitic nematodes. Smithson. Misc. Coll. (Pub. 3216) 89(6):1–63.
Filipjev, I. N. 1936. On the classification of the Tylenchinae. Proc. Helminth. Soc. Wash. 3:80–82.
Filipjev, I. N., and J. H. Schuurmans Stekhoven, Jr. 1941. A manual of agricultural helminthology. Brill, Leiden, Holland.
German, E. V. 1969. A new species of stem nematode. Vest. sel-khoz. Nauki, Alma-Ata 12(1):83–85.
Goffart, H. 1961. Unterscheidvagsmerkmale von *Ditylenchus dipsaci* (Kuhn 1857) Filipjev 1936 und *Ditylenchus destructor* Thorne 1945. Gesunde Pflanzen., Frankfurt 13(5):117–120.
Golden, A. M. 1971. Classification of the genera and higher categories of the order Tylenchida (Nematoda). *In*: B. M. Zuckerman, W. F. Mai, and R. A. Rohde, eds. Plant-parasitic nematodes, Vol. I. Morphology, anatomy, taxonomy, and ecology. Academic Press, New York and London, pp. 199–232.
Goodey, J. B. 1958. *Ditylenchus myceliophagus* n. sp. (Nematoda: Tylenchidae). Nematologica 3:91–96.
Goodey, J. B. 1959. Gall-forming nematodes of grasses in Britain. J. Sports Turf Res. Inst. 10:1–7.
Goodey, T. 1925. *Tylenchus hordei* Schoyen, a nematode parasite causing galls on the roots of barley and other Gramineae. J. Helminth. 3:193–202.
Goodey, T. 1929. On some details of comparative anatomy in *Aphelenchus, Tylenchus,* and *Heterodera*. J. Helminth. 7:223–230.
Goodey, T. 1932. The genus *Anguillulina* Gerv. and v. Ben. 1895, vel *Tylenchus* Bastian 1865. J. Helminth. 10:75–180.
Goodey, T. 1933. *Anguillulina graminophila* n. sp., a nematode causing galls on the leaves of fine bent grass. J. Helminth. 11:45–56.
Goodey, T. 1945. *Anguillulina brenani* n. sp., a nematode causing galls on the moss, *Pottia bryoides* Mitt. J. Helminth. 21:105–110.
Goodey, T. 1951. Soil and freshwater nematodes. Methuen, London.
Goodey, T. Revised by J. B. Goodey. 1963. Soil and freshwater nematodes. Methuen, London.
Gritsenko, V. P. 1971. *Ditylenchus tenuidens* n. sp. and *Aphelenchoides curiolis* n. sp. (Nematoda: Tylenchidae and Aphelenchoididae) from Kirgizia. Zool. Zh. 50(9):1402–1405.
Hirschmann, Hedwig. 1960. The Genera *Tylenchus, Psilenchus, Ditylenchus, Anguina, Tylenchorhynchus, Tetylenchus, Trophurus*, and *Macrotrophurus*. *In*: J. N. Sasser and W. R. Jenkins, eds. Nematology. Univ. North Carolina Press, Chapel Hill, pp. 171–180.
Hirschmann, Hedwig. 1962. The life cycle of *Ditylenchus triformis* (Nematoda: Tylenchida) with emphasis on postembryonic development. Proc. Helminth. Soc. Wash. 29:30–43.
Hirschmann, Hedwig, and J. N. Sasser. 1955. On the occurrence of an intersexual form in *Ditylenchus triformis* n. sp. (Nematoda: Tylenchida). Proc. Helminth. Soc. Wash. 22:115–123.
Husain, S. I., and A. M. Khan. 1967. A new subfamily, a new subgenus, and eight new species of nematodes from India belonging to superfamily Tylenchoidea. Proc. Helminth. Soc. Wash. 34:175–186.
Karimova, I. S. 1957. Eelworms of crops on the left bank of the Amu-Darya basin. *In*: A. I. Zemlyanskaya, L. V. Tikhonova, and I. S. Karimova, eds. Eelworms of agricultural crops in the Uzbek SSR Akad. Nauk Uzbek, SSR, Tashkent, pp. 133–208.

Khan, E., M. L. Chawla, and S. K. Prasad. 1969. *Tylenchus* (*Aglenchus*) *indicus* n. sp. and *Ditylenchus emus* n. sp. (Nematoda: Tylenchidae) from India. Labdev J. Sci. Technol. Ser. B. (4):311–314.
Paramonov, A. A. 1967. Problems on evolution, morphology, taxonomy, and biochemistry of nematodes of plants. Akad. Nauk, Moscow, SSR 18:78–101.
Pillai, J. K., and D. P. Taylor. 1967. Influence of ten fungus species on morphometrics of five mycophagous nematodes. Nematologica 13:149. (Abst.)
Schoyen, W. M. 1885. Bygaalen (*Tylenchus hordei* n. sp.) en ny, for Bygget skadelig Planteparasit blandt Rundormene. Forh. Vidensk. Selsk. 22:1–16.
Sher, S. A. 1970. Reclassification of the genus *Chitinotylenchus* (Micoletzky 1922) and a redescription of *C. paragracilis* (Micoletzky 1922) (Nematoda: Tylenchoidea). J. Nematol. 2:236–238.
Siddiqi, M. R. 1963. Four new species in the subfamily Tylenchinae (Nematoda) from North India. Z. Parasitenk. 23(4):397–404.
Steiner, G. 1935. Opuscula miscellanea nematologica, I. *Anguillulina gallica* n. sp. living in burls of an elm (*Ulmus* sp.) from France. Proc. Helminth. Soc. Wash. 2:41–45.
Steiner, G. 1936. *Anguillulina askenasyi* (Butschli 1873), a gall-forming nematode parasite of the common fern moss, *Thuidium delicatulum* (L.) Hedw. J. Wash. Acad. Sci. 26:410–414.
Steiner, G., and C. E. Scott. 1934. A nematosis of *Amsinckia* caused by a new variety of *Anguillulina dipsaci*. J. Agr. Res. 49:1087–1092.
Sturhan, D., and W. Friedman. 1965. *Ditylenchus convallariae* n. sp. (Nematoda: Tylenchida). Nematologica 11: 219–223.
Tarjan, A. C. 1958. A new genus, *Pseudhalenchus* (Tylenchinae: Nematoda), with descriptions of two new species. Proc. Helminth. Soc. Wash. 25:20–25.
Thorne, G. 1934. Some plant-parasitic nemas, with descriptions of three new species. J. Agr. Res. 49:755–763.
Thorne, G. 1945. *Ditylenchus destructor* n. sp., the potato-rot nematode, and *Ditylenchus dipsaci* (Kuhn 1857) Filipjev 1936, the teasel nematode (Nematoda: Tylenchidae). Proc. Helminth. Soc. Wash. 12:27–34.
Thorne, G. 1949. On the classification of the Tylenchida, new order (Nematoda: Phasmidia). Proc. Helminth. Soc. Wash. 16:37–73.
Thorne, G. 1961. Principles of nematology. McGraw-Hill, New York.
Tobar, Jimenez, A. 1964. *Ditylenchus virtudesae* n. sp. (Nematoda: Tylenchidae), habitante de los suelos granadinos. Revta iber. Parasit. 24:51–56.
Wasilewska, L. 1965. *Ditylenchus medicaginis* sp. n., a new parasitic nematode from Poland (Nematoda: Tylenchidae). Bull. Acad. Pol. Sci. Cl. II Ser. Sci. Biol. 13:167–170.
Wu, Liang-Yu. 1958. Morphology of *Ditylenchus destructor* Thorne 1945 (Nematoda: Tylenchidae), from a pure culture, with special reference to reproductive systems and esophageal glands. Can. J. Zool. 36:569–576.
Wu, Liang-Yu. 1960. Comparative study of *Ditylenchus destructor* Thorne 1945 (Nematoda: Tylenchidae), from potato, bulbous iris, and dahlia, with a discussion of De Man's ratio. Can. J. Zool. 38:1175–1187.
Wu, Liang-Yu. 1960. Further observations on the morphology of *Ditylenchus destructor* Thorne 1945 (Nematoda: Tylenchidae). Can. J. Zool. 38:47–49.
Wu, Liang-Yu. 1967. Differences of spermatheca and sperm cells in the genera *Ditylenchus* Filipjev 1936 and *Tylenchus* Bastian 1865 (Tylenchidae: Nematoda). Can. J. Zool. 45:27–30.
Yuksel, H. S. 1960. Observations on the life cycle of *Ditylenchus dipsaci* on onion seedlings. Nematologica 5:289–296.

Dolichodorus

Allen, M. W. 1957. A new species of the genus *Dolichodorus* from California (Nematoda: Tylenchida). Proc. Helminth. Soc. Wash. 24:95–98.
Christie, J. R. 1959. Plant nematodes, their bionomics and control. H. and W. B. Drew, Jacksonville, Florida.
Clark, W. C. 1963. A new species of *Dolichodorus* (Nematoda: Tylenchida) from coastal dune sands. N.Z. Jl. Sci. 6:531–534.
Cobb, N. A. 1914. The North American free-living freshwater nematodes. Trans. Am. Microscop. Soc. 33:35–99.
Fisher, J. M. 1964. *Dolichodorus adelaidensis* n. sp. and *Paralongidorus eucalypti* n. sp. from S. Australia. Nematologica 10:464–470.
Gillespie, W. H., and R. E. Adams. 1962. An awl nematode, *Dolichodorus silvestris* n. sp. from West Virginia. Nematologica 8:93–98.
Golden, A. M. 1958. *Dolichodorus similis* (Dolichodorinae), a new species of plant nematode. Proc. Helminth. Soc. Wash. 25:17–20.
Goodey, T. 1951. Soil and freshwater nematodes. Methuen, London.
Goodey, T. Revised by J. B. Goodey. 1963. Soil and freshwater nematodes. Methuen, London.
Hirschmann, Hedwig. 1960. The genera *Dolichodorus* and *Belonolaimus*. *In*: J. N. Sasser and W. R. Jenkins, eds. Nematology. Univ. North Carolina Press, Chapel Hill, pp. 191–195.
Khan, E., A. R. Seshadri, B. Weischer, and K. Mathen. 1971. Five new nematode species associated with coconut in Kerala, India. Indian J. Nematol. 1:116–127.
Loof, P. A. A. 1958. Some remarks on the status of the subfamily Dolichodorinae, with description of *Macrotrophurus arbusticola* n. g., n. sp. (Nematoda: Tylenchidae). Nematologica 3:301–307.
Luc, M. 1960. *Dolichodorus profundus* n. sp. (Nematoda: Tylenchida). Nematologica 5:1–6.
Luc, M., and F. E. Caveness. 1963. *Dolichodorus nigeriensis* n. sp. (Nematoda: Dolichodoridae). Proc. Helminth. Soc. Wash. 20:297–299.

Luc, M., and A. Dalmasso. 1971. *Dolichodorus cassati* n. sp. (Nematoda: Tylenchida). Annls. Zool. Ecol. Anim. 3(1):97–101.
Perry, V. G. 1953. The awl nematode, *Dolichodorus heterocephalus*, a devastating plant parasite. Proc. Helminth. Soc. Wash. 20:21–27.

Gracilacus

Adams, R. E., and J. J. Eichenmuller, Jr. 1962. *Gracilacus capitatus* n. sp. from Scarlet Oak in West Virginia. Nematologica 8:87–92.
Allen, M. W., and H. J. Jensen. 1950. *Cacopaurus epacris*, new species (Nematoda: Criconematidae), a nematode parasite of California black walnut roots. Proc. Helminth. Soc. Wash. 17:10–14.
Brown, Georgianna L. 1959. Three new species of the genus *Paratylenchus* from Canada (Nematoda: Criconematidae). Proc. Helminth. Soc. Wash. 26:1–8.
Cobb, N. A. 1923. Notes on *Paratylenchus*, a genus of nemas. J. Wash. Acad. Sci. 13:254–257.
Golden, A. M. 1961. *Paratylenchus steineri* (Criconematidae) a new species of plant nematode. Proc. Helminth. Soc. Wash. 28:9–11.
Golden, A. M. 1971. Classification of the genera and higher categories of the order Tylenchida (Nematoda). *In*: B. M. Zuckerman, W. F. Mai, and R. A. Rohde, eds. Plant-parasitic nematodes, Vol. I. Morphology, anatomy, taxonomy, and ecology. Academic Press, New York and London, pp. 191–232.
Jenkins, W. R. 1960. *Paratylenchus marylandicus* n. sp. (Nematoda: Criconematidae) associated with roots of pine. Nematologica 5:175–177.
Oostenbrink, M. 1953. A note on *Paratylenchus* in the Netherlands with the description of *P. goodeyi* n. sp. (Nematoda: Criconematidae). Tijdschr. Plziekt. 59:207–216.
Raski, D. J. 1962. Paratylenchidae n. fam. with descriptions of five new species of *Gracilacus* n. g. and an emendation of *Cacopaurus* Thorne 1943, *Paratylenchus* Micoletzky 1922, and Criconematidae Thorne 1943. Proc. Helminth. Soc. Wash. 29:189–207.
Siddiqi, M. R., and J. B. Goodey. 1963. The status of the genera and subfamilies of the Criconematidae (Nematoda); with a comment on the position of *Fergusobia*. Nematologica 9:363–377.
Thorne, G., and R. B. Malek. 1968. Nematodes of the northern great plains, Part I. Tylenchida (Nemata: Secernentea). S. Dak. Agr. Exp. Stn. Tech. Bull. 31:111.
Wolff Schoemaker, R. L. P. 1963. *Gracilacus peperpotti* n. sp. (Nematoda: Paratylenchidae) found in a Surinam coffee plantation soil. Nematologica 9:296–299.

Helicotylenchus

Andrassy, I. 1958. *Hoplolaimus tylenchiformis* Daday 1905 (Syn.: *H. coronatus* Cobb 1923) und die Gattungen der Unterfamilie Hoplolaiminae Filipjev 1936. Nematologica 3:44–56.
Carvalho, J. C. 1959. *Helicotylenchus elisensis* n. comb. (Nematoda: Tylenchidae). Inst. Biol., Brazil 26(7):45–48.
Cobb, N. A. 1893. Plant diseases and their remedies. Diseases of the sugarcane. Agr. Gaz., New South Wales 4:808–833.
Cobb, N. A. 1906. Free-living nematodes inhabiting the soil about the roots of cane, and their relation to root diseases. Bull. Div. Path. Physiol., Hawaiian Sug. Pltr's. Assoc. Exp. Stn. 5:163–195.
Das, V. M. 1960. Studies on the nematode parasites of plants in Hyderabad (Andhra Pradesh, India). Z. Parasitenk. 19:553–605.
Elmiligy, I. A. 1970. Three new species of the genus *Helicotylenchus* Steiner 1945 (Hoplolaiminae: Nematoda). Medelingen van de Faculteit LandbWet. Rijksuniversiteit, Gent 35:1099–1106.
Gaur, H. S., and S. K. Prasad. 1972. *Helicotylenchus teres* nomen novum for *H. thornei* Gupta and Chhabra 1967. Indian J. Nematol. 2:93–94.
Golden, A. M. 1956. Taxonomy of the spiral nematodes (*Rotylenchus* and *Helicotylenchus*), and the developmental stages and host-parasite relationships of *R. buxophilus* n. sp. attacking boxwood. Univ. Md. Bull. A-85:1–28.
Good, J. M., A. E. Steele, and T. J. Ratcliffe. 1959. Occurrence of plant-parasitic nematodes in Georgia turf nurseries. Pl. Dis. Reptr. 43:236–238.
Goodey, T. 1940. On *Anguillulina multicincta* (Cobb) and other species of *Anguillulina* associated with the roots of plants. J. Helminth. 18:21–38.
Goodey, T. 1948. A note on the presence of phasmids on the male tails of *Anguillulina multicincta, A. erythrinae, A. robusta*. J. Helminth. 22:139–140.
Goodey, T. Revised by J. B. Goodey. 1963. Soil and freshwater nematodes. Methuen, London.
Gupta, N. K., and H. K. Chhabra. 1967. *Helicotylenchus thornei* n. sp. (Nematoda: Hoplolaiminae) from a tomato-plant field in Ludhiana, Punjab. Res. Bull., Punjab Univ. Sci. 17(3/4):323–325.
Hirschmann, Hedwig, and A. C. Triantaphyllou. 1965. The question of digonic hermaphroditism in some spiral nematodes (*Helicotylenchus* spp.). Nematologica 11:39–40. (Abst.)
Ivanova, T. S. 1967. Parasitic nematodes of the subfamily Hoplolaiminae in Tadzhikistan. Izvest. Akad. Nauk Tadzhik, SSR, Otd. Biol. Nauk 1:97–100.
Jairajpuri, M. S., and Q. H. Baqri. 1973. Nematodes of high altitudes in India, I. Four new species of Tylenchida. Nematologica 19:19–30.
Khan, S. H., and M. A. Basir. 1964. Two new species of the genus *Helicotylenchus* Steiner 1945 (Nematoda: Hoplolaimidae) from India. Proc. Helminth. Soc. Wash. 31:199–202.
Loof, P. A. A. 1971. Free-living and plant-parasitic nematodes from Spitzbergen, collected by Mr. H. Van Rossen. Meded. Landbouw. Wageningen 71–7:1–86.

Lordello, L. G. E. 1955. A new nematode *Rotylenchus melancholicus* n. sp. found associated with grass roots, and its sexual dimorphism. J. Wash. Acad. Sci. 45:81–83.
Lordello, L. G. E., and A. P. L. Zamith. 1956. Novas observacoes sobre os nematodeos que parasitam a batatinha no Est. de S. Paulo. Revta Agr., Sao Paulo 31(1):45–54.
Luc, M. 1960. Trois nouvelles espèces de genre *Rotylenchoides* Whitehead 1958 (Nematoda: Tylenchida). Nematologica 5:7–17.
Perry, V. G. 1959. A note on digonic hermaphroditism in spiral nematodes (*Helicotylenchus* spp.). Nematologica 4:87–88.
Perry, V. G. 1960. The subfamily Hoplolaiminae. *In*: J. N. Sasser and W. R. Jenkins, eds. Nematology. Univ. North Carolina Press, Chapel Hill, pp. 185–190.
Perry, V. G., H. M. Darling, and G. Thorne. 1959. Anatomy, taxonomy, and control of certain spiral nematodes attacking blue grass in Wisconsin. Univ. Wis. Res. Bull. 207.
Phillips, S. P. 1971. Studies of plant and soil nematodes, 16. Eight new species of spiral nematodes (Nematoda: Tylenchoidea) from Queensland. Qd. J. Agr. Anim. Sci. 28:227–242.
Prasad, S. K., E. Khan, and M. L. Chawla. 1965. Two new species of *Helicotylenchus* from soil around maize roots in India. Indian J. Ent. 27:182–184.
Razzhivin, A. A. 1971. New species of nematodes of the family Hoplolaimidae. Zool. Zh. 50(1):133–136.
Roman, J. 1961. A new species of the genus *Helicotylenchus* (Nematoda: Hoplolaimidae) attacking sugarcane. J. Agr., Univ. P. Rico 45:300–303.
Roman, J. 1965. Nematodes of Puerto Rico. The genus *Helicotylenchus* Steiner 1945 (Nematoda: Hoplolaiminae). Tech. Paper, Univ. P. Rico Agr. Exp. Stn. 41:23 p.
Sher, S. A. 1961. Revision of the Hoplolaiminae (Nematoda), I. Classification of nominal genera and nominal species. Nematologica 6:155–169.
* Sher, S. A. 1966. Revision of the Hoplolaiminae (Nematoda), VI. Helicotylenchus Steiner 1945. Nematologica 12:1–56.
Sher, S. A. 1973. *Antarctylus humus* n. gen., n. sp. from the Subantarctic (Nematoda: Tylenchoidea). J. Nematol. 5:19–21.
Siddiqi, M. R. 1963. *Helicotylenchus mucronatus* n. sp. and *H. tunisiensis* n. sp. (Nematoda: Hoplolaiminae). Nematologica 9:386–390.
Siddiqi, M. R. 1963. Two new species of the genus *Helicotylenchus* Steiner 1945 (Nematoda: Hoplolaiminae). Z. Parasitenk. 23:239–244.
* Siddiqi, M. R. 1972. On the genus *Helicotylenchus* Steiner 1945 (Nematoda: Tylenchida), with descriptions of nine new species. Nematologica 18:74–91.
Siddiqi, M. R., and K. F. Brown. 1964. *Helicotylenchus retusus* n. sp. (Nematoda: Hoplolaiminae) found around sugar-cane roots in Negros Oriental, Philippines. Proc. Helminth. Soc. Wash. 31:209–211.
Siddiqi, M. R., and Z. Husain. 1964. Three new species of nematodes in family Hoplolaimidae found attacking citrus trees in India. Proc. Helminth. Soc. Wash. 31:211–215.
Singh, S. D. 1971. Studies on the morphology and systematics of plant and soil nematodes mainly from Andhra Pradesh, I. Tylenchoidea. J. Helminth. 45:353–369.
Steiner, G. 1945. *Helicotylenchus*, a new genus of plant-parasitic nematodes and its relationship to *Rotylenchus* Filipjev. Proc. Helminth. Soc. Wash. 12:34–38.
Swarup, G., and C. L. Sethi. 1968. Plant-parasitic nematodes of northwestern India, II. The genus *Helicotylenchus*. Bull. Ent., Loyola Coll. 9(1):76–80.
Szczygiel, A. 1969. *Tylenchorhynchus polonicus* sp. n. and *Helicotylenchus pseudodigonicus* sp. n. (Nematoda: Tylenchoidea) from Poland. Bull. Acad. Pol. Sci. Cl. II Ser. Sci. Biol. 17(11/12):685–690.
Tarjan, A. C. 1964. Two new mucronate-tailed spiral nematodes (*Helicotylenchus*: Hoplolaiminae). Nematologica 10:185–191.
Thorne, G. 1949. On the classification of the Tylenchida, new order (Nematoda: Phasmidia). Proc. Helminth. Soc. Wash. 16:37–73.
Thorne, G. 1961. Principles of nematology. McGraw-Hill, New York.
Tikyani, M. G., S. Khera, and G. C. Bhatnagar. 1969. *Helicotylenchus goodi* n. sp. from rhizosphere of great millet. Zool. Anz. 182(5/6):420–423.
Waseem, M. 1961. Two new species of the genus *Helicotylenchus* Steiner 1945 (Nematoda: Hoplolaiminae). Can. J. Zool. 39:505–509.
Williams, J. R. 1960. Studies on the nematode soil fauna of sugar-cane fields in Mauritius, 4. Tylenchoidea (partim). Occ. Paper, Maurit. Sug. Ind. Res. Inst., No. 4.
Yeates, G. W. 1967. Studies on nematodes from dune sands, I. Tylenchida. N.Z. J. Sci. 10(1):280–286.
Yuen, Pick H. 1964. Four new species of *Helicotylenchus* Steiner (Hoplolaiminae: Tylenchida) and a redescription of *H. canadensis* Waseem 1961. Nematologica 10:373–387.
Yuen, Pick H. 1964. The female gonad in the subfamily Hoplolaiminae with a note on the spermatheca of *Tylenchorhynchus*. Nematologica 10:570–580.
Yuen, Pick H. 1965. Further observations on *Helicotylenchus vulgaris* Yuen. Nematologica 11:623–637.
Zuckerman, B. M., and D. Strich-Harari. 1963. The life stages of *Helicotylenchus multicinctus* (Cobb) in banana roots. Nematologica 9:347–353.

Hemicriconemoides

Chitwood, B. G., and W. Birchfield. 1957. A new genus, *Hemicriconemoides* (Criconematidae: Tylenchina). Proc. Helminth. Soc. Wash. 24:80–86.
Choi, Y. E., and E. Geraert. 1972. Some remarkable Tylenchida from Korea. Nematologica 18:66–73.

Colbran, R. C. 1962. Studies of plant and soil nematodes, 5. Four new species of Tylenchoidea from Queensland pineapple fields. Qd. J. Agr. Anim. Sci. 19:231–239.
Dasgupta, D. R., D. J. Raski, and S. D. Van Gundy. 1969. Revision of the genus *Hemicriconemoides* Chitwood and Birchfield 1957 (Nematoda: Criconematidae). J. Nematol. 1:126–145.
Edward, J. C., and S. L. Misra. 1963. *Hemicriconemoides communis* n. sp. and *H. litchi* n. sp. (Nematoda: Criconematidae), from Uttar Pradesh, India. Nematologica 9:405–411.
* Edward, J. C., S. L. Misra, and G. R. Singh. 1965. *Hemicriconemoides birchfieldi* n. sp. (Nematoda: Criconematidae) from Allahabad, Uttar Pradesh, India; with a revision of the key to species of *Hemicriconemoides*. Nematologica 11:157–161.
Esser, R. P. 1960. Three additional species in the genus *Hemicriconemoides* Chitwood and Birchfield 1957 (Nemata: Tylenchida). Nematologica 5:64–71.
Fassuliotis, G. 1962. Life history of *Hemicriconemoides chitwoodi* Esser. Nematologica 8:110–116.
Germani, G., and M. Luc. 1970. Contribution à l'stude de genre *Hemicriconemoides* Chitwood and Birchfield 1957 (Nematoda: Criconematidae). Cah. O.R.S.T.O.M., Ser. Biol. 11:133–150.
Goodey, T. Revised by J. B. Goodey. 1963. Soil and freshwater nematodes. Methuen, London.
Grisse, A. de. 1964. *Hemicriconemoides pseudobrachyurum* n. sp. (Nematoda: Criconematidae). Nematologica 10:369–372.
Heyns, J. 1970. South African Criconematinae, Part 1. Genera *Nothocriconema, Lobocriconema, Criconemella, Xenocriconemella,* and *Discocriconemella* (Nematoda). Phytophylactica 2(1):49–56.
Heyns, J. 1970. South African Criconematinae, Part 2. Genera *Criconema, Hemicriconemoides*, and some *Macroposthonia* (Nematoda). Phytophylactica, 2(2):129–136.
Heyns, J. 1970. South African Criconematinae, Part 3. More species of *Hemicriconemoides* and *Macroposthonia* (Nematoda). Phytophylactica, 2(4):243–249.
Nakasono, K., and M. Ichinohe. 1961. *Hemicriconemoides kanayaensis* n. sp. associated with tea root in Japan (Nematoda: Criconematidae). Jap. J. Appl. Ent. Zool. 5:273–276.
Siddiqi, M. R. 1961. Studies on species of Criconematinae (Nematoda: Tylenchida) from India. Proc. Helminth. Soc. Wash. 28:19–34.
Siddiqi, M. R., and J. B. Goodey. 1963. The status of the genera and subfamilies of the Criconematidae (Nematoda); with a comment on the position of *Fergusobia*. Nematologica 9:363–377.
Suryawanshi, M. V. 1971. Studies on Tylenchida (Nematoda) from Marathwada, India, with descriptions of four new species. Nematologica 17:393–406.
Taylor, A. L. 1936. The genera and species of the Criconematinae, a subfamily of the Anguillulidae (Nematoda). Trans. Am. Microscop. Soc 55:391–421.
Thorne, G. 1961. Principles of nematology. McGraw-Hill, New York.
Whitlock, L. S., and A. E. Steele. 1960. Notes on *Hemicriconemoides gaddi* from camellias in Louisiana and Georgia. Pl. Dis. Reptr. 44:446–447.
Yokoo, T. 1963. A new ring nematode, *Hemicriconemoides ureshinoensis* n. sp., found in the soil around the root of tea-plant with some notes on its distribution in the soil. Agr. Bull., Saga Univ., No. 16, pp. 31–35.

Hemicycliophora

Brizuela, R. B. 1963. *Hemicycliophora ritteri* n. sp. (Nematoda: Criconematidae). Nematologica 9:38–40.
Brzeski, M. 1963. A new plant-parasitic nematode, *Hemicycliophora zuckermani* sp. n. (Nematoda: Criconematidae). Bull. Acad. Pol. Sci. Cl. II Ser. Sci. Biol. 11:173–176.
Brzeski, M. W., and B. M. Zuckerman. 1965. Morphological variations, life stages and emended description of *Hemicycliophora zuckermani* Brzeski (Nematoda: Criconematidae). Nematologica 11:66–72.
Carvalho, J. C., and A. de Bona. 1962. *Hemicycliophora similis* em Sao Paulo. Arq. Inst. Biol., Sao Paulo 29:227–230.
Choi, Y. E., and E. Geraert. 1971. Two new species of Tylenchida from Korea with a list of other nematodes new for this country. Nematologica 17:93–106.
Colbran, R. C. 1956. Studies of plant and soil nematodes, 1. Two new species from Queensland. Qd. J. Agr. Anim. Sci. 13:123–126.
Colbran, R. C. 1960. Studies of plant and soil nematodes, 3. *Belonolaimus hastulatus, Psilenchus tumidus*, and *Hemicycliophora labiata,* three new species from Queensland. Qd. J. Agr. Anim. Sci. 17:175–181.
Colbran, R. C. 1962. Studies of plant and soil nematodes, 5. Four new species of Tylenchoidea from Queensland pineapple fields. Qd. J. Agr. Anim. Sci. 19:231–239.
Colbran, R. C. 1963. Studies of plant and soil nematodes, 6. Two new species from citrus orchards. Qd. J. Agr. Anim. Sci. 20:469–474.
Colbran, R. C. 1969. Studies of plant and soil nematodes, 14. Five new species of *Tylenchorhynchus* Cobb, *Paratylenchus* Micoletzky, *Morulaimus* Sauer, and *Hemicycliophora* de Man (Nematoda: Tylenchoidea). Qd. J. Agr. Anim. Sci., 26:181–192.
Coomans, A. 1966. Some nematodes from Congo. Rev. Zool. Bot. Afr. 74:287–312.
Edward, J. C., and B. B. Rai. 1971. *Hemicycliophora demani* (Nematoda: Hemicycliophoridae) from Allahabad Agricultural Institute, Uttar Pradesh, India. Allahabad Fmr. 45(1):7–8.
Goffart, H. 1950. Nematoden aus unterirdischen Gewassar. Dt. zool. Z. 1:73–78.
Goodey, T. 1951. Soil and freshwater nematodes. Methuen, London.
Goodey, T. Revised by J. B. Goodey. 1963. Soil and freshwater nematodes. Methuen, London.
Heyns, J. 1962. Two new species of Criconematidae from South Africa. Nematologica 8:21–24.
Husain, S. I., and A. M. Khan. 1967. A new subfamily, a new subgenus, and eight new species of nematodes from India belonging to superfamily Tylenchoidea. Proc. Helminth. Soc. Wash. 34:175–186.

Jairajpuri, M. S., and Q. H. Baqri. 1973. Nematodes of high altitudes in India, I. Four new species of Tylenchida. Nematologica 19:19–30.
Jenkins, W. R., and J. P. Reed. 1964. Two new species of *Hemicycliophora* (Nematoda: Criconematidae) with a note on *Hemicycliophora ritteri*. Nematologica 10:111–115.
Khan, E., and M. A. Basir. 1963. Two new species of the genus *Hemicycliophora* de Man 1921 (Nematoda: Criconematidae) from North India. Nematologica 9:101–105.
Khan, E., and C. K. Nanjappa. 1972. Four new species of Criconematoidea (Nematoda) from India. Indian J. Nematol. 2:59–68.
Loof, P. A. A. 1968. Taxonomy of *Hemicycliophora* species from west and central Europe (Nematoda: Criconematoidea). Meded. LandbHoogesch. Wageningen 68(14):1–43.
Loof, P. A. A. 1970. The genera *Hemicycliophora* de Man 1921 and *Caloosia* Siddiqi and Goodey 1963 (Criconematoidea). International Nematology Symposium (10th), European Society of Nematologists, Pescara, Sept. 8–13, 1970, Summaries, p. 24. (Abst.)
Loof, P. A. A., and J. Heyns. 1969. Taxonomy of *Hemicycliophora* species from South Africa (Nematoda: Criconematiodea). Nematologica 15:464–472.
Loos, C. A. 1948. Notes on free-living and plant-parasitic nematodes of Ceylon, 3. Ceylon J. Sci. (B) 23(3):119–124.
Luc, M. 1958. Les nematodes et le fletrissement des cotonniers dans le sud-ouest de Madagascar. (Description des espèces nouvelles et precisions sur *Hemicycliophora membranifer*.) Coton Fibr. Trop. 13:239–256.
Luc, M. 1958. Trois nouvelles espèces Africaines du genre *Hemicycliophora* de Man 1921 (Nematoda: Criconematidae). Nematologica 3:15–23.
Luc, M. 1963. Recherches sur la biologie du nematode *Hemicycliophora paradoxa*. C. R. Acad. Sci., Paris 257:1794–1797.
Man, J. G. de. 1921. Nouvelles recherches sur les nematodes libres terricoles de la Hollande. Capita Zool. 1(1):3–62.
Minton, N. A., and A. M. Golden. 1965. Tail shape variability of a new species of sheath nematode (*Hemicycliophora*). Nematologica 11:44. (Abst.).
Minton, N. A., and A. M. Golden. 1966. Morphological variations of *Hemicycliophora zuckermani* Brzeski. Nematologica 12:179–180.
Oostenbrink, M. 1960. The family Criconematidae. *In*: J. N. Sasser and W. R. Jenkins, eds. Nematology. Univ. North Carolina Press, Chapel Hill, pp. 196–205.
Paetzold, D. 1958. Beobachtungen zur Stachellosigkeit der Mannchen von *Hemicycliophora typica* de Man 1921 (Criconematidae). Nematologica 3:140–142.
Raski, D. J. 1958. Four new species of *Hemicycliophora* de Man 1921, with further observations on *H. brevis* Thorne 1955 (Nematoda: Criconematidae). Proc. Helminth. Soc. Wash. 25:125–131.
Reed, J. P., and W. R. Jenkins. 1963. *Hemicycliophora vaccinium* n. sp. (Nematoda: Criconematidae) from cranberry. Proc. Helminth. Soc. Wash. 30:211–212.
Sauer, M. R. 1958. Two new species of *Hemicycliophora* (Nematoda: Tylenchida). Proc. Linn. Soc., New South Wales 83:217–221.
Siddiqi, M. R. 1961. Studies on species of Criconematinae (Nematoda: Tylenchida) from India. Proc. Helminth. Soc. Wash. 28:19–34.
Siddiqi, M. R., and J. B. Goodey. 1963. The status of the genera and subfamilies of the Criconematidae (Nematoda); with a comment on the position of *Fergusobia*. Nematologica 9:363–377.
Sofrygina, M. T. 1972. Ectoparasitic nematodes of the genus *Hemicycliophora*. Parazitologiya 6:90–94.
Tarjan, A. C. 1952. The nematode genus *Hemicycliophora* de Man 1921 (Criconematidae) with a description of a new plant-parasitic species. Proc. Helminth. Soc. Wash. 19:65–76.
Taylor, A. L. 1936. The genera and species of the Criconematinae, a subfamily of the Anguillulinidae (Nematoda). Trans. Am. Microscop. Soc. 55:391–421.
Thorne, G. 1955. Fifteen new species of the genus *Hemicycliophora* with an emended description of *H. typica* de Man (Tylenchida: Criconematidae). Proc. Helminth. Soc. Wash. 22:1–16.
Thorne, G. 1961. Principles of nematology. McGraw-Hill, New York.
Van Gundy, S. D. 1957. The first report of a species of *Hemicycliophora* attacking citrus roots. Pl. Dis. Reptr. 41:1016–1018.
Van Gundy, S. D. 1958. The pathogenicity of *Hemicycliophora arenaria* on citrus. Phytopathology 48:399. (Abst.)
Van Gundy, S. D. 1959. The life history of *Hemicycliophora arenaria* Raski (Nematoda: Criconematidae). Proc. Helminth. Soc. Wash. 26:67–72.
* Wolff Schoemaker, R. L. P. 1968. *Hemicycliophora nyanzae* n. sp. found in East Africa with a key to the species of *Hemicycliophora* de Man 1921. Nematologica 13:541–546.
Wu, Liang-Yu. 1966. Three new closely related species of *Hemicycliophora* de Man (Criconematidae: Nematoda) from Canada. Can. J. Zool. 44:225–234.
Yeates, G. W. 1967. Studies on nematodes from dune sands, 8. *Hemicycliophora halophila* n. sp. and *Ereptonema inflatum* n. sp. N.Z. Jl. Sci. 10(3):802–807.
Zuckerman, B. M., J. P. Reed, and W. R. Jenkins. 1963. Notes on *Hemicycliophora vaccinium* Reed and Jenkins. Nematologica 9:648.

Heterodera

Allen, M. W. 1952. Observations on the genus *Meloidogyne* Goeldi 1887. Proc. Helminth. Soc. Wash. 19:44–51.

Brizuela, R. B., and G. Merny. 1964. Biologie d'*Heterodera oryzae* Luc and Berdon 1961, I. Cycle du parasite et reactions histologiques de l'hote. Rev. Path. Veg. Ent. Agr. Fr. 43:43–53.

Campos Vela, Vi, A. 1968. Taxonomy, life cycle, and host range of *Heterodera mexicana* n. sp. (Nematoda: Heteroderidae). Diss. Abstr., Univ. of Wis., 28(11):4375.

Chitwood, B. G. 1949. Root-knot nematodes, Part I. A revision of the genus *Meloidogyne* Goeldi 1887. Proc. Helminth. Soc. Wash. 16:90–104.

Chitwood, B. G. 1951. The golden nematode of potatoes. Circ., U.S. Dept. Agr., No. 875.

Chitwood, B. G., and Edna M. Buhrer. 1946. Further studies on the life history of the golden nematode of potatoes (*Heterodera rostochiensis* Wollenweber), season 1945. Proc. Helminth. Soc. Wash. 13:54–56.

Chitwood, B. G., and Edna M. Buhrer. 1946. The life history of the golden nematode of potatoes, *Heterodera rostochiensis* Wollenweber, under Long Island, New York, conditions. Phytopathology 36:180–189.

Cobb, Grace S., and A. L. Taylor. 1953. *Heterodera leptonepia*, n. sp., a cyst-forming nematode found in soil with stored potatoes. Proc. Helminth. Soc. Wash. 20:13–15.

Cotten, J. 1965. Cytological investigations in the genus *Heterodera*. Nematologica 11:337–342.

Cotten, J. 1965. The pre-meiotic stages of oogenesis in *Heterodera rostochiensis* Wollenweber. Nematologica 11:335–336.

DiEdwardo, A. A., and V. G. Perry. 1963. An undescribed species of *Heterodera* pathogenic to St. Augustine grass in Florida. Phytopathology 53:874.

DiEdwardo, A. A., and V. G. Perry. 1964. *Heterodera leuceilyma* n. sp. (Nemata: Heteroderidae), a severe pathogen of St. Augustine grass in Florida. Bull. Fla. Agr. Exp. Stn. No. 687, 35 p.

Dixon, G. M., and Jane M. Ryland. 1964. Morphometric relations in the genus *Heterodera*. Nematologica 10:180–181.

Duggan, J. J., and P. A. Brennan. 1966. *Heterodera rosii* (Heteroderidae), a new species of cyst-forming nematode from curled dock (*Rumex crispus* L.). Irish J. Agr. Res. 5:113–120.

Eglitis, V. K., and Dz. K. Kaktynia. 1959. On Heteroderidae in Latvia SSR. Trudy Gel'mint Lab., Akad. Nauk, SSSR 9:403–406. [Lists n. sp. *Heterodera scleranthii*, but no diag.]

Epps, J. M., and A. M. Golden. 1966. Significance of males in reproduction of the soybean cyst nematode (*Heterodera glycines*). Proc. Helminth. Soc. Wash. 33:34.

Eroshenko, A. S., and I. P. Kazachenko. 1972. *Heterodera artemisiae* n. sp. (Nematoda: Heteroderidae), a new species of cyst-forming nematode from the Primorsk Territory. Parazitologiya 6(2):166–170.

Fenwick, D. W., and Mary T. Franklin. 1942. Identification of *Heterodera* species by larval length. Technique for estimating the constants determining the length variations within a given species. J. Helminth. 20:67–114.

Fenwick, D. W., and Mary T. Franklin. 1951. Further studies on the identification of *Heterodera* species by larval length. Estimation of the length parameters for eight species and varieties. J. Helminth. 25:57–76.

Filipjev, I. N., and J. H. Schuurmans Stekhoven, Jr. 1941. A manual of agricultural helminthology. Brill, Leiden.

Franklin, Mary T. 1940. On the specific status of the so-called biological strains of *Heterodera schachtii* Schmidt. J. Helminth. 18:193–208.

Franklin, Mary T. 1945. On *Heterodera cruciferae* n. sp. of *Brassicas*, and on a *Heterodera* strain infecting clover and dock. J. Helminth. 21:71–84.

Franklin, Mary T. 1951. The cyst-forming species of *Heterodera*. Commonwealth Agr. Bur., England.

Franklin, Mary T. 1969. *Heterodera latipons* n. sp., a cereal cyst nematode from the Mediterranean region. Nematologica 15:535–542.

Franklin, Mary T. 1971. Taxonomy of Heteroderidae. *In*: B. M. Zuckerman, W. F. Mai, and R. A. Rohde, eds. Plant-parasitic nematodes, Vol. I. Morphology, anatomy, taxonomy, and ecology. Academic Press, New York, pp. 139–162.

Franklin, Mary T., G. Thorne, and M. Oostenbrink. 1959. Proposal to stabilize the scientific name of the cereal-root eelworm (Class Nematoda). Bull. Zool. Nom. 17:76–85.

Fushtey, S. G., and P. W. Johnson. 1966. The biology of the oat-cyst nematode, *Heterodera avenae* in Canada, I. The effect of temperature on the hatchability of cysts and emergence of larvae. Nematologica 12:313–320.

Goffart, H. 1936. *Heterodera schachtii* Schmidt an gemeiner Hanfnessel (*Galeopsis tetrahit* L.) und an Kakteen. Z. Parasitenk. 8(5):528–532.

Goffart, H. 1960. Die Taxonomische Bewertung morphologisch-anatomischer Merkmale beiden Zysten der Gattung *Heterodera* (Nematoda). Mitt. Biol. Bundanst. Forstw., Berlin-Dahlem, Heft 99:24–51.

Goffart, H. 1961. Uber *Heterodera fici* Kirjanova 1954. Z. Pflkrankh. Pflpath. Pflschutz 68(10/11):597–599.

* Golden, A. M., and W. Birchfield. 1972. *Heterodera graminophila* n. sp. (Nematoda: Heteroderidae) from grass, with a key to closely related species. J. Nematol. 4:147–154.

Golden, A. M., and Grace S. Cobb. 1963. *Heterodera lespedezae* (Heteroderidae), a new species of cyst-forming nematode. Proc. Helminth. Soc. Wash. 30:281–286.

Golden, A. M., and O. J. Dickerson. 1973. *Heterodera longicolla* n. sp. (Nematoda: Heteroderidae) from Buffalo-grass (*Buchloe dactyloides*) in Kansas. J. Nematol. 5:150–154.

* Golden, A. M., and D. M. S. Ellington. 1972. Redescription of *Heterodera rostochiensis* (Nematoda: Heteroderidae) with a key and notes on closely related species. Proc. Helminth. Soc. Wash. 39:64–78.

Golden, A. M., and J. M. Epps. 1965. Morphological variations in the soybean-cyst nematode. Nematologica 11:38. (Abst.)

Golden, A. M., G. J. Rau, and Grace S. Cobb. 1962. *Heterodera cyperi* (Heteroderidae), a new species of cyst-forming nematode. Proc. Helminth. Soc. Wash. 29:168–173.

Goodey, T. 1929. On some details of comparative anatomy in *Aphelenchus, Tylenchus*, and *Heterodera*. J. Helminth. 7:223–230.
Goodey, T. 1951. Soil and freshwater nematodes. Methuen, London.
Goodey, T. Revised by J. B. Goodey. 1963. Soil and freshwater nematodes. Methuen, London.
Granek, I. 1955. Additional morphological differences between the cysts of *Heterodera rostochiensis* and *Heterodera tabacum*. Pl. Dis. Reptr. 39:716–718.
Grisse, A. de, and A. Gillard. 1963. Morphology and biology of hop-cyst eelworm (*Heterodera humuli* Filipjev 1934). Nematologica 9:41–48.
Hesling, J. J. 1959. The identification of *Heterodera* cysts. Plant Nematology, Bull. Minist. Agr. Fish. Fd., London 7:64–70 and chart.
Hijner, J. A., M. Oostenbrink, and H. den Ouden. 1953. Morphological differences between the most important *Heterodera* species in the Netherlands. Tijdschr. Plziekt. 59:245–251.
Hirschmann, Hedwig. 1956. A morphological comparison of two cyst nematodes, *Heterodera glycines* and *H. trifolii*. Phytopathology 46:15. (Abst.)
Hirschmann, Hedwig. 1956. Comparative morphological studies on the soybean cyst nematode, *Heterodera glycines* and the clover cyst nematode, *H. trifolii* (Nematoda: Heteroderidae). Proc. Helminth. Soc. Wash. 23:140–151.
Hirschmann, Hedwig, and R. D. Riggs. 1969. *Heterodera betulae* n. sp. (Heteroderidae), a cyst-forming nematode from river birch. J. Nematol. 1:169–179.
Hirschmann, Hedwig, and A. C. Triantaphyllou. 1965. Comparative cytological and morphometric studies in some parthenogenetic *Heterodera* species. Phytopathology 55:1061. (Abst.)
Horne, C. W., Jr. 1966. The taxonomic status, morphology and biology of a cyst nematode (Nematoda: Heteroderidae) found attacking *Poa annua* L. Diss. Abstr., Texas A. and M. Univ., 26:3573.
Ichinohe, M. 1952. On the soybean nematode *Heterodera glycines* n. sp., from Japan. Mag. Appl. Zool., Tokyo 17(1–2):4 p.
Ichinohe, M. 1955. Studies on the morphology and ecology of the soybean nematode, *Heterodera glycines*, in Japan. Rep. Hokkaido Natl. Agr. Expt. Stn. No. 48.
Jones, F. G. W. 1950. A new species of root eelworm attacking carrots. Nature, London 165(4185):81.
Kampfe, L. 1955. Missbildungen und Veranderungen an Larven von *Heterodera schachtii* Schmidt und *H. rostochiensis* Wollenweber (Nematodes). Zool. Anz. 155(3/4):91–100.
Kampfe, L. 1960. Uber den wert von schwanzform und korpermassen fur die artdiagnose der nematoden (dargestellt an der gattung *Heterodera* Schm.) Nematologica Suppl. II, 1960:112–122. [On the value of tail form and body measurements in the diagnosis of nematode species, illustrated by the genus *Heterodera* Schm.]
Khan, A. M., and S. I. Husain. 1965. *Heterodera mothi* n. sp. (Tylenchida: Heteroderidae) parasitizing *Cyperus rotundus* L. at Aligarh, U.P., India, Nematologica 11:167–172.
Kirjanova, E. S. 1954. Progress and prospective development of phytonematology in USSR. [In Russian.] Trudy probl. temat. Soveshch. Zool. Inst. 3:9–47.
Kirjanova, E. S. 1959. On the systematics of the nematodes belonging to the genus *Heterodera* Schmidt 1871. [In Russian.] Zool. Zh. 38(11):1620–1626.
Kirjanova, E. S. 1962. *Heterodera oxiana* n. sp. (Nematoda: Heteroderidae) from Kara-Kalpak USSR. Trudy vses. Soveshch. Fitogelmint. (5th), Samarkand, pp. 122–131.
Kirjanova, E. S. 1969. The structure of the subcrystalline layer in *Heterodera* (Nematoda: Heteroderidae) with descriptions of two new species. Parazitologiya. 3(1):81–91.
Kirjanova, E. S., and T. S. Ivanova. 1967. First finding of root nematodes of the genera *Heterodera* and *Meloidodera* (Nematoda: Heteroderidae) in Tadzhikistan. *In*: B. M. Zuckerman, M. W. Brzeski, and K. H. Deubert, eds. English translation of selected East European papers in nematology. East Wareham, Mass., Univ. of Mass., pp. 97–100.
Kirjanova, E. S., and T. S. Ivanova. 1969. A cyst-forming nematode, *Heterodera cardiolata* n. sp. (Nematoda: Heteroderidae) from Dushanbe, Tadzhikistan. Dokl. Akad. Nauk Tadzhik, SSR 12(12):59–62.
Kirjanova, E., and E. Krall. 1963. The Estonian cyst-forming nematode *Heterodera estonica* n. sp. (Nematoda: Heteroderidae). (Acad. Sci. Estonian SSR, Inst. Zool. Bot.) Eesti NSV Tead. Akad. Toim. 12: 219–223.
Kirjanova, E., and E. Krall. 1965. The milfoil cyst nematode—*Heterodera millefolii* n. sp. (Nematodes: Heteroderidae). Eesti NSV Tead. Akad. Toim., Seer. Biol. 3:325–328.
Kirjanova, E. S., and L. M. Shagalina. 1965. *Heterodera turcomanica* n. sp. (Nematoda: Heteroderidae). Izvest. Akad. Nauk Turkmen, SSR, Seer. Biol. 6:73–75.
Koshy, P. K. 1967. A new species of *Heterodera* from India. Indian Phytopath. 20(3):272–274.
Koshy, P. K., G. Swarup, and C. L. Sethi. 1971. *Heterodera zeae* n. sp. (Nematoda: Heteroderidae), a cyst-forming nematode on *Zea mays*. Nematologica 16:511–516.
Lownsbery, B. F., and J. W. Lownsbery. 1954. *Heterodera tabacum* n. sp., a parasite of solanaceous plants in Connecticut. Proc. Helminth. Soc. Wash. 21:42–47.
Luc, M., and R. B. Brizuela. 1961. *Heterodera oryzae* n. sp. (Nematoda: Tylenchoidea) parasite du riz en Cote d'Ivoire. Nematologica 6:272–279.
Luc, M., and G. Merny. 1963. *Heterodera sacchari* n. sp. (Nematoda: Tylenchoidea) parasite de la canne à sucre au Congo-Brazzaville. Nematologica 9:31–37.
Mackintosh, G. M. 1960. The morphology of the brassica root eelworm *Heterodera cruciferae* Franklin 1945. Nematologica 5:158–165.
Mathews, H. J. P. 1971. Two new species of cyst nematode, *Heterodera mani* n. sp. and *H. iti* n. sp., from Northern Ireland. Nematologica 17:553–565.

Miller, L. I. 1966. Variation in development of two morphologically different isolates of *Heterodera glycines* obtained from the same field. Phytopathology 56:585. (Abst.)

Miller, L. I., and P. L. Duke. 1967. Morphological variation of eleven isolates of *Heterodera glycines* in the United States. Nematologica 13:145–146. (Abst.)

Miller, L. I., and Betty J. Gray. 1968. Horsenettle cyst nematode, *Heterodera virginiae* n. sp., a parasite of solanaceous plants. Nematologica 14:535–543.

Miller, L. I., and Betty J. Gray. 1972. *Heterodera solanacearum* n. sp., a parasite of solanaceous plants. Nematologica 18:404–413.

Miller, L. I., M. B. Harrison, and A. F. Schindler. 1962. Horsenettle and Osborne's cyst nematodes—two undescribed nematodes occurring in Virginia. Phytopathology 52:743. (Abst.)

Moriarty, F. 1965. The development of the beet eelworm *Heterodera schachtii* Schm. in monoxenic culture. Parasitology 55:719–722.

Mulvey, R. H. 1957. Taxonomic value of the cone top and the underbridge in the cyst-forming nematodes *Heterodera schachtii*, *H. schachtii* var. *trifolii*, and *H. avenae* (Nematoda: Heteroderidae). Can. J. Zool. 35:421–423.

Mulvey, R. H. 1960. The value of cone top and underbridge structures in the separation of some cyst-forming nematodes. *In*: J. N. Sasser and W. R. Jenkins, eds. Nematology. Univ. North Carolina Press, Chapel Hill, pp. 212–215.

* Mulvey, R. H. 1972. Identification of *Heterodera* cysts by terminal and cone top structures. Can. J. Zool. 50:1277–1292.

Oostenbrink, M. 1949. Enkele algemene beschouwingen over de *Heterodera*-Ziekten in de Landbouw. Maandbl. LandbVoorlDienst. 6(3–4):164–166.

Oostenbrink, M. 1951. Het Erwtencystenaaltje, *Heterodera gottingiana* Liebscher, in Nederland. Tijdschr. Plziekt. 57:52–64.

Oostenbrink, M. 1960. The genus *Heterodera*. *In*: J. N. Sasser and W. R. Jenkins, eds. Nematology. Univ. of North Carolina Press, Chapel Hill, pp. 206–211.

Oostenbrink, M., and H. den Ouden. 1953. Het koolcystenaaltje, *Heterodera cruciferae* Franklin 1945, in Nederland. Tijdschr. Plziekt. 59:95–100.

Oostenbrink, M., and H. den Ouden. 1954. The structure of the cone top as a taxonomical character in the *Heterodera* species with lemon-shaped cysts. Tijdschr. Plziekt. 60:146–151.

Pogossian, E. E. 1961. *Heterodera rumicis* n. sp. (Nematoda: Heteroderidae) from the Armenian SSR. Dokl. Akad. Nauk Armyan., SSR 32(3):171–175.

Potter, J. W., and J. A. Fox. 1965. Hybridization of *Heterodera schachtii* and *H. glycines*. Phytopathology 55:800–801.

Raski, D. J. 1950. The life history and morphology of the sugar-beet nematode *Heterodera schachtii* Schmidt. Phytopathology 40:135–152.

Schindler, A. F., and A. M. Golden. 1965. Significance of males in reproduction of *Heterodera cyperi*, a cyst-nematode parasite of nut-grass. Pl. Dis. Reptr. 49:5–6.

Schmidt, A. 1871. Uber den Ruben-Nematoden (*Heterodera schachtii* A.S.). Z. Ver. Rubenzuckerindust. Zollverein 21:1–19.

Seidel, M. 1972. *Heterodera longicaudata* n. sp., a cyst nematode on Gramineae in grassland soils of northern DDR. [*Heterodera longicaudata* n. sp., ein an Gramineen vorkommendes Zystenalchen von Grunlandflachen im Norden der DDR.] Nematologica 18:31–37.

Sen, A. K., and H. J. Jensen. 1967. An amended description of larvae and males of *Heterodera humuli* Filipjev 1934. Nematologica 13:378–384.

Sengbusch, R. 1927. Beitrag zur Biologie des Rubennematoden, *Heterodera schachtii*. Z. Pflkrankh. Pflpath. Pflschutz 37:86–102.

Skarbilovich, Tatiana S. 1959. On the structure of systematics of nematodes of the order Tylenchida Thorne 1949. Acta Polonica 7:117–132.

Skarvilovich, Tatiana S. 1959. The family Heteroderidae Skarbilovich 1947, and its place in the zoological system. Vsesoiuzn, Inst. Gel'mint. imeni I. I. Skriabina, Trudy 6:387–394.

Smart, G. C., Jr. 1965. The aperture and lumen of the spear in three members of Tylenchoidea (Nemata). Nematologica 11:45–46. (Abst.)

Steiner, G. 1949. Plant nematodes the grower should know. Proc. Soil. Sci. Soc. Fla. Bull. 131:72–117.

Stynes, B. A. 1971. *Heterodera graminis* n. sp., a cyst nematode from grass in Australia. Nematologica 17:213–218.

Tarjan, A. C. 1951. An explanation of the revision of the root-knot nematodes, *Meloidogyne* spp. Pl. Dis. Reptr. 35:216.

Thorne, G. 1928. *Heterodera punctata* n. sp., a nematode parasitic on wheat roots from Saskatchewan. Sci. Agr. 8:707–711.

Thorne, G. 1949. On the classification of the Tylenchida, new order (Nematoda: Phasmidia). Proc. Helminth. Soc. Wash. 16:37–73.

Thorne, G. 1961. Principles of nematology. McGraw-Hill, New York.

Tobar Jimenez, A. 1963. Contribution al conocimiento de la *Heterodera fici* Kirjanova 1954 (Heteroderidae: Nematoda) y su diferenci action morfologica de la *H. humuli* Filipjev 1934. Revta iber. Parasit. 23: 341–345.

Tobar Jimenez, A. 1964. Especies del genero *Heterodera* A. Schmidt 1871 (Heteroderidae: Nematoda) de la provincia de Granada, II. La *Heterodera schachtii* A. Schmidt 1871 y sus diferencias morfologicas con la *H. trifolii* Goffart 1932. Revta iber. Parasit. 24:105–109.

Triantaphyllou, A. C., and Hedwig Hirschmann. 1962. Oogenesis and mode of reproduction in the soybean cyst nematode, *Heterodera glycines*. Nematologica 7:235–241.
Triffitt, Marjorie J. 1928. On the morphology of *Heterodera schachtii* with special reference to the potato strain. J. Helminth. 6:39–50.
Wollenweber, H. W. 1923. Krankheiten und Beschadigungen der Kartoffel. Arb. Forsch. Inst. Kartoff., Berlin 7:1–56.
Wouts, W. M. 1972. A revision of the family Heteroderidae (Nematoda: Tylenchoidea), I. The family Heteroderidae and its subfamilies. Nematologica 18:439–446.
Wouts, W. M. 1973. A revision of the family Heteroderidae (Nematoda: Tylenchoidea), II. The subfamily Meloidoderinae. Nematologica 19:218–235.

Hirschmanniella

Breda de Haan, J. van. 1902. Een aoltjes-ziekte der rijst, "omo mentek" of "omo bambang." Meded. Pl. Tuin., Batavia 53:1–65.
Das, V. M. 1960. Studies on the nematode parasites of plants in Hyderabad (Andhra Pradesh, India). Z. Parasitenk. 19:553–605.
Goodey, T. 1936. On *Anguillulina oryzae* (v. Breda de Haan 1902) Goodey 1932, a nematode parasite of the roots of rice, *Oryza sativa* L. J. Helminth. 14:107–112.
Hirschmann, Hedwig. 1955. *Radopholus gracilis* (de Man 1880) n. comb. (Syn.: *Tylenchorhynchus gracilis* [de Man 1880] Filipjev 1936.) Proc. Helminth. Soc. Wash. 22:57–63.
Hirschmann, Hedwig. 1955. *Tylenchorhynchus gracilis* (de Man 1880) Filipjev 1936 *Radopholus gracilis* (de Man 1880) n. c. und seine Synonyme. Zool. Anz. 154:288–301.
Luc, M. 1957. *Radopholus lavabri* n. sp. (Nematoda: Tylenchidae) parasite du riz au Cameroun francais. Nematologica 2:144–148.
Luc, M., and J. B. Goodey. 1962. *Hirschmannia* n. g. differentiated from *Radopholus* Thorne 1949 (Nematoda: Tylenchoidea). Nematologica 7:197–202.
Luc, M., and J. B. Goodey. 1963. *Hirschmanniella* nom. nov. for *Hirschmannia*. Nematologica 9:471.
Man, J. G. de. 1880. Die Einheimischen, frei in der reinen Erde und im Sussen Wasser lebende Nematoden. Vorlaufiger Bericht und descriptiv-systematischer Theil. Tijdschr. Nederl. Dierk Vereen. 5:1–104.
Merny, G. 1972. Les nematodes phyto-parasites des rizieres inondees de Cote d'Ivoire, III. Etudes sur la dynamique de populations de deux endoparasites: *Hirschmanniella spinicaudata* et *Heterodera oryzae*. Cah. O.R.S.T.R.O.M., Ser. Biol. 16:31–87.
Sanwal, K. C. 1957. The morphology of the nematode *Radopholus gracilis* (de Man 1880) Hirschmann 1955, parasitic in roots of wild rice, *Zizania aquatica* L. Can. J. Zool. 35:75–92.
Schuurmans Stekhoven, J. H. 1944. Nematodes libres d'eau douce. Exploration of the Albert National Park (1935–1936) 9:31.
Sher, S. A. 1968. Revision of the genus *Hirschmanniella* Luc and Goodey 1963 (Nematoda: Tylenchoidea). Nematologica 14:243–275.
Siddiqi, M. R. 1963. On the classification of the Pratylenchidae (Thorne 1949) nov. grad. (Nematoda: Tylenchida), with a description of *Zygotylenchus browni* nov. gen. et nov. sp. Z. Parasitenk. 23(4): 390–396.
Siddiqi, M. R. 1966. *Hirschmanniella nana* n. sp. and *H. magna* n. sp. (Nematoda: Pratylenchidae) from India. Proc. Helminth. Soc. Wash. 33:173–177.
Thorne, G. 1949. On the classification of the Tylenchida, new order (Nematoda: Phasmidia). Proc. Helminth. Soc. Wash. 16:37–73.

Histotylenchus

Jairajpuri, M. S., and Q. H. Baqri. 1968. *Tylenchorhynchus hexincisus* n. sp. and *Telotylenchus historicus* n. sp. (Tylenchida) from India. Nematologica 14:217–222.
Netscher, C., and G. Germani. 1969. *Telotylenchus baoulensis* n. sp. et *Trichotylenchus rectangularis* n. sp. (Nematoda: Tylenchoidea). Nematologica 15:347–352.
Siddiqi, M. R. 1971. On the plant-parasitic nematode genera *Histotylenchus* gen. n. and *Telotylenchoides* gen. n. (Telotylenchinae), with observations on the genus *Paratrophurus* Arias (Trophurinae). Nematologica 17:190–200.

Hoplolaimus

Andrassy, I. 1954. Uber einige von Daday beschriebene Nematoden-Arten. Zool. Anz. 152(5–6):138–144.
Andrassy, I. 1958. *Hoplolaimus tylenchiformis* Daday 1905 (Syn.: *H. coronatus* Cobb 1923) und die Gattungen der Unterfamilie Hoplolaiminae Filipjev 1936. Nematologica 3:44–56.
Christie, J. R. 1959. Plant nematodes, their bionomics and control. H. and W. B. Drew, Jacksonville, Florida.
Cobb, N. A. 1923. An emendation of *Hoplolaimus* Daday 1905, nec auctores. J. Wash. Acad. Sci. 13:211–214.
Coomans, A. 1963. Observations on the variability of morphological structures in *Hoplolaimus pararobustus*. Nematologica 9:241–254.
Daday, J. 1905. Untersuchungen uber die Susswasser-Mikrofauna Paraguays. Zoologica Stuttg. 44:1–374.
Golden, A. M., and N. A. Minton. 1970. Description and larval heteromorphism of *Hoplolaimus concaudajuvencus* n. sp. (Nematoda: Hoplolaimidae). J. Nematol. 2:161–166.

Good, J. M., A. E. Steele, and T. J. Ratcliffe. 1959. Occurrence of plant-parasitic nematodes in Georgia turf nurseries. Pl. Dis. Reptr. 43:236–238.
Goodey, J. B. 1957. *Hoplolaimus proporicus* n. sp. (Hoplolaiminae: Tylenchida). Nematologica 2:108–113.
Goodey, T. Revised by J. B. Goodey. 1963. Soil and freshwater nematodes. Methuen, London.
Guiran, D. de. 1963. Un nematode nouveau associée aux cultures d'agrumes au Maroc. Comptes Rendus des Séances de l'Academie d'Agriculture de France 49(5):392–394.
Hirschmann, Hedwig. 1959. Histological studies on the anterior region of *Heterodera glycines* and *Hoplolaimus tylenchiformis* (Nematoda: Tylenchida). Proc. Helminth. Soc. Wash. 26:73–90.
Jairajpuri, M. S., and Q. H. Baqri. 1973. Nematodes of high altitudes in India, I. Four new species of Tylenchida. Nematologica 19:19–30.
Krueger, H. J., and M. B. Linford. 1957. Sex differences in the cephalic region of *Hoplolaimus coronatus*. Proc. Helminth. Soc. Wash. 24:20–23.
Luc, M. 1958. Nematodes and wilting in cotton in southwestern Madagascar. Coton Fibr. Trop. 13:239–256. [In French.]
Perry, V. G. 1960. The subfamily Hoplolaiminae. *In:* J. N. Sasser and W. R. Jenkins, eds. Nematology. Univ. North Carolina Press, Chapel Hill, pp. 185–190.
Perry, V. G., H. M. Darling, and G. Thorne. 1959. Anatomy, taxonomy and control of certain spiral nematodes attacking blue grass in Wisconsin. Univ. Wis. Res. Bull. 207.
Ramirez, C. T. 1964. *Hoplolaimus puertoricensis* n. sp. (Nematoda: Hoplolaimidae). J. Agr., Univ. P. Rico 48:127–130.
Sher, S. A. 1961. Revision of the Hoplolaiminae (Nematoda), I. Classification of nominal genera and nominal species. Nematologica 6:155–169.
Sher, S. A. 1963. Revision of the Hoplolaiminae (Nematoda), II. *Hoplolaimus* Daday 1905 and *Aorolaimus* n. gen. Nematologica 9:267–295.
Siddiqi, M. R. 1972. Two new species of *Scutellonema* from cultivated soils in Africa with a description of *Hoplolaimus aorolaimoides* sp. n. from Portugal (Nematoda: Hoplolaiminae). Proc. Helminth. Soc. Wash. 39:7–13.
Smart, G. C., Jr. 1965. The aperture and lumen of the spear in three members of Tylenchoidea (Nemata). Nematologica 11:45–46. (Abst.)
Steiner, G. 1928. *Tylenchus pratensis* and various other nemas attacking plants. J. Agr. Res. 35:961–981.
Suryawanshi, M. V. 1971. Studies on Tylenchida (Nematoda) from Marathwada, India, with descriptions of four new species. Nematologica 17:393–406.
Thames, W. H., Jr. 1959. Plant-parasitic nematode populations of some Florida soils under cultivated and natural conditions. Diss. Abstr., Univ. of Fla., Gainsville, Fla. 20:1109–1110.
Thorne, G. 1949. On the classification of the Tylenchida, new order (Nematoda: Phasmidia). Proc. Helminth. Soc. Wash. 16:37–73.
Thorne, G. 1961. Principles of nematology. McGraw-Hill, New York.
Timm, R. W. 1956. Nematode parasites of rice in East Pakistan. Pakist. Rev. Agr. 2:115–118.
Tulaganov, A. T. 1941. On the description of a new species of nematode of the genus *Hoplolaimus*, *Hoplolaimus zavadskii*. Uzbek State Univ., Trudy, Nov. Ser. No. 16, Biologiia No. 11:21–22.
Van Den Berg, E., and J. Heyns. 1970. South African Hoplolaiminae, I. The genus *Hoplolaimus* Daday 1905. Phytophylactica 2(4):221–226.
Whitehead, A. G. 1959. East African Agriculture and Forestry Research Organization, Kenya. *Hoplolaimus aberrans* n. sp. (Hoplolaiminae: Tylenchida). Nematologica 4:268–271.
Whitehead, A. G. 1959. *Hoplolaimus angustalatus* n. sp. (Hoplolaiminae: Tylenchida). Nematologica 4:99–105.
Yuen, Pick H. 1964. The female gonad in the subfamily Hoplolaiminae, with a note on the spermatheca of *Tylenchorhynchus*. Nematologica 10:570–580.

Hoplotylus

Goodey, T. Revised by J. B. Goodey. 1963. Soil and freshwater nematodes. Methuen, London.
s'Jacob, J. J. 1959. *Hoplotylus femina* n. g., n. sp. (Pratylenchinae: Tylenchida), associated with ornamental trees. Nematologica 4:317–321.

Longidorus

Aboul-Eid, H. Z. 1970. Systematic notes on *Longidorus* and *Paralongidorus*. Nematologica 16:159–179.
Aboul-Eid, H. Z., and A. Coomans. 1966. Intersexuality in *Longidorus macrosoma*. Nematologica 12: 343–345.
Allen, M. W. 1960. The genera *Xiphinema*, *Longidorus*, and *Trichodorus*. *In*: J. N. Sasser and W. R. Jenkins, eds. Nematology. Univ. North Carolina Press, Chapel Hill, pp. 227–228.
Altherr, E. 1953. Nematodes du sol du Jura vaudois et francais. Bull. Soc. Vaudoise Sci. Nat. (284)65: 429–460.
Christie, J. R. 1959. Plant nematodes, their bionomics and control. H. and W. B. Drew, Jacksonville, Florida.
Clark, W. C. 1963. A new species of *Longidorus* (Micol.) (Dorylaimida: Nematoda). N.Z. Jl. Sci. 6:607–611.
Coomans, A. 1964. Structure of the female gonads in members of the Dorylaimina. Nematologica 10: 601–622.

Corbett, D. C. M. 1964. *Longidorus utriculoides* n. sp. (Nematoda: Dorylaimidae) from Nyasaland. Nematologica 10:496–499.
Dalmasso, A. 1969. Etude anatomique et taxonomique des genres *Xiphinema*, *Longidorus*, et *Paralongidorus* (Nematoda: Dorylaimidae). Mem. Mus. Natn. Hist. Nat., Paris, Serie A, Zool., 61(2):33–82.
Edward, J. C., S. L. Misra, and G. R. Singh. 1964. *Longidorus pisi* n. sp. (Nematoda: Dorylaimoidea) associated with the rhizosphere of *Pisum sativum*, from Uttar Pradesh, India. Jap. J. Appl. Ent. Zool. 8(4):310–312.
Goodey, J. B., and D. J. Hooper. 1963. The nerve rings of *Longidorus* and *Xiphinema*. Nematologica 9:303–304.
Goodey, T. 1951. Soil and freshwater nematodes. Methuen, London.
Goodey, T. Revised by J. B. Goodey. 1963. Soil and freshwater nematodes. Methuen, London.
Harrison, B. D., W. P. Mowat, and C. E. Taylor. 1961. Transmission of a strain of tomato black-ring virus by *Longidorus elongatus* (Nematoda). Virology 14:480–485.
Heyns, J. 1965. New species of the genera *Paralongidorus* and *Longidorus* (Nematoda: Dorylaimoidea) from South Africa. S. Afr. J. Agr. Sci. 8:863–874.
Heyns, J. 1966. Further studies on South African Longidoridae (Nematoda). S. Afr. J. Agr. Sci. 9(4):927–943.
Heyns, J. 1967. *Paralongidorus capensis* n. sp. and *Longidorus belondiroides* n. sp., with a note on *L. taniwha* Clark 1963 (Nematoda: Longidoridae). Nematologica 12:568–574.
Heyns, J. 1969. *Longidorus cohni* n. sp., a nematode parasite of alfalfa and Rhodes grass in Israel. Israel J. Agr. Res. 19(4):179–183.
Hooper, D. J. 1961. A redescription of *Longidorus elongatus* (de Man 1876) Thorne and Swanger 1936 (Nematoda: Dorylaimidae) and descriptions of five new species of *Longidorus* from Great Britain. Nematologica 6:237–257.
Hooper, D. J. 1965. *Longidorus profundorum* n. sp. (Nematoda: Dorylaimidae). Nematologica 11:489–495.
Jensen, H. J., and C. E. Horner. 1957. Peppermint decline caused by *Longidorus sylphus* can be controlled by soil fumigation. Phytopathology 47:18. (Abst.)
Khan, E. 1964. *Longidorus afzali* n. sp., and *Xiphinema arcum* n. sp. (Nematoda: Longidoridae) from India. Nematologica 10:313–318.
Khan, E., A. R. Seshadri, B. Weischer, and K. Mathen. 1971. Five new species associated with coconut in Kerala, India. Indian J. Nematol. 1:116–127.
Konicek, D. E., and H. J. Jensen. 1961. *Longidorus menthasolanus*, a new plant parasite from Oregon (Nemata: Dorylaimoidea). Proc. Helminth. Soc. Wash. 28:216–218.
Man, J. G. de. 1876. Onderzoekingen over vrijinde aarde levende nematoden. Tijdschr. Nederl. Dierk Vereen. 2:78–196.
Merny, G. 1966. Nematodes d'Afrique tropicale: un nouveau *Paratylenchus* (Criconematidae), deux nouveaux *Longidorus*, et observations sur *Longidorus laevicapitatus* Williams 1959 (Dorylaimidae). Nematologica 12:385–395.
Merzheevskaya, O. I. 1951. New species of nematodes. [In Russian.] Sb. nauch. Trudy, Akad. Nauk Belorussk., SSR, Inst. Biol. 2:112–120.
Micoletzky, H. 1922. Die freilebender Erd-Nematoden mit besonderer Berücksichtigung der Steiermark und der Bukowina, zugleich mit einer Revision samtlicher nicht mariner, freilebender Nematoden in Form von Genus-Beschreibungen und Bestimmungsschlusseln. Arch. Naturg., Berlin (1921), Abt. A. 87(8/9):1–650.
Raina, R. 1966. *Longidorus reneyii* sp. nov. (Nematoda: Longidoridae) from Srinagar, Kashmir. Indian J. Ent. 28(4):438–441.
Schuurmans Stekhoven, J. H., and R. J. H. Teunissen. 1938. Nematodes libres terrestres. Fasc. 22, Mission (de Witte) (1933–35), Exploration of the Albert National Park.
Siddiqi, M. R. 1959. Studies on *Xiphinema* spp. (Nematoda: Dorylaimoidea) from Aligarh (North India), with comments on the genus *Longidorus* Micoletzky 1922. Proc. Helminth. Soc. Wash. 26:151–163.
Siddiqi, M. R. 1962. *Longidorus tarjani* n. sp. found around oak roots in Florida. Nematologica 8:152–156.
Siddiqi, M. R. 1962. Studies on the genus *Longidorus* Micoletzky 1922 (Nematoda: Dorylaimoidea), with descriptions of three new species. Proc. Helminth. Soc. Wash. 29:177–188.
* Siddiqi, M. R. 1965. *Longidorus nirulai* n. sp., a parasite of potato plants in Shillong, India, with a key to species of *Longidorus* (Nematoda: Dorylaimoidea). Proc. Helminth. Soc. Wash. 32:95–99.
Sturhan, D. 1963. Beitrag zur Systematik der Gattung *Longidorus*. Nematologica 9:131–142.
Sturhan, D., and B. Weischer. 1964. *Longidorus vineacola* n. sp. (Nematoda: Dorylaimidae). Nematologica 10:335–341.
Thorne, G. 1939. A monograph of the nematodes of the superfamily Dorylaimoidea. Capita Zool. 8(5):261.
Thorne, G. 1961. Principles of nematology. McGraw-Hill, New York.
Thorne, G., and Helen H. Swanger. 1936. A. monograph of the nematode genera *Dorylaimus* Dujardin, *Aporcelaimus* n. g., *Dorylaimoides* n. g., and *Pugentus* n. g. Capita Zool., 6(4):225.
Tulaganow, D. A. 1937. Nematoden der Tomate und des sie umgebenden Bodens. Zool. Anz., Leipzig 118(9–10):283–285.
Tulagnov, A. T. 1938. The fauna of nematodes of cotton and surrounding soil in Katta-Kurgan district of the Uzbek SSR. Trudy Uzbek, Gos. Univ. 12(2):1–25.
Williams, J. R. 1959. Studies on the nematode soil fauna of sugar-cane fields in Mauritius, 3. Dorylaimidae (Dorylaimoidea: Enoplida). Occ. Paper, Maurit. Sug. Ind. Res. Inst., No. 3.
Williams, T. D. 1961. The female reproductive system of *Longidorus elongatus* (de Man 1876) Thorne and Swanger 1936 (Nematoda: Longidorinae), with a note on the genera *Longidorus* (Micoletzky 1922) Thorne and Swanger 1936, and *Xiphinema* Cobb 1913. Can. J. Zool. 39:413–418.

Macrotrophurus

Goodey, T. Revised by J. B. Goodey. 1963. Soil and freshwater nematodes. Methuen, London.
Hirschmann, Hedwig. 1960. The genera *Tylenchus*, *Psilenchus*, *Ditylenchus*, *Anguina*, *Tylenchorhynchus*, *Tetylenchus*, *Trophurus*, and *Macrotrophurus*. *In*: J. N. Sasser and W. R. Jenkins, eds. Nematology. Univ. North Carolina Press, Chapel Hill, pp. 171–180.
Loof, P. A. A. 1958. Some remarks on the status of the subfamily Dolichodorinae, with description of *Macrotrophurus arbusticola* n. g., n. sp. (Nematoda: Tylenchidae). Nematologica 3:301–307.
Thorne, G. 1961. Principles of nematology. McGraw-Hill, New York.

Meloidodera

Chitwood, B. G., and W. Birchfield. 1956. Nematodes, their kinds and characteristics. Bull. State Plant Board, Fla. 2(9).
Chitwood, B. G., C. I. Hannon, and R. P. Esser. 1956. A new nematode genus, *Meloidodera*, linking the genera *Heterodera* and *Meloidogyne*. Phytopathology 46:264–266.
Franklin, Mary T. 1971. Taxonomy of Heteroderidae. *In*: B. M. Zuckerman, W. F. Mai, and R. A. Rohde, eds. Plant-parasitic nematodes, Vol. I. Morphology, anatomy, taxonomy, and ecology. Academic Press, New York, pp. 139–162.
Goodey, T. Revised by J. B. Goodey. 1963. Soil and freshwater nematodes. Methuen, London.
Hopper, B. E. 1958. Plant-parasitic nematodes in the soils of Southern forest nurseries. Pl. Dis. Reptr. 42:308–314.
Hopper, B. E. 1960. Contributions to the knowledge of the genus *Meloidodera* (Nematoda: Tylenchida), with a description of *M. charis* n. sp. Can. J. Zool. 38:939–947.
Hutchinson, M. T., and J. P. Reed. 1959. The pine cystoid nematode in New Jersey. Pl. Dis. Reptr. 43: 801–802.
Kirjanova, E. S., and T. S. Ivanova. 1967. First finding of root nematodes of the genera *Heterodera* and *Meloidodera* (Nematoda: Heteroderidae) in Tadzhikistan. *In*: B. M. Zuckerman, M. W. Brzeski, and K. H. Deubert, eds. English translation of selected East European papers in nematology. East Wareham, Mass., Univ. of Mass., pp. 97–100.
Pogosian, E. E. 1960. *Meloidodera armeniaca* n. sp. (Nematoda: Heteroderidae) from the Armenian SSR Dokl. Akad. Nauk Armyan., SSR 31:311–313.
Sasser, J. N. 1960. The genera *Meloidogyne* and *Meloidodera*. *In*: J. N. Sasser and W. R. Jenkins, eds. Nematology. Univ. North Carolina Press, Chapel Hill, pp. 216–219.
Thorne, G. 1961. Principles of nematology. McGraw-Hill, New York.
Wouts, W. M. 1972. A revision of the family Heteroderidae (Nematoda: Tylenchoidea), I. The family Heteroderidae and its subfamilies. Nematologica 18:439–446.
Wouts, W. M. 1973. A revision of the family Heteroderidae (Nematoda: Tylenchoidea), II. The subfamily Meloidoderinae. Nematologica 19:218–235.

Meloidogyne

Allen, M. W. 1952. Observations on the genus *Meloidogyne* Goeldi 1887. Proc. Helminth. Soc. Wash. 19:44–51.
Birchfield, W. 1965. Host-parasite relations and host-range studies of a new *Meloidogyne* species in southern U.S.A. Phytopathology 55:1359–1361.
Bird, A. F., and G. E. Rogers. 1965. Ultrastructure of the cuticle and its formation in *Meloidogyne javanica*. Nematologica 11:224–230.
Bird, A. F., and G. E. Rogers. 1965. Ultrastructural and histochemical studies of the cells producing the gelatinous matrix in *Meloidogyne*. Nematologica 11:231–238.
Chitwood, B. G. 1949. Root-knot Nematodes, Part I. A revision of the genus *Meloidogyne* Goeldi 1887. Proc. Helminth. Soc. Wash. 16:90–104.
Chitwood, B. G. 1962. Type specimens of pyroid nemic taxa with comments on variability and host range. Nematologica 7:11–12. (Abst.)
Chitwood, B. G., and B. A. Oteifa. 1952. Nematodes parasitic on plants. Ann. Rev. Microbiol. 6:151–184.
Christie, J. R. 1959. Plant nematodes, their bionomics and control. H. and W. B. Drew, Jacksonville, Florida.
Coetzee, Victoria. 1956. *Meloidogyne acronea*, a new species of root-knot nematode. Nature, London 177(4515):899–900.
Davide, R. G., and A. C. Triantaphyllou. 1967. Influence of the environment on development and sex differentiation of root-knot nematodes, I. Effect of infection density, age of the host plant, and soil temperature. Nematologica 13:102–110.
Davide, R. G., and A. C. Triantaphyllou. 1967. Influence of the environment on development and sex differentiation of root-knot nematodes, I. Effect of host nutrition. Nematologica 13:111–117.
Elmiligy, I. A. 1968. Three new species of the genus *Meloidogyne* Goeldi 1887 (Nematoda: Heteroderidae). Nematologica 14:577–590.
Fox, J. A. 1967. Biological studies of the blueberry root-knot nematode (*Meloidogyne carolinensis* n. sp.). Diss. Abstr., N. Car. Univ., Raleigh, N. Car. 28(4):1311–1312.
Franklin, Mary T. 1961. A British root-knot nematode, *Meloidogyne artiella* n. sp. J. Helminth., R. T. Leiper Suppl., pp. 85–92.
Franklin, Mary T. 1962. Preparation of posterior cuticular patterns of *Meloidogyne* spp. for identification. Nematologica 7:336–337.

Franklin, Mary T. 1965. A root-knot nematode, *Meloidogyne naasi* n. sp., on field crops in England and Wales. Nematologica 11:79–86.

Franklin, Mary T. 1971. Taxonomy of Heteroderidae. *In*: B. M. Zuckerman, W. F. Mai, and R. A. Rohde, eds. Plant-parasitic nematodes, Vol. I. Morphology, anatomy, taxonomy, and ecology. Academic Press, New York, pp. 139–162.

Geraert, E. 1965. The head structures of some Tylenchs with special attention to the amphidial apertures. Nematologica 11: 131–136.

Goeldi, E. A. 1887. Relatorio sobre a molestia do cafeeiro na provincia do Rio de Janeiro. Arch. Mus. Nac., Rio de Janeiro 8:1–121.

Goeldi, E. A. 1889. Biologische Miscellen aus Brasilien, VII. Der Kaffeenematode Brasiliens (*Meloidogyne exigua* G.) Zool. Jb. f. syst. 4:262–267.

Goffart, H. 1957. Bemerkungen zu einigen Arten der Gattung *Meloidogyne*. Nematologica 2:177–184.

Golden, A. M., and W. Birchfield. 1965. *Meloidogyne graminicola* (Heteroderidae) a new species of root-knot nematode from grass. Proc. Helminth. Soc. Wash. 32:228–231.

Goodey, T. Revised by J. B. Goodey. 1963. Soil and freshwater nematodes. Methuen, London.

de Grisse, A. 1960. *Meloidogyne kikuyensis* n. sp., a parasite of kikuyu grass (*Pennisetum clandestinum*) in Kenya. Mematologica 5:303–308.

Ibrahim, I. K. A., J. P. Hollis, and W. Birchfield. 1966. Ultrastructure of the body wall of *Meloidogyne hapla*. Phytopathology 56:883. (Abst.)

Ishibashi, N. 1965. The increase in male adults by gamma-ray irradiation in the root-knot nematode, *Meloidogyne incognita* Chitwood. Nematologica 11:361–369.

Itoh, Y., Y. Ohshima, and M. Ichinohe. 1969. A root-knot nematode, *Meloidogyne mali* n. sp., on apple tree from Japan (Tylenchida: Heteroderidae). Jap. J. Appl. Ent. Zool. 4(4):194–202.

Jimenez Millan, F. et al. 1964. Morfologia de las especies del genero *Meloidogyne* (Nematoda) de varios focos de infeccion de cultivos espanoles. Boln. R. Soc. esp. Hist. Nat. (B) 62:143–153.

Kirjanova, E. S., and T. S. Ivanova. 1965. Eelworm fauna of *Pelargonium roseum* L. in the Tadzhik SSR Izvest. Akad. Nauk, Tadzhik, SSR, Otd. Biol. Nauk 1(18):24–31.

Kirjanova, E. S., and T. S. Ivanova. 1967. Survey of nematodes associated with *Pelargonium roseum* L. in Tadzhikistan. *In*: B. M. Zuckerman, M. W. Brzeski, and K. H. Deubert, eds. English translation of selected East European papers in nematology. East Wareham, Mass., Univ. of Mass., pp. 89–96.

Lee, S. H. 1965. Attempts to use immunodiffusion for species identification of *Meloidogyne*. Nematologica 11:41. (Abst.)

Loos, C. A. 1953. *Meloidogyne brevicauda*, n. sp., a cause of root-knot of mature tea in Ceylon. Proc. Helminth. Soc. Wash. 20:83–91.

Lordello, L. G. E. 1956. *Meloidogyne inornata* sp. n., a serious pest of soybean in the state of Sao Paulo, Brazil. Rev. Bras. Biol. 16:65–70.

Lordello, L. G. E. 1956. Nematoides que parasitam a soja na regiao de bauru. Bragantia 15(6):55–64.

Lordello, L. G. E. 1958. *Meloidogyne incognita*, a nematode pest of fig orchards at the Valinhos Region (State of S. Paulo, Brazil). Rev. Bras. Biol. 18:375–379.

Lordello, L. G. E., and A. P. L. Zamith. 1958. On the morphology of the coffee root-knot nematode, *Meloidogyne exigua* Goeldi 1887. Proc. Helminth. Soc. Wash. 25:133–137.

Lordello, L. G. E., and A. P. L. Zamith. 1960. *Meloidogyne coffeicola* sp. n., a pest of coffee trees in the state of Parana, Brazil (Nematoda: Heteroderidae). Rev. Bras. Biol. 20:375–379.

Maggenti, A. R., and M. W. Allen. 1960. The origin of the gelatinous matrix in *Meloidogyne*. Proc. Helminth. Soc. Wash. 27:4–10.

Minton, N. A. 1965. Rectal matrix gland in *Meloidogyne arenaria*. Proc. Helminth. Soc. Wash. 32:163.

Pogosian, E. E. 1960. *Meloidogyne armeniaca* n. sp. (Nematoda: Heteroderidae) from the Armenian SSR Dokl. Akad. Nauk Armyan. SSR 31(5):311–313.

Pogosian, E. E. 1961. A root-knot nematode new for USSR from Armenia. [In Russian; Armenian summary, p. 97.] Izvest. Akad. Nauk Armyan., SSR 14(1):95–97.

Pogosian, E. E. 1966. A new nematode genus and species of the family Heteroderidae (Nematoda) in the Armenian SSR. [In Russian.] Dokl. Akad. Nauk Armyan., SSR 42:117–123.

Ponte, J. J. Da. 1969. *Meloidogyne lordelloi* sp. n., a nematode parasite of *Cereus macrogonus* Salm-Dick. Bolm Soc. Cearense Agron. 10:59–63.

Riffle, J. W. 1963. *Meloidogyne ovalis* (Nematoda: Heteroderidae), a new species of root-knot nematode. Proc. Helminth. Soc. Wash. 30:287–292.

Santos, M. S. N. De A. 1968. *Meloidogyne ardenensis* n. sp. (Nematoda: Heteroderidae), a new British species of root-knot nematode. Nematologica 13:593–598.

Sasser, J. N. 1954. Identification and host-parasite relationships of certain root-knot nematodes (*Meloidogyne* spp.). Univ. Md. Agr. Exp. Stn. Tech. Bull. A-77.

Sasser, J. N. 1960. The genera *Meloidogyne* and *Meloidodera*. *In*: J. N. Sasser and W. R. Jenkins, eds. Nematology. Univ. North Carolina Press, Chapel Hill, pp. 216–219.

Sasser, J. N. 1963. Variation within and among species of *Meloidogyne*. Phytopathology 53:887–888. (Abst.)

Singh, S. P. 1969. A new plant-parasitic nematode *Meloidogyne lucknowica* n. sp. from the root galls of *Luffa cylindrica* (sponge-gourd) in India. Zool. Anz. 182(3/4):259–270.

Sledge, E. B., and A. M. Golden. 1964. *Hypsoperine graminis* (Nematoda: Heteroderidae), a new genus and species of plant-parasitic nematode. Proc. Helminth. Soc. Wash. 31:83–88.

Tarjan, A. C. 1951. An explanation of the revision of the root-knot nematodes *Meloidogyne* spp. Pl. Dis. Reptr. 35:216.

Tarjan, A. C. 1952. Life histories of the root-knot nematodes. Phytopathology 42:20. (Abst.)
Taylor, A. L., V. H. Dropkin, and G. C. Martin. 1955. Perineal patterns of root-knot nematodes. Phytopathology 45:26–34.
Terenteva, T. G. 1965. *Meloidogyne kirjanovae* n. sp. (Nematoda: Heteroderidae). Mater. nauch. Kong. vses. Obshch. Gel'mint., Part 4, pp. 277–281.
Thorne, G. 1961. Principles of nematology. McGraw-Hill, New York.
Triantaphyllou, A. C. 1960. Sex determination in *Meloidogyne incognita* Chitwood 1949 and intersexuality in *M. javanica* (Treub 1885) Chitwood 1949. Ann. Inst. Phytopathology 3:12–31.
Triantaphyllou, A. C., and Hedwig Hirschmann. 1960. Postinfection development of *Meloidogyne incognita* Chitwood 1949 (Nematoda: Heteroderidae). Ann. Inst. Phytopathology 3:3–11.
Triantaphyllou, A. C., and J. N. Sasser. 1960. Variation in perineal patterns and host specificity of *Meloidogyne incognita*. Phytopathology 50:724–735.
Whitehead, A. G. 1959. The root-knot nematodes of East Africa, I. *Meloidogyne africana* n. sp., a parasite from Arabica coffee (*Coffea arabica* L.). Nematologica 4:272–278.
* Whitehead, A. G. 1968. Taxonomy of *Meloidogyne* (Nematodea: Heteroderidae) with descriptions of four new species. Trans. Zool. Soc. London 31(3):263–401.
Wouts, W. M. 1972. A revision of the family Heteroderidae (Nematoda: Tylenchoidea), I. The family Heteroderidae and its subfamilies. Nematologica 18:439–446.
Wouts, W. M. 1973. A revision of the family Heteroderidae (Nematoda: Tylenchoidea), II. The subfamily Meloidoderinae. Nematologica 19:218–235.
Wouts, W. M., and S. A. Sher. 1971. The genera of the subfamily Heteroderinae (Nematoda: Tylenchoidea) with a description of two new genera. J. Nematol. 3:129–144.

Morulaimus

Colbran, R. C. 1960. Studies of plant and soil nematodes, 3. *Belonolaimus hastulatus*, *Psilenchus tumidus*, and *Hemicycliophora labiata*, three new species from Queensland. Qd. J. Agr. Anim. Sci. 17:175–181.
Colbran, R. C. 1969. Studies of plant and soil nematodes, 14. Five new species of *Tylenchorhynchus* Cobb, *Paratylenchus* Micoletzky, *Morulaimus* Sauer, and *Hemicycliophora* DeMan (Nematoda: Tylenchoidea). Qd. J. Agr. Anim. Sci. 26:181–192.
Coomans, A., and A. de Grisse. 1963. Observations on *Trichotylenchus falciformis* Whitehead 1959. Nematologica 9:320–326.
Fisher, J. M. 1964. *Telotylenchus whitei* n. sp. from S. Australia, with observations on *Telotylenchus hastulatus* (Colbran 1960) n. comb. Nematologica 10:563–569.
Sauer, M. R. 1965. *Morulaimus*, a new genus of the Belonolaiminae. Nematologica 11:609–618.
Whitehead, A. G. 1959. *Trichotylenchus falciformis* n. g., n. sp. (Belonolaiminae n. subfam.: Tylenchida Thorne 1949), an associate of grass roots (*Hyparrhenia* sp.) in Southern Tanganyika. Nematologica 4:279–285.

Nacobbus

Allen, M. W. 1960. The genera *Pratylenchus*, *Radopholus*, *Pratylenchoides*, *Rotylenchulus*, and *Nacobbus*; *Tylenchulus*, *Trophotylenchulus*, *Trophonema*, and *Sphaeronema*. *In*: J. N. Sasser and W. R. Jenkins, eds. Nematology. Univ. North Carolina Press, Chapel Hill, pp. 181–184.
Christie, J. R. 1959. Plant nematodes, their bionomics and control. H. and W. B. Drew, Jacksonville, Florida.
Clark, Sybil A. 1967. The development and life history of the false root-knot nematode, *Nacobbus serendipiticus*. Nematologica 13:91–101.
Franklin, Mary T. 1959. *Nacobbus serendipiticus* n. sp., a root-galling nematode from tomatoes in England. Nematologica 4:286–293.
Goodey, T. 1951. Soil and freshwater nematodes. Methuen, London.
Goodey, T. Revised by J. B. Goodey. 1963. Soil and freshwater nematodes. Methuen, London.
Lordello, L. G. E., A. P. L. Zamith, and O. J. Boock. 1961. Two nematodes found attacking potato in Cochabamba, Bolivia. Anals Acad. Bras. Cienc. 33:209–215.
Sher, S. A. 1970. Revision of the genus *Nacobbus* Thorne and Allen 1944 (Nematoda: Tylenchoidea). J. Nematol. 2:228–235.
Thorne, G., and M. W. Allen. 1944. *Nacobbus dorsalis* nov. gen. nov. spec. (Nematoda: Tylenchidae), producing galls on the roots of alfileria, *Erodium cicutarium* (L.) L'Her. Proc. Helminth. Soc. Wash. 11:27–31.
Thorne, G., and M. L. Schuster. 1956. *Nacobbus batatiformis* n. sp. (Nematoda: Tylenchidae), producing galls on the roots of sugar beets and other plants. Proc. Helminth. Soc. Wash. 23:128–134.

Nothanguina

Goodey, T. 1934. *Anguillulina cecidoplastes* n. sp., a nematode causing galls on the grass, *Andropogon pertusus* Willd. J. Helminth. 12:225–236.
Goodey, T. Revised by J. B. Goodey. 1963. Soil and freshwater nematodes. Methuen, London.
Thorne, G. 1961. Principles of nematology. McGraw-Hill, New York.
Venkatarayan, S. V. 1932. *Tylenchus* sp., forming leaf-galls on *Andropogon pertusus* Willd. J. Indian Bot. Soc. 11:243–247.
Whitehead, A. G. 1959. *Nothanguina cecidoplastes* n. comb., syn. *Anguina cecidoplastes* (Goodey 1934) Filipjev 1936 (Nothotylenchinae: Tylenchida). Nematologica 4:70–75.

Nothotylenchus

Andrassy, I. 1958. Erd- und Susswassernematoden aus Bulgarien. Acta. Zool. 4(1–2):1–88.
Das, V. M. 1960. Studies on the nematode parasites of plants in Hyderabad (Andhra Pradesh, India). Z. Parasitenk. 19(6):553–605.
Goodey, T. Revised by J. B. Goodey. 1963. Soil and freshwater nematodes. Methuen, London.
Khan, A. M., and M. R. Siddiqi. 1968. Three new species of *Nothotylenchus* (Nematoda: Neotylenchidae) from North India. Nematologica 14:369–376.
Khan, S. H. 1965. *Nothotylenchus acutus* n. sp. and *N. basiri* n. sp. (Nematoda: Nothotylenchinae) from North India. Proc. Helminth. Soc. Wash. 32:90–93.
* Kheiri, A. 1971. Two new species of *Nothotylenchus* Thorne 1941 from Iran and a redescription of *N. affinis* Thorne 1941 (Nematoda: Neotylenchidae), with a key to the species of the genus. Nematologica 16: 591–600.
Nishizawa, T., and K. Iyatomi. 1955. *Nothotylenchus acris* Thorne, as a parasitic nematode of strawberry plant. Jap. J. Appl. Zool. 20(1–2):48–55.
Thorne, G. 1941. Some nematodes of the family Tylenchidae which do not possess a valvular median esophageal bulb. The Great Basin Naturalist 2:37–85.
Thorne, G. 1961. Principles of nematology. McGraw-Hill, New York.
Tikyani, M. G., and S. Khera. 1969. *Nothotylenchus bhatnagari* n. sp. from the rhizosphere of great millet (*Sorghum vulgare* Per.). Zool. Anz. 182(1/2):87–91.

Paralongidorus

Aboul-Eid, H. Z. 1970. Systematic notes on *Longidorus* and *Paralongidorus*. Nematologica 16:159–179.
Dalmasso, A. 1969. Etude anatomique et taxonomique des genres *Xiphinema*, *Longidorus*, et *Paralongidorus* (Nematoda: Dorylaimidae). Mem. Mus. Natn. Hist. Nat., Paris, Ser. A, Zool. 61(2):33–82.
Edward, J. C., S. L. Misra, and S. L. Singh. 1964. A new species of *Paralongidorus* (Nematoda: Dorylaimoidea) from Allahabad, Uttar Pradesh, India. Jap. J. Appl. Ent. Zool. 8(4):313–316.
Fisher, J. M. 1964. *Dolichodorus adelaidensis* n. sp. and *Paralongidorus eucalypti* n. sp. from S. Australia. Nematologica 10:464–470.
Goodey, J. B., and D. J. Hooper. 1963. The nerve rings of *Longidorus* and *Xiphinema*. Nematologica 9:303–304.
Heyns, J. 1965. New species of the genera *Paralongidorus* and *Longidorus* (Nematoda: Dorylaimoidea) from South Africa. S. Afr. J. Agr. Sci. 8:863–874.
Heyns, J. 1966. *Paralongidorus capensis* n. sp. and *Longidorus belondiroides* n. sp., with a note on *L. taniwha* Clark 1963 (Nematoda: Longidoridae). Nematologica 12:568–574.
Khan, E., M. Saha, and A. R. Seshadri. 1972. Plant parasitic nematodes from Kumaon Hills, India, I. Two new species of *Paralongidorus* (Nematoda: Longidoridae). Nematologica 18:38–43.
Khan, E., A. R. Seshadri, B. Weischer, and K. Mathen. 1971. Five new nematode species associated with coconut in Kerala, India. Indian J. Nematol. 1:116–127.
Siddiqi, M. R. 1959. Studies on *Xiphinema* spp. (Nematoda: Dorylaimoidea) from Aligarh (North India), with comments on the genus *Longidorus* Micoletzky 1922. Proc. Helminth. Soc. Wash. 26:151–163.
* Siddiqi, M. R. 1964. *Xiphinema conurum* n. sp. and *Paralongidorus microlaimus* n. sp. with a key to the species of *Paralongidorus* (Nematoda: Longidoridae). Proc. Helminth. Soc. Wash. 31:133–137.
Siddiqi, M. R., D. J. Hooper, and E. Khan. 1963. A new nematode genus *Paralongidorus* (Nematoda: Dorylaimoidea) with descriptions of two new species and observations on *Paralongidorus citri* (Siddiqi 1959) n. comb. Nematologica 9:7–14.
Siddiqi, M. R., and Z. Husain. 1965. *Paralongidorus beryllus* n. sp. (Nematoda: Dorylaimoidea) from India. Proc. Helminth. Soc. Wash. 32:243–245.
Thorne, G. 1961. Principles of nematology. McGraw-Hill, New York.

Paratrophurus

Arias, M. 1970. *Paratrophurus loofi* n. gen., n. sp. (Tylenchidae) from Spain. Nematologica 16:47–50.
Loof, P. A. A. 1958. Some remarks on the status of the subfamily Dolichodorinae, with description of *Macrotrophurus arbusticola* n. g., n. sp. (Nematoda: Tylenchidae). Nematologica 3:301–307.
Loof, P. A. A., and A. M. Yassin. 1970. Three new plant-parasitic nematodes from the Sudan, with notes on *Xiphinema basiri* Siddiqi 1959. Nematologica 16:537–546.
Paramonov, A. A. 1967. Problems on evolution, morphology, taxonomy, and biochemistry of nematodes of plants. Akad. Nauk Moscow, SSR 18:78–101.
Siddiqi, M. R. 1970. On the plant-parasitic nematode genera *Merlinius* gen. n. and *Tylenchorhynchus* Cobb and the classification of the families Dolichodoridae and Belonolaimidae n. rank. Proc. Helminth. Soc. Wash. 37:68–77.
Siddiqi, M. R. 1971. On the plant-parasitic nematode genera *Histotylenchus* gen. n. and *Telotylenchoides* gen. n. (Telotylenchinae) with observations on the genus *Paratrophurus* Arias (Trophurinae). Nematologica 17:190–200.

Paratylenchus

Adams, R. E., and J. J. Eichenmuller, Jr. 1962. *Gracilacus capitatus* n. sp. from scarlet oak in West Virginia. Nematologica 8:87–92.
Allen, M. W., and H. J. Jensen. 1950. *Cacopaurus epacris* n. sp. (Nematoda: Criconematidae), a nematode parasite of California black walnut roots. Proc. Helminth. Soc. Wash. 17:10–14.

Allen, M. W., and S. A. Sher. 1967. Taxonomic problems concerning the phytoparasitic nematodes. *In*: Annual Review of Phytopathology, Vol. 5. Annual Reviews, Palo Alto, California. pp. 247–264.
Andrassy, I. 1959. Neue und wenig bekannte Nematoden aus Jugoslawien. Annls. Hist., Nat. Mus. Natn. Hungary (Ser. Nov.) 51:259–275.
Bally, W., and G. A. Reydon. 1931. The present status of the question of the nematode [sic] diseases of coffee. Arch. Koffiecult. Nederl.-Indie. 5:23–216.
Brown, Georgianna L. 1959. Three new species of the genus *Paratylenchus* from Canada (Nematoda: Criconematidae). Proc. Helminth. Soc. Wash. 26:1–8.
Brzeski, M. 1963. *Paratylenchus macrodorus* n. sp. (Nematoda: Paratylenchidae), a new plant-parasitic nematode from Poland. Bull. Acad. Pol. Sci. Cl. II Ser. Sci. Biol. 11(6):277–280.
Brzeski, M. W., and A. Szczygiel. 1963. Studies on the nematodes of the genus *Paratylenchus* Micoletzky (Nematoda: Paratylenchinae) in Poland. Nematologica 9:613–625.
Christie, J. R. 1959. Plant nematodes, their bionomics and control. H. and W. B. Drew, Jacksonville, Florida.
Cobb, N. A. 1923. Notes on *Paratylenchus*, a genus of nemas. J. Wash. Acad. Sci. 13:254–257.
Colbran, R. C. 1965. Studies of plant and soil nematodes, 10. *Paratylenchus coronatus* n. sp. (Nematoda: Criconematidae), a pin nematode associated with citrus. Qd. J. Agr. Anim. Sci. 22:277–279.
Colbran, R. C. 1969. Studies of plant and soil nematodes, 14. Five new species of *Tylenchorhynchus* Cobb, *Paratylenchus* Micoletzky, *Morulaimus* Sauer, and *Hemicycliophora* de Man (Nematoda: Tylenchoidea). Qd. J. Agr. Anim. Sci. 26:181–192.
Corbett, D. C. M. 1966. Central African nematodes, II. *Paratylenchus crenatus* n. sp. (Nematoda: Criconematidae) from Malawi. Nematologica 12:101–104.
Edward, J. C., and S. L. Misra. 1963. *Paratylenchus nainianus* n. sp. (Nematoda: Criconematidae) from Uttar Pradesh, India. Nematologica 9:215–217.
Edward, J. C., S. L. Misra, and G. R. Singh. 1967. *Paratylenchus micoletzkyi* n. sp., with the description of the allotype of *P. nainianus* Edward and Misra 1936. Nematologica 13:347–352.
Fisher, J. M. 1965. Studies on *Paratylenchus nanus*, I. Effects of variation in environment on several morphometric characters of adults. Nematologica 11:269–279.
Fisher, J. M. 1966. Observations on moulting of fourth-stage larvae of *Paratylenchus nanus*. Aust. J. Biol. Sci. 19:1073–1079.
Geraert, E. 1965. The genus *Paratylenchus*. Nematologica 11:301–334.
Golden, A. M. 1961. *Paratylenchus steineri* (Criconematidae) a new species of plant nematode. Proc. Helminth. Soc. Wash. 28:9–11.
Goodey, T. 1934. Observations on *Paratylenchus macrophallus* (de Man 1880). J. Helminth. 12:79–88.
Goodey, T. 1951. Soil and freshwater nematodes. Methuen, London.
Goodey, T. Revised by J. B. Goodey. 1963. Soil and freshwater nematodes. Methuen, London.
Grisse, A. de. 1962. *Paratylenchus vandenbrandei* n. sp. (Nematoda: Criconematidae), nouvelle espèce de *Paratylenchus*, associée aux racines d'Agave au Kenya. Nematologica 8:229–232.
* Jenkins, W. R. 1956. *Paratylenchus projectus*, new species (Nematoda: Criconematidae), with a key to the species of *Paratylenchus*. J. Wash. Acad. Sci. 46:296–298.
Jenkins, W. R. 1960. *Paratylenchus marylandicus* n. sp. (Nematoda: Criconematidae) associated with roots of pine. Nematologica 5:175–177.
Jenkins, W. R., and D. P. Taylor. 1956. *Paratylenchus dianthus* n. sp. (Nematoda: Criconematidae), a parasite of carnation. Proc. Helminth. Soc. Wash. 23:124–127.
Jenkins, W. R., and D. P. Taylor. 1967. Plant nematology. Reinhold, New York.
Khan, E., S. K. Prasad, and V. K. Mathur. 1967. Two new species of the genus *Paratylenchus* Micoletzky 1922 (Nematoda: Criconematidae) from India. Nematologica 13:79–84.
Linford, M. B., Juliette M. Oliveira, and M. Ishii. 1949. *Paratylenchus minutus* n. sp., a nematode parasitic on roots. Pacif. Sci. 3:111–119.
Loof, P. A. A., and M. Oostenbrink. 1968. Redescription of *Paratylenchus bukowinensis* Micoletzky 1922 (Criconematoidea). Nematologica 14:152–154.
Luc, M., and G. de Guiran. 1962. Deux nouveaux *Paratylenchus* (Nematoda: Criconematidae) de Cote d'Ivoire. Nematologica 7:133–138.
Mathur, V. K., E. Khan, and S. K. Prasad. 1967. *Paratylenchus neonanus* n. sp. (Nematoda: Criconematidae) from India. Labdev. J. Sci. Technol. 5(2):146–147.
Merny, G. 1966. Nematodes d'Afrique tropicale: un nouveau *Paratylenchus* (Criconematidae), deux nouveaux *Longidorus* et observations sur *Longidorus laevicapitatus* Williams 1959 (Dorylaimidae). Nematologica 12:385–395.
Micoletzky, H. 1922. Die freilebenden Erd-Nematoden. Arch. Naturg. 87:1–650.
Raski, D. J. 1962. Paratylenchidae n. fam. with descriptions of five new species of *Gracilacus* n. g. and an emendation of *Cacopaurus* Thorne 1949, *Paratylenchus* Micoletzky 1922 and Criconematidae Thorne 1943. Proc. Helminth. Soc. Wash. 29:189–207.
Rhoades, H. L., and M. B. Linford. 1961. Biological studies on some members of the genus *Paratylenchus*. Proc. Helminth. Soc. Wash. 28:51–59.
Siddiqi, M. R., and J. B. Goodey. 1963. The status of the genera and subfamilies of the Criconematidae (Nematoda); with a comment on the position of *Fergusobia*. Nematologica 9:363–377.
Thorne, G., and R. B. Malek. 1968. Nematodes of the great plains, Part 1. Tylenchida (Nemata: Secernentea). S. Dak. Agr. Exp. Stn. Tech. Bull. 31, 111 p.
van der Linde, W. J. 1938. A contribution to the study of nematodes. Ent. Memoirs, Dept. Agr. For., Union South Afr. 2(3):1–40.

Wolff Schoemaker, R. L. P. 1963. *Gracilacus peperpotti* n. sp. (Nematoda: Paratylenchidae) found in a Surinam coffee plantation soil. Nematologica 9:296–299.
Wouts, W. M. 1966. *Paratylenchus halophilus* (Nematoda: Criconematidae), a new species from New Zealand. N.Z. Jl. Sci. 9:281–286.
Wu, Liang-Yu. 1961. *Paratylenchus tenuicaudatus* n. sp. (Nematoda: Criconematidae). Can. J. Zool. 29:163–165.
Wu, Liang-Yu. 1962. *Paratylenchus brevihastus* n. sp. (Criconematidae: Nematoda). Can. J. Zool. 40: 391–393.
Wu, Liang-Yu. 1962. *Paratylenchus veruculatus* n. sp. (Criconematidae: Nematoda) from Scotland. Can. J. Zool. 40:773–775.
Wu, Liang-Yu, and J. L. Townshend. 1973. *Paratylenchus tateae* n. sp. (Paratylenchinae: Nematoda). Can. J. Zool. 51:109–111.

Peltamigratus

Andrassy, I. 1958. *Hoplolaimus tylenchiformis* Daday 1905 (Syn.: *H. coronatus* Cobb 1923) und die Gattungen der Unterfamilie Hoplolaiminae Filipjev 1936. Nematologica 3:44–56.
Golden, A. M., and A. L. Taylor. 1956. *Rotylenchus christiei* n. sp., a new spiral nematode species associated with roots of turf. Proc. Helminth. Soc. Wash. 23:109–112.
Khan, S. H., and Zakiuddin. 1968. A new species of the genus *Peltamigratus* Sher 1963 (Nematoda: Hoplolaiminae) from Trinidad, West Indies. Annls. Zool. Ecol. Anim. 1(4):495–498.
* Knobloch, N. A. 1969. *Peltamigratus thornei* sp. n. (Nematoda: Hoplolaimidae) from soil in Central America. Proc. Helminth. Soc. Wash. 36:208–210.
Loof, P. A. A. 1964. Free-living and plant-parasitic nematodes from Venezuela. Nematologica 10:201–300.
Sher, S. A. 1963. Revision of the Hoplolaiminae (Nematoda), II. *Hoplolaimus* Daday 1905 and *Aorolaimus* n. gen. Nematologica 9:267–295.
Sher, S. A. 1963. Revision of the Hoplolaiminae (Nematoda), III. *Scutellonema* Andrassy 1958. Nematologica 9:421–443.
Sher, S. A. 1963. Revision of the Hoplolaiminae (Nematoda), IV. *Peltamigratus* n. gen. Nematologica 9:455–467.
Smit, J. J. 1971. Deux nouvelles espèces africaines d'Hoplolaiminae (Nematoda: Tylenchoidea); *Peltamigratus striatus* n. sp. et *Scutellonema africanum* n. sp. Nematologica 17:113–126.
Yuen, Pick H. 1964. The female gonad in the subfamily Hoplolaiminae with a note on the spermatheca of *Tylenchorhynchus*. Nematologica 10:570–580.

Pratylenchoides

Allen, M. W. 1960. The genera *Pratylenchus, Radopholus, Pratylenchoides, Rotylenchulus* and *Nacobbus*; *Tylenchulus, Trophotylenchulus, Trophonema,* and *Sphaeronema*. *In*: J. N. Sasser and W. R. Jenkins, eds. Nematology. Univ. North Carolina Press, Chapel Hill, pp. 181–184.
Bor, N. A., and J. J. s'Jacob. 1966. *Pratylenchoides maritimus*, a new nematode species from the Boschplaat, Terschelling. Nematologica 12:462–466.
Braun, A. L., and P. A. A. Loof. 1966. *Pratylenchoides laticauda* n. sp., a new endoparasitic phytonematode. Neth. J. Pl. Path. 72:241–245.
Cobb, N. A. 1893. Nematode worms found attacking sugar-cane. *In*: Plant diseases and their remedies. Agr. Gaz., New South Wales 4:808–833.
Goodey, T. 1932. The genus *Anguillulina* Gerv. and v. Ben. 1859, vel *Tylenchus* Bastian 1865. J. Helminth. 10:75–180.
Goodey, T. 1940. On *Anguillulina multicincta* (Cobb) and other species of *Anguillulina* associated with the roots of plants. J. Helminth. 18:21–38.
Goodey, T. Revised by J. B. Goodey. 1963. Soil and freshwater nematodes. Methuen, London.
Guiran, G. de. 1963. *Mesotylus*: nouveau genre de Pratylenchinae (Nematoda: Tylenchoidea). Nematologica 9:567–575.
Guiran, G. de, and M. R. Siddiqi. 1967. Characters differentiating the genera *Zygotylenchus* Siddiqi 1963 (Syn.: *Mesotylus* de Guiran 1964) and *Pratylenchoides* Winslow 1958 (Nematoda: Pratylenchinae). Nematologica 13:235–240.
Jairajpuri, M. S. 1964. On *Pratylenchoides crenicauda* Winslow 1958 (Nematoda: Pratylenchinae) from Srinagar (Kashmir), India. Current Science 33:339.
Sher, S. A. 1968. Revision of the genus *Radopholus* Thorne 1949 (Nematoda: Tylenchoidea). Proc. Helminth. Soc. Wash. 35:219–237.
Sher, S. A. 1970. Revision of the genus *Pratylenchoides* Winslow 1958 (Nematoda: Tylenchoidea). Proc. Helminth. Soc. Wash. 37:154–166.
Siddiqi, M. R. 1963. On the classification of the Pratylenchidae (Thorne 1949) nov. grad. (Nematoda: Tylenchida) with a description of *Zygotylenchus browni* nov. gen. et nov. sp. Z. Parasitenk. 23:390–396.
Tarjan, A. C., and B. Weischer. 1965. Observations on some Pratylenchinae (Nemata), with additional data on *Pratylenchoides guevarai* Tobar Jimenez 1963 (Syn.: *Zygotylenchus browni* Siddiqi 1963 and *Mesotylus gallicus* de Guiran 1964). Nematologica 11:432–440.
Tobar Jimenez, A. 1963. *Pratylenchoides guevarai* n. sp., nuevo nematode Tylenchido relacionado con el cipres (*Cupressus sempervirens* L.). Revta iber. Parasit. 23:27–36.
Winslow, R. D. 1958. The taxonomic position of *Anguillulina obtusa* Goodey 1932 and 1940. Nematologica 3:136–139.

Pratylenchus

Allen, M. W. 1960. The Genera *Pratylenchus*, *Radopholus*, *Pratylenchoides*, *Rotylenchulus*, and *Nacobbus*; *Tylenchulus*, *Trophotylenchulus*, *Trophonema*, and *Sphaeronema*. *In*: J. N. Sasser and W. R. Jenkins, eds. Nematology. Univ. North Carolina Press, Chapel Hill, pp. 181–184.

Allen, M. W., and H. J. Jensen. 1951. *Pratylenchus vulnus* n. sp. (Nematoda: Pratylenchinae), a parasite of trees and vines in California. Proc. Helminth. Soc. Wash. 18:47–50.

Baranovskaya, I. A., and M. M. Haque. 1968. Description of *Pratylenchus clavicaudatus* n. sp. (Nematoda: Pratylenchinae Thorne 1949). Zoo. Zh. 47(5):759–761.

Christie, J. R. 1959. Plant nematodes, their bionomics and control. H. and W. B. Drew, Jacksonville, Florida.

Cobb, N. A. 1917. A new parasitic nema found infesting cotton and potatoes. J. Agr. Res. 11(1):27–33.

* Corbett, D. C. M. 1969. *Pratylenchus pinguicaudatus* n. sp. (Pratylenchinae: Nematoda) with a key to the genus *Pratylenchus*. Nematologica 15:550–556.

Das, V. M. 1960. Studies on the nematode parasites of plants in Hyderabad (Andhra Pradesh, India). Z. Parasitenk. 19:553–605.

Edward, J. C., S. L. Misra, B. B. Rai, and E. Peter. 1969. Association of *Pratylenchus chrysanthus* n. sp. with chrysanthemum root rot. Allahabad Fmr. 43(3):175–179.

Ferris, Virginia R. 1961. A new species of *Pratylenchus* (Nemata: Tylenchida) from roots of soybeans. Proc. Helminth. Soc. Wash. 28:109–111.

Filipjev, I. N. 1934. The classification of the free-living nematodes and their relations to the parasitic nematodes. Smithson. Misc. Coll. (Pub. 3216) 89(6):1–63.

Filipjev, I. N. 1936. On the classification of the Tylenchinae. Proc. Helminth. Soc. Wash. 3:80–82.

Filipjev, I. N., and J. H. Schuurmans Stekhoven, Jr. 1941. A manual of agricultural helminthology. Brill, Leiden.

Godfrey, G. H. 1929. A destructive root disease of pineapples and other plants due to *Tylenchus brachyurus* n. sp. Phytopathology 19:611–629.

Goodey, T. 1932. The genus *Anguillulina* Gerv. and v. Ben. 1859, vel *Tylenchus* Bastian 1865. J. Helminth. 10:75–180.

Goodey, T. 1951. Soil and freshwater nematodes. Methuen, London.

Goodey, T. Revised by J. B. Goodey. 1963. Soil and freshwater nematodes. Methuen, London.

Graham, T. W. 1951. Nematode root rot of tobacco and other plants. S.Car. Agr. Exp. Stn. Bull. 390.

Haque, M. M. 1966. Contribution to the knowledge of the genus *Pratylenchus* Filipjev 1934 (Nematoda: Pratylenchinae Thorne 1949). Zool. Zh. 45:342–344.

Hastings, R. J. 1939. The biology of the meadow nematode *Pratylenchus pratensis* (de Man) Filipjev 1936. Can. J. Res. Sect. D 17:39–44.

Ivanova, T. S. 1968. Nematodes of cereals from the Zeravshan Valley of Tadzhikistan. Dushanbe, Izdatelstvo "Donish."

Loof, P. A. A. 1959. Ueber das Vorkommen von Endotokia matricida bei Tylenchida. Nematologica 4: 238–240.

Loof, P. A. A. 1960. Taxonomic studies on the genus *Pratylenchus* (Nematoda). Tijdschr. Plziekt. 66:29–90.

Loof, P. A. A., and A. M. Yassin. 1971. Three new plant-parasitic nematodes from the Sudan, with notes on *Xiphinema basiri* Siddiqi 1959. Nematologica 16:537–546.

Lordello, L. G. E. 1956. Sobre um nematodeo do genero *Pratylenchus*, parasito das raizes de *Allium cepa*. Revta Agr., Sao Paulo 31(3):181–188.

Lordello, L. G. E., A. P. L. Zamith, and O. J. Boock. 1954. Novo nematodeo parasito da batatinha. Bragantia, Campinas 13:141–149.

Luc, M. 1958. Nematodes and wilting in cotton in southwestern Madagascar. Coton Fibr. Trop. 13:239–256.

Man, J. G. de. 1880. Die Einheimischen, frei in der reinen Erde und im Sussen Wasser lebenden Nematoden. Volaufiger Bericht und descriptivsystematischer Theil. Tijdschr. Nederl. Dierk Vereen. 5(1–2):1–104.

Merzheevskaya, O. I. 1951. New species of nematodes. [In Russian.] Sb. nauch. Trudy, Akad. Nauk Belorussk., SSR, Inst. Biol. 2:112–120.

Meyl, A. 1961. Die freilebenden Erd- und Susswassernematoden (Fadenwurmer). *In*: Bohamer, Ehrmann, and Ulmer. Die Tierwelt Mitteleuropas. Quelle and Meyer, Leipzig, pp. 1–273.

Nandakumar, C., and S. Khera. 1970. A new nematode species, *Pratylenchus mulchandi*, from millets of Rajasthan. Indian Phytopath. 22:359–363.

Romaniko, V. I. 1960. Novoe v izuchenii biologii i ekologii vozbuditelya pratilenkhoza rastenii iz semeistva bobovykh (Leguminosae)—*Pratylenchus globulicola* Romaniko nov. sp. (Nematodes: Pratylenchidae). *In*: Materialy k pyatomu vsesoiuznomu soveshschaniyu po izucheniyu nematod. Samarkand, Gosud Univ. imeni Alishera Navoi, Soviet Min., Usbekskoi, SSR, pp. 85–87.

Romaniko, V. I. 1966. Two new species of plant nematodes from wheat. Zool. Zh. 45(6):929–931.

Seinhorst, J. W. 1959. Two new species of *Pratylenchus*. Nematologica 4:83–86.

Seinhorst, J. W. 1968. Three new *Pratylenchus* species with a discussion of the structure of the cephalic framework and of the spermatheca in this genus. Nematologica 14:497–510.

Sher, S. A., and M. W. Allen. 1953. Revision of the genus *Pratylenchus* (Nematoda: Tylenchidae). Univ. Calif. Publns. Zool. 57(6):441–470.

Sherbakoff, C. D., and W. W. Stanley. 1943. The more important diseases and insect pests of crops in Tennessee. Tenn. Agr. Exp. Stn. Bull. 186.

Siddiqi, M. R. 1963. On the classification of the Pratylenchidae (Thorne 1949) nov. grad. (Nematoda: Tylenchida), with a description of *Zygotylenchus browni* nov. gen. et nov. sp. Z. Parasitenk. 23(4): 390–396.

Steiner, G. 1927. *Tylenchus pratensis* and various other nemas attacking plants. J. Agr. Res. 35:961–981.
Taylor, D. P., and W. R. Jenkins. 1957. Variation within the nematode genus *Pratylenchus*, with the descriptions of *P. hexincisus* n. sp. and *P. subpenetrans* n. sp. Nematologica 2:159–174.
Thorne, G. 1949. On the classification of the Tylenchida new order (Nematoda: Phasmidia). Proc. Helminth. Soc. Wash. 16:37–73.
Thorne, G. 1961. Principles of nematology. McGraw-Hill, New York.
Wu, Liang-Yu. 1971. *Pratylenchus macrostylus* n. sp. (Pratylenchinae: Nematoda). Can. J. Zool. 49: 487–489.
Zyubin, B. N. 1966. *Pratylenchus montanus* n. sp. (Nematoda: Pratylenchidae) on a crop of the opium poppy in Kirgizia. *In*: M. M. Tokobaev, ed. Helminths of animals in Kirgizia and adjacent territories. Frunze, Izdatelstvo "ILIM," pp. 147–151.

Psilenchus

Andrassy, I. 1952. Freilebende Nematoden aus dem Bukk-Gebirge. Annls. Hist., Nat. Mus. Natn. Hungary (Ser. Nov.) 2:13–65.
Andrassy, I. 1962. Neue nematoden-arten aus ungarn, I. Zehn neue arten der unterklasse Secernentea (Phasmidia). Acta Zool., Budapest 8:1–23.
Andrassy, I. 1962. Zwei neue nematoden-arten aus dem Uberschwemmungsgebiet der Donau (Danubialia Hungarica, XIII). Opusc. Zool., Budapest 4(2–4):3–8.
Colbran, R. C. 1960. Studies of plant and soil nematodes, 3. *Belonolaimus hastulatatus*, *Psilenchus tumidus*, and *Hemicycliophora labiata*, three new species from Queensland. Qd. J. Agr. Anim. Sci. 17:175–181.
Geraert, E. 1965. The head structures of some Tylenchs with special attention to the amphidial apertures. Nematologica 11:131–136.
Goodey, T. 1932. The genus *Anguillulina* Gerv. and v. Ben. 1859, vel *Tylenchus* Bastian 1865. J. Helminth. 10:75–180.
Goodey, T. 1951. Soil and freshwater nematodes. Methuen, London.
Goodey, T. Revised by J. B. Goodey. 1963. Soil and freshwater nematodes. Methuen, London.
Hagemeyer, Joyce W., and M. W. Allen. 1952. *Psilenchus duplexus* n. sp. and *Psilenchus terextremus* n. sp., two additions to the nematode genus *Psilenchus* de Man 1921. Proc. Helminth. Soc. Wash. 19: 51–54.
Hirschmann, Hedwig. 1960. The genera *Tylenchus*, *Psilenchus*, *Ditylenchus*, *Anguina*, *Tylenchorhynchus*, *Tetylenchus*, *Trophurus* and *Macrotrophurus*. *In*: J. N. Sasser and W. R. Jenkins, eds. Nematology. Univ. North Carolina Press, Chapel Hill, pp. 171–181.
Jairajpuri, M. S. 1965. A redefinition of *Psilenchus* de Man 1921 and *Tylenchus* subgenus *Filenchus* Andrassy 1954 with the erection of *Clavilenchus* n. subgenus under *Tylenchus* Bastian 1865. Nematologica 11:619–622.
Jairajpuri, M. S., and A. H. Siddiqi. 1963. On *Psilenchus neoformis* n. sp. (Nematoda: Tylenchida) from Solon (H.P.), North India. [Correspondence.] Current Science, Bangalore 32(7):318–319.
* Kheiri, A. 1970. Two new species in the family Tylenchidae (Nematoda) from Iran, with a key to *Psilenchus* de Man 1921. Nematologica 16:359–368.
Man, J. G. de. 1921. Nouvelles recherches sur les nematodes libres terricoles de la Hollande. Capita Zool. 1:3–62.
Siddiqi, M. R. 1963. On the diagnosis of the nematode genera *Psilenchus* de Man 1921, and *Basiria* Siddiqi 1959, with a description of *Psilenchus hilarus* n. sp. Z. Parasitenk. 23(2):164–169.
Thorne, G. 1949. On the classification of the Tylenchida, new order (Nematoda: Phasmidia). Proc. Helminth. Soc. Wash. 16:37–73.
Thorne, G. 1961. Principles of nematology. McGraw-Hill, New York.

Radopholoides

Filipjev, I. N. 1934. The classification of the free-living nematodes and their relations to the parasitic nematodes. Smithson. Misc. Coll. 89(6):1–63.
Guiran, G. de. 1967. Description de *Radopholoides litoralis* n. g., n. sp. (Nematoda: Pratylenchinae). Nematologica 13:231–234.
s'Jacob, J. J. 1959. *Hoplotylus femina* n. g., n. sp. (Pratylenchinae: Tylenchida) associated with ornamental trees. Nematologica 4:317–321.
Siddiqi, M. R. 1963. On the classification of the Pratylenchidae (Thorne 1949). nov. grad. (Nematoda: Tylenchida), with a description of *Zygotylenchus browni* nov. gen. et. nov. sp. Z. Parasitenk. 23:390–396.
Thorne, G. 1949. On the classification of the Tylenchida, new order (Nematoda: Phasmidia). Proc. Helminth. Soc. Wash. 16:37–73.

Radopholus

Allen, M. W. 1960. The genera *Pratylenchus*, *Radopholus*, *Pratylenchoides*, *Rotylenchulus* and *Nacobbus*; *Tylenchulus*, *Trophotylenchulus*, *Trophonema*, and *Sphaeronema*. *In*: J. N. Sasser and W. R. Jenkins, eds. Nemtology. Univ. North Carolina Press, Chapel Hill, pp. 181–184.
Allen, M. W., and S. A. Sher. 1967. Taxonomic problems concerning the phytoparasitic nematodes. Ann. Rev. Phytopath. 5:247–264.
Christie, J. R. 1959. Plant nematodes, their bionomics and control. H. and W. B. Drew, Jacksonville, Florida.

Cobb, N. A. 1893. Nematode worms found attacking sugar-cane. *In*: Plant diseases and their remedies. Agr. Gaz., New South Wales 4:808–833.
Cobb, N. A. 1915. *Tylenchus similis*, the cause of a root disease of sugar cane and banana. J. Agr. Res. 4:561–568.
Colbran, R. C. 1971. Studies of plant and soil nematodes, 15. Eleven new species of *Radopholus* Thorne and a new species of *Radopholoides* de Guiran (Nematoda: Tylenchoidea) from Australia. Qd. J. Agr. Anim. Sci. 27:437–460.
Das, V. M. 1960. Studies on the nematode parasites of plants in Hyderabad (Andhra Pradesh, India). Z. Parasitenk. 19:553–605.
Egunjobi, O. A. 1968. Three new species of nematodes from New Zealand. N.Z. Jl. Sci. 11(3):488–497.
Goodey, T. 1932. The genus *Anguillulina* Gerv. and v. Ben. 1859, vel *Tylenchus* Bastian 1865. J. Helminth. 10:75–180.
Goodey, T. 1951. Soil and freshwater nematodes. Methuen, London.
Goodey, T. Revised by J. B. Goodey. 1963. Soil and freshwater nematodes. Methuen, London.
Luc, M. 1957. *Radopholus lavabri* n. sp. (Nematoda: Tylenchidae) parasite du riz au Cameroun francais. Nematologica 2:144–148.
Luc, M., and J. B. Goodey. 1962. *Hirschmannia* n. g. differentiated from *Radopholus* Thorne 1949 (Nematoda: Tylenchoidea). Nematologica 7:197–202.
Sauer, M. R. 1968. *Hoplolaimus gracilidens*, *Radopholus inaequalis* and *Radopholus neosimilis*, three new tylenchs native to Victoria, Australia. Nematologica 3:97–107.
* Sher, S. A. 1968. Revision of the genus *Radopholus* Thorne 1949 (Nematoda: Tylenchoidea). Proc. Helminth. Soc. Wash. 35:219–237.
Siddiqi, M. R. 1963. On the classification of the Pratylenchidae (Thorne 1949) nov. grad. (Nematoda: Tylenchida), with a description of *Zygotylenchus browni* nov. gen. et nov. sp. Z. Parasitenk. 23(4): 390–396.
Siddiqi, M. R. 1964. *Radopholus williamsi* n. sp. (Nematoda: Pratylenchidae), a parasite of sugarcane roots at L'Etoile, Mauritius. Indian J. Ent. 26:207–208.
Taylor, A. L. 1969. The Fiji banana-root nematode, *Radopholus similis*. Proc. Helminth. Soc. Wash. 36: 157–163.
Thorne, G. 1949. On the classification of the Tylenchida, new order (Nematoda: Phasmidia). Proc. Helminth. Soc. Wash. 16:37–73.
Thorne, G. 1961. Principles of nematology. McGraw-Hill, New York.
Weerdt, L. G. van. 1958. Studies on the biology of *Radopholus similis* (Cobb 1893) Thorne 1949, Part II. Morphological variation within the progenies of single females. Nematologica 3:184–196.
Weerdt, L. G. van. 1960. Studies on the biology of *Radopholus similis* (Cobb 1893) Thorne 1949, Part III. Embryology and postembryonic development. Nematologica 5:43–52.
Winslow, R. D. 1960. Some aspects of the ecology of free-living and plant-parasitic nematodes. *In*: J. N. Sasser and W. R. Jenkins, eds. Nematology. Univ. North Carolina Press, Chapel Hill, pp. 341–415.

Rhadinaphelenchus

Cobb, N. A. 1919. A newly discovered nematode (*Aphelenchus cocophilus* n. sp.) connected with a serious disease of the coconut palm. W. Indian Bull., Barbados 17:203–210.
Corbett, M. K. 1959. Diseases of the coconut palm, III. Red ring. Principles 3:83–86.
Fenwick, D. W. 1963. On the distribution of *Rhadinaphelenchus cocophilus* (Cobb 1919) Goodey 1960, in coconut palms suffering from red-ring disease. J. Helminth. 37:15–20.
Fenwick, D. W. 1963. Recovery of *Rhadinaphelenchus cocophilus* (Cobb 1919) Goodey 1960 from coconut tissues. J. Helminth. 37:11–14.
Fenwick, D. W., and S. B. Maharaj. 1960. Presence of *Aphelenchoides cocophilus* in the roots of *Cocos nucifera*, the coconut palm. Nature, London 185(4708):259–260.
Fenwick, D. W., and S. B. Maharaj. 1963. Observations on the course of red-ring disease of coconuts caused by the nematode *Rhadinaphelenchus cocophilus* (Cobb 1919) Goodey 1960 in naturally infected trees. J. Helminth. 37:21–26.
Fenwick, D. W., and S. B. Maharaj. 1963. *Rhadinaphelenchus cocophilus* (Cobb 1919) Goodey 1960 as a root parasite of coconuts. J. Helminth. 37:27–38.
Goodey, J. B. 1960. *Rhadinaphelenchus cocophilus* (Cobb 1919) n. comb., the nematode associated with "Red-Ring" disease of coconut. Nematologica 5:98–102.
Goodey, J. B. 1960. The classification of the *Aphelenchoides* Fuchs 1937. Nematologica 5:111–126.
Goodey, T. Revised by J. B. Goodey. 1963. Soil and freshwater nematodes. Methuen, London.
Husain, S. I., and A. M. Khan. 1967. On the status of the genera of the superfamily Aphelenchoidea (Fuchs 1937) Thorne 1949, with the descriptions of six new species of nematodes from India. Proc. Helminth. Soc. Wash. 34:167–174.
Lordello, L. G. E., and A. P. L. Zamith. 1954. Constatacao da moletia de "anel vermelho" do coqueiro no Estado do Rio de Janeiro. Redescricao do agente caudator *Aphelenchoides cocophilus* (Cobb 1919) Goodey 1933 (Nematoda: Aphelenchidae). Anais Esc. sup. Agr. "Luiz Queiroz" 11:125–132.
Thorne, G. 1961. Principles of nematology. McGraw-Hill, New York.

Rotylenchoides

Golden, A. M. 1971. Classification of the genera and higher categories of the order Tylenchida (Nematoda). *In*: B. M. Zuckerman, W. F. Mai, R. A. Rohde, eds. Plant-parasitic nematodes, Vol. I. Morphology, anatomy, taxonomy and ecology. Academic Press, New York and London, pp. 191–232.

Goodey, T. Revised by J. B. Goodey. 1963. Soil and freshwater nematodes. Methuen, London.
Luc, M. 1960. Trois nouvelles espèces du genre *Rotylenchoides* Whitehead 1958 (Nematoda: Tylenchida). Nematologica 5:7–17.
Thorne, G. 1949. On the classification of the Tylenchida, new order (Nematoda: Phasmidia). Proc. Helminth. Soc. Wash. 16:37–73.
Whitehead, A. G. 1958. *Rotylenchoides brevis* n. g., n. sp. Rotylenchoidinae n. subfam: Tylenchida. Nematologica 3:327–331.

Rotylenchulus

Allen, M. W. 1960. The genera *Pratylenchus*, *Radopholus*, *Pratylenchoides*, *Rotylenchulus* and *Nacobbus*; *Tylenchulus*, *Trophotylenchulus*, *Trophonema*, and *Sphaeronema*. *In*: J. N. Sasser and W. R. Jenkins, eds. Nematology. Univ. North Carolina Press, Chapel Hill, pp. 181–184.
Allen, M. W., and S. A. Sher. 1967. Taxonomic problems concerning the phytoparasitic nematodes. *In*: Annual Review of Phytopathology, Vol. 5. Annual Reviews, Inc., Palo Alto, California, pp. 247–264.
Birchfield, W. 1968. Comparative studies of a new *Rotylenchulus* species with *R. reniformis* in Louisiana. Phytopathology 58:1043–1044. (Abst.)
Christie, J. R. 1959. Plant nematodes, their bionomics and control. H. and W. B. Drew, Jacksonville, Florida.
Dasgupta, D. R. 1968. A revision of the genus *Rotylenchulus* Linford and Oliveira 1940 and notes on the biology of *R. parvus* (Williams) Sher. Diss. Abstr. Univ. of Calif. at Davis, Davis, Calif. 29(4):1395.
Dasgupta, D. R., D. J. Raski, and S. A. Sher. 1968. A revision of the genus *Rotylenchulus* Linford and Oliveira 1940 (Nematoda: Tylenchidae). Proc. Helminth. Soc. Wash. 35:169–192.
Goodey, T. 1951. Soil and freshwater nematodes. Methuen, London.
Goodey, T. Revised by J. B. Goodey. 1963. Soil and freshwater nematodes. Methuen, London.
Husain, S. I., and A. M. Khan. 1965. On *Rotylenchulus stakmani* n. sp. with a key to the species of the genus (Nematoda: Tylenchida). Proc. Helminth. Soc. Wash. 32:21–23.
Husain, S. I., and A. M. Khan. 1967. A new subfamily, a new subgenus and eight new species of nematodes from India belonging to superfamily Tylenchoidea. Proc. Helminth. Soc. Wash. 34:175–186.
Linford, M. B., and Juliette M. Oliveira. 1940. *Rotylenchulus reniformis* nov. gen., n. sp., a nematode parasite of roots. Proc. Helminth. Soc. Wash. 7:35–42.
Linford, M. B., and F. Yap. 1940. Some host plants of the reniform nematode in Hawaii. Proc. Helminth. Soc. Wash. 7:42–44.
* Loof, P. A. A., and M. Oostenbrink. 1962. *Rotylenchulus borealis* n. sp. with a key to the species of *Rotylenchulus*. Nematologica 7:83–90.
Nakasono, K. 1966. An instance of male and young female development of the reniform nematode (*Rotylenchulus* sp.) within the egg-shell (Tylenchida: Hoplolaimidae). Jap. J. Appl. Ent. Zool. 1:49–50.
Thorne, G. 1961. Principles of nematology. McGraw-Hill, New York.

Rotylenchus

Colbran, R. C. 1962. Studies of plant and soil nematodes, 5. Four new species of Tylenchoidea from Queensland pineapple fields. Qd. J. Agr. Anim. Sci. 19:231–239.
Coomans, A. 1962. Morphological observations on *Rotylenchus goodeyi* Loof and Oostenbrink 1958, I. Redescription and variability. Nematologica 7:203–215.
Coomans, A. 1962. Morphological observations on *Rotylenchus goodeyi* Loof and Oostenbrink 1958, II. Detailed Morphology. Nematologica 7:242–250.
Das, V. M. 1960. Studies on the nematode parasites of plants in Hyderabad (Andhra Pradesh, India). Z. Parasitenk. 19:553–605.
Filipjev, I. N. 1936. On the classification of the Tylenchinae. Proc. Helminth. Soc. Wash. 3:80–82.
Golden, A. M. 1956. Taxonomy of the spiral nematodes (*Rotylenchus* and *Helicotylenchus*), and the developmental stages and host-parasite relationships of *R. buxophilus* n. sp. attacking boxwood. Univ. Md. Bull. A-85:1–28.
Golden, A. M. 1971. Classification of the genera and higher categories of the order Tylenchida (Nematoda). *In*: B. M. Zuckerman, W. F. Mai, and R. A. Rhode, eds. Plant-parasitic nematodes, Vol. I. Morphology, anatomy, taxonomy and ecology. Academic Press, New York and London, pp. 191–232.
Golden, A. M., and A. L. Taylor. 1956. *Rotylenchus christei* n. sp., a new spiral nematode species associated with roots of turf. Proc. Helminth. Soc. Wash. 23:109–112.
Good, J. M., A. E. Steele, and T. J. Ratcliffe. 1959. Occurrence of plant-parasitic nematodes in Georgia turf nurseries. Pl. Dis. Reptr. 43:236–238.
Goodey, J. B., and J. W. Seinhorst. 1960. Further observations and comments on the identity of *Rotylenchus robustus* (de Man 1876) Filipjev 1934 with a redescription of a proposed neotype and a new definition of *Rotylenchus goodeyi*. Nematologica 5:136–148.
Goodey, T. Revised by J. B. Goodey. 1963. Soil and freshwater nematodes. Methuen, London.
Hopper, B. E. 1959. Three new species of the genus *Tylenchorhynchus* (Nematoda: Tylenchida). Nematologica 4:23–30.
Husain, S. I., and A. M. Khan. 1967. A new subfamily, a new subgenus, and eight new species of nematodes from India belonging to superfamily Tylenchoidea. Proc. Helminth. Soc. Wash. 34:175–186.
Jairajpuri, M. S. 1963. *Rotylenchus sheri* n. sp. (Nematoda: Tylenchida) from North India. Nematologica 9:378–380.
Jairajpuri, M. S., and Q. H. Baqri. 1973. Nematodes of high altitudes in India, I. Four new species of Tylenchida. Nematologica 19:19–30.

Loof, P. A. A., and M. Oostenbrink. 1958. Die identitat von *Tylenchus robustus* de Man. Nematologica 3:34–43.
Lordello, L. G. E. 1955. A new nematode, *Rotylenchus melancholicus* n. sp., found associated with grass roots, and its sexual dimorphism. J. Wash. Acad. Sci. 45:81–83.
Lordello, L. G. E. 1957. A note on nematode parasites of red anthurium (*Anthurium andraeanum* Lind.), with a description of *Rotylenchus boocki* n. sp. Nematologica 2:273–276.
Mamiya, Y. 1968. *Rotylenchus pini* n. sp. (Nematoda: Hoplolaimidae) from forest nurseries in Japan. Proc. Helminth. Soc. Wash. 35:38–40.
Man, J. G. de. 1876. Onderzoekingen over vrjin de aarde levende nematoden. Tijdschr. Nederl. Dierk Vereen. 2:78–196.
Perry, V. G. 1960. The subfamily Hoplolaiminae. *In*: J. N. Sasser and W. R. Jenkins, eds. Nematology. Univ. North Carolina Press, Chapel Hill, pp. 185–190.
Perry, V. G., H. M. Darling, and G. Thorne. 1959. Anatomy, taxonomy and control of certain spiral nematodes attacking blue grass in Wisconsin. Univ. Wis. Res. Bull. 207.
Sauer, M. R. 1958. *Hoplolaimus gracilidens*, *Radopholus inaequalis*, and *Radopholus neosimilis*, three new tylenchs native to Victoria, Australia. Nematologica 3:97–107.
Sher, S. A. 1961. Revision of the Hoplolaiminae (Nematoda), I. Classification of nominal genera and nominal species. Nematologica 6:155–169.
* Sher, S. A. 1965. Revision of the Hoplolaiminae (Nematoda), V. *Rotylenchus* Filipjev 1936. Nematologica 11:173–198.
Siddiqi, M. R. 1964. *Rotylenchus eximius* n. sp. (Nematoda: Hoplolaiminae) found around almond roots in Tunisia. Nematologica 10:101–104.
Siddiqi, M. R., and Z. Husain. 1964. Three new species of nematodes in the family Hoplolaimidae found attacking citrus trees in India. Proc. Helminth. Soc. Wash. 31:211–215.
Steiner, G. 1945. *Helicotylenchus*, a new genus of plant-parasitic nematodes and its relationship to *Rotylenchus* Filipjev. Proc. Helminth. Soc. Wash. 12:34–38.
Thorne, G. 1949. On the classification of the Tylenchida, new order (Nematoda: Phasmidia). Proc. Helminth. Soc. Wash. 16:37–73.
Thorne, G. 1961. Principles of nematology. McGraw-Hill, New York.
Yuen, Pick H. 1964. The female gonad in the subfamily Hoplolaiminae with a note on the spermatheca of *Tylenchorhynchus*. Nematologica 10:570–580.

Sarisodera

Wouts, W. M., and S. A. Sher. 1971. The genera of the subfamily Heteroderinae (Nematoda: Tylenchoidea) with a description of two new genera. J. Nematol. 3:129–144.

Scutellonema

Andrassy, I. 1958. *Hoplolaimus tylenchiformis* Daday 1905 (Syn.: *H. coronatus* Cobb 1923) und die Gattungen der Unterfamilie Hoplolaiminae Filipjev 1936. Nematologica 3:44–56.
Carvalho, J. C. 1959. Descricao do Macho de *Scutellonema boocki* (Nematoda: Tylenchidae). Inst. Biol., Brazil 26(6):41–44.
Edward, J. C., and B. B. Rai. 1970. Plant-parasitic nematodes association with hill orange (*Citrus reticulata* Blanco) in Sikkim. Allahabad Fmr. 44(4):251–254.
Golden, A. M. 1956. Taxonomy of the spiral nematodes (*Rotylenchus* and *Helicotylenchus*), and the developmental stages and host-parasite relationships of *R. buxophilus* n. sp. attacking boxwood. Univ. Md. Bull. A-85:1–28.
Golden, A. M., and A. L. Taylor. 1956. *Rotylenchus christei* n. sp., a new spiral nematode species associated with roots of turf. Proc. Helminth. Soc. wash. 23:109–112.
Goodey, J. B. 1952. *Rotylenchus coheni* n. sp. (Nematoda: Tylenchida) parasitic on the roots of *Hippeastrum* sp. J. Helminth. 26:91–96.
Goodey, T. 1935. Observations on a nematode disease of yams. J. Helminth. 13:173–190.
Goodey, T. Revised by J. B. Goodey. 1963. Soil and freshwater nematodes. Methuen, London.
Khan, S. H., and M. A. Basir. 1965. *Scutellonema mangiferae* n. sp. (Nematoda: Hoplolaimidae) from India. Proc. Helminth. Soc. Wash. 32:136–138.
Lordello, L. G. E. 1957. A note on nematode parasites of red Anthurium (*Anthurium andraeanum* Lind.), with a description of *Rotylenchus boocki* n. sp. Nematologica 2:273–276.
Lordello, L. G. E. 1959. A nematosis of yam in Pernambuco, Brazil, caused by a new species of the genus "*Scutellonema*." Revta. Bras. Biol. 19(1):35–41.
Perry, V. G. 1960. The subfamily Hoplolaiminae. *In*: J. N. Sasser and W. R. Jenkins, eds. Nematology. Univ. North Carolina Press, Chapel Hill, pp. 185–190.
Phillips, S. P. 1971. Studies of plant and soil nematodes, 16. Eight new species of spiral nematodes (Nematoda: Tylenchoidea) from Queensland. Qd. J. Agr. Anim. Sci. 28:227–242.
Razzhivin, A. A. 1971. New species of nematodes of the family Hoplolaimidae. Zool. Zh. 50(1):133–136.
Sher, S. A. 1961. Revision of the Hoplolaiminae (Nematoda), I. Classification of nominal genera and nominal species. Nematologica 6:155–169.
Sher, S. A. 1963. Revision of the Hoplolaiminae (Nematoda), III. *Scutellonema* Andrassy 1958. Nematologica 9:421–443.
* Sher, S. A. 1964. Revised key to the *Scutellonema* Andrassy 1958 (Hoplolaiminae: Nematoda). Nematologica 10:648.

Siddiqi, M. R. 1972. Two new species of *Scutellonema* from cultivated soils in Africa with a description of *Hoplolaimus aorolaimoides* sp. n. from Portugal (Nematoda: Hoplolaiminae). Proc. Helminth. Soc. Wash. 39:7–13.
Smit, J. J. 1971. Deux nouvelles espèces africaines d'Hoplolaiminae (Nematoda: Tylenchoidea): *Peltamigratus striatus* n. sp. et *Scutellonema africanum* n. sp. Nematologica 17:113–126.
Steiner, G. 1938. Nematodes infesting red spiderlillies. J. Agr. Res. 56:1–8.
Steiner, G., and Rowena R. LeHew. 1933. *Hoplolaimus bradys* n. sp. (Tylenchidae: Nematodes), the cause of a disease of yam (*Dioscorea* sp.). Zool. Anz. 101(9–10):260–264.
Taylor, D. P. 1959. The male of *Scutellonema brachyurum* (Steiner 1938) Andrassy 1958. Proc. Helminth. Soc. Wash. 26:51–53.
Thorne, G. 1961. Principles of nematology. McGraw-Hill, New York.
Timm, R. W. 1965. *Scutellonema siamense* n. sp. (Tylenchida: Hoplolaiminae) from Thailand. Nematologica 11:370–372.
Whitehead, A. G. 1959. *Hoplolaimus aberrans* n. sp. (Hoplolaiminae: Tylenchida). Nematologica 4:268–271.
Whitehead, A. G. 1959. *Scutellonema clathricaudatum* n. sp. (Hoplolaiminae: Tylenchida), a suspected ectoparasite of the roots of the cotton plant (*Gossypium hirsutum* L. var. UK 51). Nematologica 4:56–59.
Yeates, G. W. 1967. Studies on nematodes from dune sands, I. Tylenchida. N.Z. Jl. Sci. 10(1):280–286.
Yuen, Pick H. 1964. The female gonad in the subfamily Hoplolaiminae with a note on the spermatheca of *Tylenchorhynchus*. Nematologica 10:570–580.

Sphaeronema

Allen, M. W. 1960. The genera *Pratylenchus, Radopholus, Pratylenchoides, Rotylenchulus* and *Nacobbus; Tylenchulus, Trophotylenchulus, Trophonema*, and *Sphaeronema. In*: J. N. Sasser and W. R. Jenkins, eds. Nematology. Univ. North Carolina Press, Chapel Hill, pp. 181–184.
Goodey, J. B. 1958. *Sphaeronema minutissimum* n. sp. (Sphaeronematinae: Tylenchulidae). Nematologica 3:169–172.
Goodey, T. Revised by J. B. Goodey. 1963. Soil and freshwater nematodes. Methuen, London.
Kirjanova, E. S. 1970. *Sphaeronema rumicis* n. sp. (Nematoda: Sphaeronematidae). Parazitologiya 4(5):489–493.
Raski, D. J. 1956. *Sphaeronema arenarium*, n. sp. (Nematoda: Criconematidae), a nematode parasite of salt rush, *Juncus leseurii* Boland. Proc. Helminth. Soc. Wash. 23:75–77.
Raski, D. J., and S. A. Sher. 1952. *Sphaeronema californicum* nov. gen., nov. spec. (Criconematidae: Sphaeronematinae, nov. subfam.) an endoparasite of the roots of certain plants. Proc. Helminth. Soc. Wash. 19:77–80.
Sledge, E. B., and J. R. Christie. 1962. *Sphaeronema whittoni* n. sp. (Nematoda: Criconematidae). Nematologica 8:11–14.

Subanguina

Filipjev, I. N. 1936. On the classification of the Tylenchinae. Proc. Helminth. Soc. Wash. 3:80–82.
Golden, A. M. 1971. Classification of the genera and higher categories of the order Tylenchida (Nematoda). *In*: B. M. Zuckerman, W. F. Mai, and R. A. Rohde, eds. Plant parasitic nematodes, Vol. I. Morphology, anatomy, taxonomy and ecology. Academic Press, New York and London, pp. 191–232.
Goodey, T. 1932. On the nomenclature of the root-gall nematodes. J. Helminth. 10:21–28.
Goodey, T. 1932. Some observations on the biology of the root-gall nematode, *Anguillulina radicicola* Greeff (1872). J. Helminth. 10:33–44.
Greeff, R. 1872. Uber nematoden in Wurtzelanschwellungen (Golden) verschiedener Pflanzen. Sitzber. Ges. Naturw. Marburg., pp. 169–174.
Paramonov, A. A. 1967. Problems on evolution, morphology, taxonomy, and biochemistry of nematodes of plants. Akad. Nauk Moscow, SSR 18:78–101.
Thorne, G. 1961. Principles of nematology. McGraw-Hill, New York.

Telotylenchoides

Loof, P. A. A., and A. M. Yassin. 1970. Three new plant-parasitic nematodes from the Sudan, with notes on *Xiphinema basiri* Siddiqi 1959. Nematologica 16:537–546.
Raski, D. J., S. K. Prasad, and G. Swarup. 1964. *Telotylenchus housei*, a new nematode species from Mysore State, India (Tylenchidae: Nematoda). Nematologica 10:83–86.
Siddiqi, M. R. 1971. On the plant-parasitic nematode genera *Histotylenchus* gen. n. and *Telotylenchoides* gen. n. (Telotylenchinae), with observations on the genus *Paratrophurus* Arias (Trophurinae). Nematologica 17:190–200.

Telotylenchus

Colbran, R. C. 1960. Studies of plant and soil nematodes, 3. *Belonolaimus hastulatus, Psilenchus tumidus*, and *Hemicycliophora labiata*, three new species from Queensland. Qd. J. Agr. Anim. Sci. 17:175–181.
Fisher, J. M. 1964. *Telotylenchus whitei* n. sp. from S. Australia with observations on *Telotylenchus hastulatus* (Colbran 1960) n. comb. Nematologica 10:563–569.
Geraert, E. 1966. On some Tylenchidae and Neotylenchidae from Belgium with the description of a new species, *Tylenchorhynchus microdorus*. Nematologica 12:409–416.
Goodey, T. Revised by J. B. Goodey. 1963. Soil and freshwater nematodes. Methuen, London.

Husain, S. I., and A. M. Khan. 1967. A new subfamily, a new subgenus, and eight new species of nematodes from India belonging to superfamily Tylenchoidea. Proc. Helminth. Soc. Wash. 34:175–186.
Jairajpuri, M. S. 1963. On the status of the subfamilies Rotylenchoidinae Whitehead 1958, and Telotylenchinae Siddiqi 1960. Z. Parasitenk. 23(4):320–323.
Jairajpuri, M. S. 1971. On the synonymy of *Telotylenchus* Siddiqi 1960 with *Trichotylenchus* Whitehead 1959 (Nematoda: Tylenchida). Indian J. Nematol. 1:3–6.
Jairapuri, M. S., and Q. H. Baqri. 1968. *Tylenchorhynchus hexincisus* n. sp. and *Telotylenchus historicus* n. sp. (Tylenchida) from India. Nematologica 14:217–222.
Loof, P. A. A. 1963. A new species of *Telotylenchus* (Nematoda: Tylenchida). Nematologica 9:76–80.
Netscher, C., and G. Germani. 1969. *Telotylenchus baoulensis* n. sp. et *Trichotylenchus rectangularis* n. sp. (Nematoda: Tylenchoidea). Nematologica 15:347–352.
Raski, D. J., S. K. Prasad, and G. Swarup. 1964. *Telotylenchus housei*, a new nematode species from Mysore State, India (Tylenchidae: Nematoda). Nematologica 10:83–86.
Siddiqi, M. R. 1960. *Telotylenchus,* a new nematode genus from North India (Tylenchida: Telotylenchinae n. sub-fam.). Nematologica 5:73–77.
Tarjan, A. C. 1973. A synopsis of the genera and species in the Tylenchorhynchinae (Tylenchoidea: Nematoda). Proc. Helminth. Soc. Wash. 40:123–144.
Tikyani, M. G., and S. Khera. 1970. A new species of *Telotylenchus* (Nematoda: Tylenchida). Labdev. J. Sci. Technol. Ser. B. 8(1):27–29.

Tetylenchus

Ferris, Virginia R., and J. M. Ferris. 1967. Morphological variant of *Tetylenchus joctus* Thorne (Nemata: Tylenchida) associated with cultivated blueberries in Indiana. Proc. Helminth. Soc. Wash. 34:30–32.
Filipjev, I. N. 1936. On the classification of the Tylenchinae. Proc. Helminth. Soc. Wash. 3:80–82.
Goodey, T. 1932. The genus *Anguillulina* Gerv. and v. Ben. 1859, vel *Tylenchus* Bastian 1865. J. Helminth. 10:75–180.
Goodey, T. 1951. Soil and freshwater nematodes. Methuen, London.
Goodey, T. Revised by J. B. Goodey. 1963. Soil and freshwater nematodes. Methuen, London.
Hirschmann, Hedwig. 1960. The Genera *Tylenchus, Psilenchus, Ditylenchus, Anguina, Tylenchorhynchus, Tetylenchus, Trophurus* and *Macrotrophurus*. *In*: J. N. Sasser and W. R. Jenkins, eds. Nematology. Univ. North Carolina Press, Chapel Hill, pp. 171–180.
Merny, G. 1964. Un nouveau Tylenchida d'Afrique tropicale: *Tetylenchus annulatus* n. sp. Nematologica 10:425–430.
Micoletzky, H. 1922. Die freilebender Erd-Nematoden mit besonderer Berücksichtigung freilebender Nematoden in Form von Genus-Beschreibungen und Bestimmungsschlusseln. Arch. Naturg., Berlin (1921), abt. A. 87:1–320.
* Tarjan, A. C. 1973. A synopsis of the genera and species in the Tylenchorhynchinae (Tylenchoidea: Nematoda). Proc. Helminth. Soc. Wash. 40:123–144.
Thorne, G. 1949. On the classification of the Tylenchida, new order (Nematoda: Phasmidia). Proc. Helminth. Soc. Wash. 16:37–73.
Thorne, G. 1961. Principles of nematology. McGraw-Hill, New York.
Wilski, A. 1965. The nematode plant-parasitic fauna of glasshouse soils in Poland. Pr. Nauk. Inst. Ochr. Rosl., 1964, 6:5–59.
Zuckerman, B. M. 1960. Parasitism of cranberry roots by *Tetylenchus joctus* Thorne. Nematologica 5:253–254.

Trichodorus

Allen, M. W. 1957. A review of the nematode genus *Trichodorus* with descriptions of ten new species. Nematologica 2:32–62.
Allen, M. W. 1960. The genera *Xiphinema, Longidorus*, and *Trichodorus*. *In*: J. N. Sasser and W. R. Jenkins, eds. Nematology. Univ. North Carolina Press, Chapel Hill, pp. 227–228.
Arias Delgado, M., F. Jimenez Millan, and J. M. Lopez Pedregal. 1965. Tres nuevas especies de nematodos posibles fitoparasitos en suelos espanoles. Publnes. Inst. Biol. Apl., Barcelona 38:47–58.
Bird, G. W. 1966. Influence of host and geographical origin on populations of *Trichodorus christiei*. Nematologica 12:88–89. (Abst.)
Bird, G. W. 1967. *Trichodorus acutus* n. sp. (Nematodea: Diphtherophoridea) and a discussion of allometry. Can. J. Zool. 45:1201–1204.
Bird, G. W., and W. F. Mai. 1965. Plant species in relation to morphometric variation of the New York population of *Trichodorus christiei*. Nematologica 11:34. (Abst.)
Bird, G. W., and W. F. Mai. 1968. A numerical study of the growth and development of *Trichodorus christiei*. Can. J. Zool. 46:109–114.
Christie, J. R. 1959. Plant nematodes, their bionomics and control. H. and W. B. Drew, Jacksonville, Florida.
Clark, W. C. 1961. A revised classification of the order Enoplida (Nematoda). N.Z. Jl. Sci. 4:123–150.
Clark, W. C. 1963. A new species of *Trichodorus* (Nematoda: Enoplida) from Westland, New Zealand. N.Z. Jl. Sci. 6(3):414–417.
Cobb, N. A. 1913. New nematode genera found inhabiting fresh water and nonbrackish soils. J. Wash. Acad. Sci. 3:432–444.

Colbran, R. C. 1956. Studies of plant and soil nematodes, I. Two new species from Queensland. Qd. J. Agr. Anim. Sci. 13:123–126.

Colbran, R. C. 1965. Studies of plant and soil nematodes, 9. *Trichodorus lobatus* n. sp. (Nematoda: Trichodoridae), a stubby-root nematode associated with citrus and peach trees. Qd. J. Agr. Anim. Sci. 22:273–276.

Edward, J. C., and S. L. Misra. 1970. Two new species of *Trichodorus* from Uttar Pradesh, India. Allahabad Fmr. 44(4):167–171.

Goodey, T. 1951. Soil and fresh water nematodes. Methuen, London.

Goodey, T. Revised by J. B. Goodey. 1963. Soil and freshwater nematodes. Methuen, London.

Harrison, B. D., W. P. Mowat, and C. E. Taylor. 1961. Transmission of a strain of tomato black-ring virus by *Longidorus elongatus* (Nematoda). Virology 14:480–485.

Hooper, D. J. 1962. Three new species of *Trichodorus* (Nematoda: Dorylaimoidea) and observations on *T. minor* Colbran 1956. Nematologica 7:273–280.

Hooper, D. J. 1963. *Trichodorus viruliferus* n. sp. (Nematoda: Dorylaimida). Nematologica 9:200–204.

Hooper, D. J. 1972. Two new species of *Trichodorus* (Nematoda: Dorylaimida) from England. Nematologica 18:59–65.

Hooper, D. J., K. Kuiper, and P. A. A. Loof. 1963. Observations on the identity of *Trichodorus teres* Hooper 1962 and *T. flevensis* Kuiper and Loof 1962. Nematologica 9:646.

Jensen, H. J. 1963. *Trichodorus allius,* a new species of stubby-root nematode from Oregon (Nemata: Dorylaimoidea). Proc. Helminth. Soc. Wash. 30:157–159.

Kuiper, K., and P. A. A. Loof. 1961. *Trichodorus flevensis* n. sp. (Nematoda: Enoplida) a plant nematode from new polder soil. Versl. Meded. Plziektenk. Dienst Wageningen 136:192–200.

Kuiper, K., and P. A. A. Loof. 1964. Observations on some *Trichodorus* species. Nematologica 10:77. (Abst.)

Loof, P. A. A. 1965. *Trichodorus anemones* n. sp. with a note on *T. teres* Hooper 1962 (Nematoda: Enoplida). Versl. Meded. Plziektenk. Dienst Wageningen 142:132–136.

Lordello, L. G. E., and A. P. L. Zamith. 1958. Nota sobre o genero *Trichodorus* Cobb 1913, com descricao de *Trichodorus bucrius* sp. n. (Nematoda: Dorylaimoidea). Anals Acad. Bras. Cienc. 30:103–105.

Man, J. G. de. 1880. Die Einheimischen, frei in der reinen Erde und im sussen wasser lebenden Nematoden. Volaufiger Bericht und descriptivsystematischer Theil. Tijdschr. Nederl. Dierk Vereen. 5(1–2):1–104.

Micoletzky, H. 1922. Die freilebender Erd-Nematoden mit besonderer Berücksichtigung der Steiermark und der Bukowina, zugleich mit einer Revision samtlicher nicht mariner, freilebender Nematoden in Form von Genus-Beschreibungen und Bestimmungschlusseln. Arch. Naturg., Berlin (1921), Abt. A 87(8/9):1–650.

Seinhorst, J. W. 1954. On *Trichodorus pachydermus* n. sp. (Nematoda: Enoplida). J. Helminth. 28:111–114.

Seinhorst, J. W. 1963. A redescription of the male of *Trichodorus primitivus* (de Man), and the description of a new species, *T. similis*. Nematologica 9:125–130.

Siddiqi, M. R. 1960. Two new species of the genus *Trichodorus* (Nematoda: Dorylaimoidea) from India. Proc. Helminth. Soc. Wash. 27:22–27.

Siddiqi, M. R. 1962. *Trichodorus pakistanensis* n. sp. (Nematoda: Trichodoridae), with observations on *T. porosus* Allen 1957, *T. mirzai* Siddiqi 1960 and *T. minor* Colbran 1956, from India. Nematologica 8:193–200.

Siddiqi, M. R. 1963. *Trichodorus* spp. (Nematoda: Trichodoridae) from Tunisia and Nicaragua. Nematologica 9:69–75.

Siddiqi, M. R., and K. F. Brown. 1965. *Trichodorus rhodesiensis* and *Amphidelus trichurus*, two new nematode species from cultivated soils of Africa. Proc. Helminth. Soc. Wash. 32:239–242.

Szczygiel, A. 1968. *Trichodorus sparsus* sp. n. (Nematoda: Trichodoridae). Bull. Acad. Pol. Sci. Cl. II Ser. Sci. Biol. 16(11):695–698.

Thorne, G. 1939. A monograph of the nematodes of the superfamily Dorylaimoidea. Capita Zool. 8(5):261 p.

Thorne, G. 1961. Principles of nematology. McGraw-Hill, New York.

Yeates, G. W. 1969. The status of *Trichodorus clarki* Yeates 1967, *Ereptonema inflatum* Yeates 1967 and *Mononchoides potohikus* Yeates 1969. Nematologica 15:430.

Yokoo, T. 1964. On the stubby-root nematodes from the western Japan. Agr. Bull., Saga Univ., No. 20, pp. 57–62.

Yokoo, T. 1966. On a new stubby root nematode (*Trichodorus kurumeensis* n. sp.) from Kyushum Japan. Agr. Bull., Saga Univ., No. 23, pp. 1–6.

Trichotylenchus

Cooman, A., and A. de Grisse. 1963. Observations on *Trichotylenchus falciformis* Whitehead 1959. Nematologica 9:320–326.

Golden, A. M. 1971. Classification of the genera and higher categories of the order of Tylenchida (Nematoda). *In*: B. M. Zuckerman, W. F. Mai, and R. A. Rohde, eds. Plant-parasitic nematodes, Vol. I. Morphology, anatomy, taxonomy, and ecology. Academic Press, New York and London, pp. 191–232.

Jairajpuri, M. S. 1969. On the identity of *Telotylenchus* Siddiqi 1960 and *Trichotylenchus* Whitehead 1959 with remarks on their systematic position. All-India Nemat. Symp., IARI, New Delhi. (Abst., p. 25.)

Jairajpuri, M. S. 1971. On the synonymy of *Telotylenchus* Siddiqi 1960 with *Trichotylenchus* Whitehead 1959 (Nematoda: Tylenchida). Indian J. Nematol. 1:3–6.

Khan, E., and C. K. Nanjappa. 1971. *Trophurus similis* sp. n. and *Trichotylenchus astriatus* sp. n. (Nematoda: Tylenchoidea) from Mysore, India. Indian J. Nematol. 1:75–79.

Netscher, C., and G. Germani. 1969. *Telotylenchus baoulensis* n. sp. et *Trichotylenchus rectangularis* n. sp. (Nematoda: Tylenchoidea). Nematologica 15:347–352.

Seinhorst, J. W. 1968. *Trichotylenchus rhopalocercus* (Seinhorst 1963) n. comb. (Syn.: *Tylenchorhynchus rhopalocercus* Seinhorst 1963) and *Tylenchorhynchus clavicauda* nom. nov. (Syn.: *T. clavicaudatus* Seinhorst 1963). Nematologica 14:596.

Seinhorst, J. W. 1971. On the genera *Trichotylenchus* and *Telotylenchus*. Nematologica 17:413–416.

Siddiqi, M. R. 1970. On the plant-parasitic nematode genera *Merlinius* gen. n. and *Tylenchorhynchus* Cobb and the classification of the families Dolichodoridae and Belonolaimidae n. rank. Proc. Helminth. Soc. Wash. 37:68–77.

Singh, S. D. 1971. Studies on the morphology and systematics of plant and soil nematodes mainly from Andhra Pradesh, I. Tylenchoidea. J. Helminth. Soc. Wash. 45:353–369.

Tarjan, A. C. 1973. A synopsis of the genera and species in the Tylenchorhychinae (Tylenchoidea: Nematoda). Proc. Helminth. Soc. Wash. 40:123–144.

Whitehead, A. G. 1959. *Trichotylenchus falciformis* n. g., n. sp. (Belonolaiminae n. subfam.: Tylenchida Thorne 1949) an associate of grass roots (*Hyparrhenia* sp.) in Southern Tanganyika. Nematologica 4:279–285.

Trophonema

Allen, M. W. 1960. The genera *Pratylenchus, Radopholus, Pratylenchoides, Rotylenchulus,* and *Nacobbus; Tylenchulus, Trophotylenchulus, Trophonema*, and *Spaeronema*. *In*: J. N. Sasser and W. R. Jenkins, eds. Nematology. Univ. North Carolina Press, Chapel Hill, pp. 181–184.

Goodey, T. Revised by J. B. Goodey. 1963. Soil and freshwater nematodes. Methuen, London.

Raski, D. J. 1956. *Sphaeronema arenarium* n. sp. (Nematoda: Criconematidae), a nematode parasite of salt rush, *Juncus leseurii* Boland. Proc. Helminth. Soc. Wash. 23:75–77.

Raski, D. J. 1957. *Trophotylenchulus* and *Trophonema*, two new genera of Tylenchulidae n. fam. (Nematoda). Nematologica 2:85–90.

Thorne, G. 1961. Principles of nematology. McGraw Hill, New York.

Trophotylenchulus

Allen, M. W. 1960. The genera *Pratylenchus, Radopholus, Pratylenchoides, Rotylenchulus* and *Nacobbus; Tylenchulus, Trophotylenchulus, Trophonema* and *Sphaeronema*. *In*: J. N. Sasser and W. R. Jenkins, eds. Nematology. Univ. North Carolina Press, Chapel Hill, pp. 181–184.

Goodey, T. Revised by J. B. Goodey. 1963. Soil and freshwater nematodes. Methuen, London.

Luc, M. 1957. *Tylenchulus mangenoti*, n. sp. (Nematoda: Tylenchulidae). Nematologica 2:329–334.

Raski, D. J. 1957. *Trophotylenchulus* and *Trophonema*, two new genera of Tylenchulidae n. fam. (Nematoda). Nematologica 2:85–90.

Thorne, G. 1961. Principles of nematology. McGraw-Hill, New York.

Trophurus

Caveness, F. E. 1958. *Clavaurotylenchus minnesotensis*, n. gen., n. sp. (Tylenchinae: Nematoda) from Minnesota. Proc. Helminth. Soc. Wash. 25:122–124.

Caveness, F. E. 1959. *Trophurus minnesotensis* (Caveness 1958), n. comb. Proc. Helminth. Soc. Wash. 26:64.

Chawla, M. L., B. L. Bhamburkar, E. Khan, and S. K. Prasad. 1968. One new genus and seven new species of nematodes from India. Labdev. J. Sci. Technol., Ser. B. 6(2):86–100.

Goodey, T. Revised by J. B. Goodey. 1963. Soil and freshwater nematodes. Methuen, London.

Hirschmann, Hedwig. 1960. The genera *Tylenchus, Psilenchus, Ditylenchus, Anguina, Tylenchorhynchus, Tetylenchus, Trophurus*, and *Macrotrophurus*. *In*: J. N. Sasser and W. R. Jenkins, eds. Nematology. Univ. North Carolina Press, Chapel Hill, pp. 171–180.

Khan, E., and C. K. Nanjappa. 1971. *Trophurus similis* sp. n., and *Trichotylenchus astriatus* sp. n. (Nematoda: Tylenchoidea) from Mysore, India. Indian J. Nematol. 1:75–79.

Loof, P. A. A. 1956. *Trophurus*, a new Tylenchid genus (Nematoda). Versl. Meded. Plziektenk. Dienst Wageningen 129:191–195.

Roman, J. 1962. *Trophurus longimarginatus* n. sp. (Tylenchida: Nematoda) from Puerto Rico. J. Agr., Univ. P. Rico 46(4):269–271.

Suryawanshi, M. V. 1971. Studies on Tylenchida (Nematoda) from Marathwada, India, with descriptions of four new species. Nematologica 17:393–406.

Thorne, G. 1961. Principles of nematology. McGraw-Hill, New York.

Tylenchorhynchus

Allen, M. W. 1955. A review of the nematode genus *Tylenchorhynchus*. Univ. Calif. Publns. Zool. 61(3):129–165.

Andrassy, I. 1962. Neue nematoden-arten aus ungarn, I. Zehn neue arten der unterklasse Secernentea (Phasmidia). Acta Zool., Budapest 8(1–2):1–23.

Andrassy, I. 1969. The scientific results of the Hungarian Soil Zoological Expedition to the Brazzaville-Congo, 40. Vier neue Bodennematoden-Arten. Opusc. Zool., Budapest 9(1):15–29.

Arias Delgado, M., F. Jimenez Millan, and J. M. Lopez Pedregal. 1965. Tres nuevas especies de nematodos posibles fitoparasitos en suelos espanoles. Publnes. Inst. Biol. Apl., Barcelona 38:47–58.

* Baqri, Q. H., and M. S. Jairajpuri. 1970. On the intraspecific variations of *Tylenchorhynchus mashhodi*

Siddiqi and Basir 1959, and an emended key to species of *Tylenchorhynchus* Cobb 1913 (Nematoda). Revta Bras. Biol. 30(1):61–68.
Brown, Georgiana L. 1956. *Tylenchorhynchus lenorus* n. sp., (Nematoda: Tylenchida), associated with the roots of wheat. Proc. Helminth. Soc. Wash. 23:152–153.
Chawla, M. L., B. L. Bhamburkar, E. Khan, and S. K. Prasad. 1968. One new genus and seven new species of nematodes from India. Labdev. J. Sci. Technol., Ser. B. 6(2):86–100.
Christie, J. R. 1959. Plant nematodes, their bionomics and control. H. and W. B. Drew, Jacksonville, Florida.
Cobb, N. A. 1913. New nematode genera found inhabiting fresh water and nonbrackish soils. J. Wash. Acad. Sci. 3:432–444.
Das, V. M. 1960. Studies on the nematode parasites of plants in Hyderabad (Andhra Pradesh, India). Z. Parasitenk. 19:553–605.
Eliava, I. Y. 1964. The position of the genus *Tylenchorhynchus* within the Tylenchoidea (Nematoda: Tylenchida). Soobshch. Akad. Nauk Gruz, SSR 34:669–673.
Elmiligy, I. A. 1969. Redescription of *Tylenchorhynchus clarus* Allen 1955. Nematologica 15:288–290.
Ferris, Virginia R. 1963. *Tylenchorhynchus silvaticus* n. sp. and *Tylenchorhynchus agri* n. sp. (Nematoda: Tylenchida). Proc. Helminth. Soc. Wash. 30:165–168.
Fielding, M. J. 1956. *Tylenchorhynchus martini*, a new nematode species found in the sugarcane and rice fields of Louisiana and Texas. Proc. Helminth. Soc. Wash. 23:47–48.
Filipjev, I. N. 1936. On the classification of the Tylenchinae. Proc. Helminth. Soc. Wash. 3:80–82.
Geraert, E. 1966. On some Tylenchidae and Neotylenchidae from Belgium with the description of a new species, *Tylenchorhynchus microdorus*. Nematologica 12:409–416.
Goodey, J. B. 1952. *Tylenchorhynchus tessellatus* n. sp. (Nematoda: Tylenchida). J. Helminth. 26:87–90.
Goodey, T. 1951. Soil and freshwater nematodes. Methuen, London.
Goodey, T. Revised by J. B. Goodey. 1963. Soil and freshwater nematodes. Methuen, London.
Graham, T. W. 1954. The tobacco stunt nematode in South Carolina. Phytopathology 44:332. (Abst.)
Guiran, G. de. 1967. Description de deux espèces nouvelles du genre *Tylenchoryhnchus* Cobb 1913 (Nematoda: Tylenchinae) accompagnée d'une clé des femelles, et précisions sur *T. mamillatus* Tobar Jimenez 1966. Nematologica 13:117–230.
Hasbrouck, E. R., and W. R. Jenkins. 1960. Morphological variations in *Tylenchorhynchus claytoni*. Phytopathology 50:571. (Abst.)
Hirschmann, Hedwig. 1960. The genera *Tylenchus, Psilenchus, Ditylenchus, Anguina, Tylenchorhynchus, Tetylenchus, Trophurus*, and *Macrotrophurus*. *In*: J. N. Sasser and W. R. Jenkins, eds. Nematology. Univ. North Carolina Press, Chapel Hill, pp. 171–180.
Hollis, J. P., and L. S. Whitlock. 1959. Variants of *Tylenchorhynchus martini* and *T. ewingi*. Phytopathology 49:541. (Abst.)
Hopper, B. E. 1959. Three new species of the genus *Tylenchorhynchus* (Nematoda: Tylenchida). Nematologica 4:23–30.
Husain, S. I., and A. M. Khan. 1967. A new subfamily, a new subgenus, and eight new species of nematodes from India belonging to superfamily Tylenchoidea. Proc. Helminth. Soc. Wash. 34:175–186.
Ibrahim, I. K. A. 1965. Structure of the cuticle of *Tylenchorhynchus martini* (Nematoda: Tylenchida). Phytopathology 55:1062–1063. (Abst.)
Ibrahim, I. K. A. 1967. Morphological differences between the cuticle of swarming and nonswarming *Tylenchorhynchus martini*. Proc. Helminth. Soc. Wash. 34:18–20.
Ivanova, T. S. 1968. Nematodes of cereals from the Zeravshan Valley of Tadzhikistan. Dushanbe, Izdatelstvo "Donish."
Izatullaeva, R. I. 1967. New nematode species from ornamental flowering plants in Kazakhstan. Izvest. Akad. Nauk Kazakh., SSR. Ser. Biol. 5(2):45–50.
Jairajpuri, M. S. 1971. On *Scutylenchus mamillatus* (Tobar Jimenez 1966) n. comb. Nat. Acad. Sci. India, 40th Session, p. 18. (Abst.)
Jairajpuri, M. S., and Q. H. Baqri. 1968. *Tylenchorhynchus hexincisus* n. sp. and *Telotylenchus historicus* n. sp. (Tylenchida) from India. Nematologica 14:217–222.
Khak, M. M. 1967. Description of *Tylenchorhynchus undyferrus* n. sp. (Nematoda: Hoplolaimidae [Filipjev 1934] Weiser 1953). Zool. Zh., 46(1):132–134.
Knobloch, N. A. 1971. Emendation of the description of *Tylenchorhynchus longus* Wu 1969. Nematologica 17:602.
Kozlowska, J. 1966. The hitherto unknown male *Tylenchorhynchus judithae* Andrassy 1962 (Nematoda: Tylenchidae). Bull. Acad. Pol. Sci. Cl. II Ser. Sci. Biol. 14:491–493.
Krall, E. 1959. New and little known tylenchs (Nematoda: Tylenchida) including a description of gyandromorphism in the genus *Aphelenchoides*. Eesti NSV Tead. Akad. Toim, Seer. Biol. 8(3):190–198.
Krusberg, L. R. 1959. Investigations on the life cycle, reproduction, feeding habits and host range of *Tylenchorhynchus claytoni* Steiner. Nematologica 4:187–197.
Litvinova, N. F. 1946. Four new species of *Tylenchorhynchus* (Nematoda) from Kazakhstan. Proc. Zool. Soc., London 116:120–128.
Loof, P. A. A. 1959. Miscellaneous notes on genus *Tylenchorhynchus*. (Tylenchinae: Nematoda). Nematologica 4:294–306.
Loof, P. A. A. 1964. Free-living and plant-parasitic nematodes from Venezuela. Nematologica 10:201–300.
Marinari, Anna. 1962. *Tylenchorhynchus goodeyi* n. sp. (Tylenchinae: Nematoda). Redia 47:119–122.
Merny, G., and G. Germani. 1968. *Tylenchorhynchus palustris* n. sp. (Nematoda: Tylenchinae), hote des rizieres de Cote d'Ivoire. Annls. Epiphyt. 19(4):601–603.

Muchina, T. I. 1970. A new species of *Tylenchorhynchus* Cobb 1913 (Nematoda: Tylenchoidea). Parazitologiya 4(4):342–344.
Mulk, M. M., and M. S. Jairajpuri. 1972. A redescription of *Tylenchorhynchus phaseoli* Sethi and Swarup 1968. Indian J. Nematol. 2:54–58.
Mulvey, R. H. 1969. Nematodes of the genus *Tylenchorhynchus* (Tylenchoidea: Nematoda) from the Canadian High Arctic. Can. J. Zool. 47:1245–1248.
Paetzold, D. 1958. Beitrage zur Nematodenfauna mitteldeutscher Salzstellen im Raum von Halle. Wiss. Z. Martin-Luther, Univ. Halle, Wittenb. 8:17–48.
Sabova, M. 1967. Two new soil-inhabiting nematode species (*Tylenchorhynchus tatrensis* and *Alaimus andrassyi* n. spp.) from Czechoslovakia. Opusc. Zool., Budapest 7(2):237–240.
Schuurmans Stekhoven. J. H. 1944. Nematodes libres d'eau douce. Explanation of the Albert National Park (1935–36) 9:31.
Seinhorst, J. W. 1963. Five new *Tylenchorhynchus* species from West Africa. Nematologica 9:173–180.
Seshadri, A. R., T. S. Muthukrishnan, and S. Shunmugam. 1967. A new species of *Tylenchorhynchus* (Tylenchidae: Nematoda) from Madras State, India. [Correspondence.] Current Science 36(20):551–553.
Sethi, C. L., and G. Swarup. 1968. Plant-parasitic nematodes of northwestern India, I. The genus *Tylenchorhynchus*. Nematologica 14:77–88.
Siddiqi, M. R. 1961. Studies on *Tylenchorhynchus* spp. (Nematoda: Tylenchida) from India. Z. Parasitenk. 21:46–64.
Siddiqi, M. R. 1963. On the diagnosis of the nematode genera *Psilenchus* de Man 1921, and *Basiria* Siddiqi 1959, with a description of *Psilenchus hilarus* n. sp. Z. Parasitenk. 23(2):164–169.
Siddiqi, M. R. 1970. On the plant-parasitic nematode genera *Merlinius* gen. n. and *Tylenchorhynchus* Cobb and the classification of the families Dolichodoridae and Belonolaimidae n. rank. Proc. Helminth. Soc. Wash. 37:68–77.
Siddiqi, M. R. 1971. Structure of the esophagus in the classification of the superfamily Tylenchoidea (Nematoda). Indian J. Nematol. 1:25–43.
Siddiqi, M. R., and M. A. Basir. 1959. On some plant-parasitic nematodes occurring in South India, with the description of two new species of the genus *Tylenchorhynchus* Cobb 1913. Proc. 46th Meet. Indian Sci. Congr., Pt. 4, 35. (Abst.)
Singh, S. D. 1971. Studies on the morphology and systematics of plant and soil nematodes mainly from Andhra Pradesh, I. Tylenchoidea. J. Helminth. 45:353–369.
Steiner, G. 1937. Opuscula miscellanea nematologica, V. Proc. Helminth. Soc. Wash. 4:33–38.
Sturhan, D. 1966. Uber Verbreitung, Pathogenitat und Taxonomie der Nematoden-gattung *Tylenchorhynchus*. Mitt. Biol. Bundnst. Forstw., Berlin-Dahlem, Heft 118:82–99.
Szczygiel, A. 1969. *Tylenchorhynchus polonicus* sp. n. and *Helicotylenchus pseudodigonicus* n. sp. (Nematoda: Tylenchoidea) from Poland. Bull. Acad. Pol. Sci. Cl. II Ser. Sci. Biol. 17(11/12):685–690.
Tarjan, A. C. 1964. A compendium of the genus *Tylenchorhynchus* (Tylenchidae: Nematoda). Proc. Helminth. Soc. Wash. 31:270–280.
* Tarjan, A. C. 1973. A synopsis of the genera and species in the Tylenchorhynchinae (Tylenchoidea: Nematoda). Proc. Helminth. Soc. Wash. 40:123–144.
Thorne, G. 1949. On the classification of the Tylenchida, new order (Nematoda: Phasmidia). Proc. Helminth. Soc. Wash. 16:37–73.
Thorne, G. 1961. Principles of nematology. McGraw-Hill, New York.
Thorne, G., and R. B. Malek. 1968. Nematodes of the northern great plains, Part 1. Tylenchida (Nemata: Secernentea) S. Dak. Agr. Exp. Stn. Tech. Bull. 31:111.
Timm, R. W. 1963. *Tylenchorhynchus trilineatus* n. sp. from West Pakistan, with notes on *T. nudus* and *T. martini*. Nematologica 9:262–266.
Tobar Jimenez, A. 1966. *Tylenchorhynchus mamillatus* n. sp. (Nematoda: Tylenchidae), componente de la microfauna de los suelos andaluces. Revta. iber. Parasit. 26(2/3):163–169.
Tobar, Jimenez, A. 1969. Description del *Tylenchorhynchus ventrosignatus* n. sp. (Nematoda: Tylenchidae). Revta iber. Parasit. 29(4):399–403.
Tobar Jimenez, A. 1970. Descripcion de dos nuevas especies del genero *Tylenchorhynchus* Cobb, 1913 (Nematoda: Tylenchidae), conalgunos datos adicionales sobre el *T. sulcatus* de Guiran 1967. Revta iber. Parasit. 30(2/3):257–270.
Tulaganov, A. T. 1954. Results and future investigations of nematodes of plants of Uzbekistan. Trudy probl. temat. Soveshch. Zool. Inst. (3):161–170.
Williams, J. R. 1960. Studies on the nematode soil fauna of sugar-cane fields in Mauritius, 4. Tylenchoidea (partim). Occ. Paper, Maurit. Sug. Ind. Res. Inst., No. 4.
Wouts, W. M. 1966. The identity of New Zealand populations of *Tylenchorhynchus capitatus* Allen 1955, with a description of an intersex. N.Z. Jl. Sci. 9:878–881.
Wu, Liang-Yu. 1969. Three new species of the genus *Tylenchorhynchus* Cobb 1913 (Tylenchidae: Nematoda) from Canada. Can. J. Zool. 47:563–567.
Yuen, Pick H. 1964. The female gonad in the subfamily Hoplolaiminae with a note on the spermetheca of *Tylenchorhynchus*. Nematologica 10:570–580.

Tylenchulus

Allen, M. W. 1960. The genera *Pratylenchus, Radopholus, Pratylenchoides, Rotylenchulus* and *Nacobbus; Tylenchulus, Trophotylenchulus, Trophonema* and *Sphaeronema*. *In*: J. N. Sasser and W. R. Jenkins, eds. Nematology. Univ. North Carolina Press, Chapel Hill, pp. 181–184.

Arias, M., et al. 1964. Estudio bioestadistico de *Tylenchulus semipenetrans* Cobb (Nematoda), parasito de algunas especies de citrus espanoles. Revta iber. Parasit. 24:91–103.
Christie, J. R. 1959. Plant nematodes, their bionomics and control. H. and W. B. Drew, Jacksonville, Florida.
Cobb, N. A. 1913. Notes on *Mononchus* and *Tylenchulus*. J. Wash. Acad. Sci. 3:287–288.
Cobb, N. A. 1914, Citrus-root nematode. J. Agr. Res. 2:217–230.
Colbran, R. C. 1961. Studies of plant and soil nematodes, 4. *Tylenchulus obscurus* n. sp. (Nematoda: Tylenchulidae). Qd. J. Agr. Anim. Sci. 18:203–207.
Colbran, R. C. 1966. Studies of plant and soil nematodes, 13. *Tylenchulus clavicaudatus* n. sp. (Nematoda: Tylenchulidae), a parasite of the liana *Deeringia arborescens* (R. Br.) Druce. Qd. J. Agr. Anim. Sci. 23:423–427.
Filipjev, I. N., and J. H. Schuurmans Stekhoven, Jr. 1941. A manual of agricultural helminthology. Brill, Leiden.
Goodey, T. 1951. Soil and freshwater nematodes. Methuen, London.
Goodey, T. Revised by J. B. Goodey. 1963. Soil and freshwater nematodes. Methuen, London.
Karimova, I. S. 1957. Eelworms of crops on the left bank of the Amu-Darya basin. *In*: A. I. Zemlyanskaya, L. V. Tikhinova, and I. S. Karimova, eds. Eelworms of agricultural crops in the Uzbek SSR Akad. Nauk Uzbek, SSR, Tashkent, pp. 133–208.
Luc, M. 1957. *Tylenchulus mangenoti* n. sp. (Nematoda: Tylenchulidae). Nematologica 2:329–334.
Maggenti, A. R. 1962. The production of the gelatinous matrix and its taxonomic significance in *Tylenchulus* (Nematoda: Tylenchulinae). Proc. Helminth. Soc. Wash. 29:139–144.
Thomas, E. E. 1923. The citrus nematode, *Tylenchulus semipenetrans*. Univ. Calif. Agr. Exp. Stn. Tech. Paper 2:1–35.
Thorne, G. 1961. Principles of nematology. McGraw-Hill, New York.
Van Gundy, S. D. 1958. The life history of the citrus nematode, *Tylenchulus semipenetrans* Cobb. Nematologica 3:283–294.

Tylenchus

Andrassy, I. 1952. Freilebende nematoden aus dem Bukk-Gebirge. Annls. Hist., Nat. Mus. Natn. Hungary (Ser. Nov.) 2:13–65.
Andrassy, I. 1954. Revision der Gattung *Tylenchus* Bastian, 1865 (Tylenchidae: Nematoda). Acta Zool., Budapest 1:5–42.
Andrassy, I. 1959. Freilebende Nematoden aus Rumanien. Ann. Univ. Sci., Budapest 2:3–27.
Andrassy, I. 1963. The zoological results of Gy. Topal's collectings in South Argentina, 2. Nematoda. Neve und einige seltene Nematoden-Arten aus Argentinien. Annls. Hist., Nat. Mus. Natn. Hungary 55:243–273.
Andrassy, I. 1968. Fauna paraguayensis, 2. Nematoden aus den Galeriewaldern des Acaray-Flusses. Opusc. Zool., Budapest 8(2):167–315.
Baqri, Q. H., and M. S. Jairajpuri. 1969. Two known and three new species of nematodes associated with fibrous crops in India. Annls. Zool. Ecol. Anim. 1(3):327–337.
Bastian, H. C. 1865. Monograph on the Anguillulidae or free nematoids, marine, land, and freshwater; with descriptions of 100 new species. Trans. Linn. Soc., London 25:73–184.
Bello, A., and E. Geraert. 1972. Redescription of eight species belonging to the superfamily Tylenchoidea (Nematoda: Tylenchida). Nematologica 18:190–200.
Brzeski, M. W. 1963. On the taxonomic status of *Tylenchus filiformis* Butschli 1873, and description of *T. vulgaris* sp. n. (Nematoda: Tylenchidae). Bull. Acad. Pol. Sci. Cl. II. Ser. Sci. Biol. 11(11):531–535.
Brzeski, M. W. 1963. *Tylenchus ditissimus* sp. n., a new nematode from Poland (Nematoda: Tylenchidae). Bull. Acad. Pol. Sci. Cl. II. Ser. Sci. Biol. 11(11):537–540.
Chawla, M. L., S. K. Prasad, E. Khan, and S. Nand. 1969. Two new species of the genus *Tylenchus* Bastian 1865 (Nematoda: Tylenchidae) from Uttar Pradesh, India. Labdev. J. Sci. Technol., Ser. B. 7(4):291–294.
Das, V. M. 1960. Studies on the nematode parasites of plants in Hyderabad (Andhra Pradesh, India). Z. Parasitenk. 19:553–605.
Egunjobi, O. A. 1967. Four new species of the genus *Tylenchus* Bastian 1865 (Nematoda: Tylenchida). Nematologica 13:417–424.
Egunjobi, O. A. 1968. Three new species of nematodes from New Zealand. N.Z. Jl. Sci. 11(3):488–497.
Elmiligy, I. A. 1971. Two new species of Tylenchidae, *Basiroides nortoni* n. sp. and *Tylenchus hageneri* n. sp. (Nematoda: Tylenchida). J. Nematol. 3:108–112.
Filipjev, I. N., and J. H. Schuurmans Stekhoven, Jr. 1941. A manual of agricultural helminthology. Brill, Leiden.
Geraert, E. 1965. The head structures of some Tylenchs with special attention to the amphidial apertures. Nematologica 11:131–136.
Geraert, E., and J. B. Goodey. 1963. The priority of *Tylenchus hexalineatus* over *T. megacephalus*. Nematologica 9:471.
Golden, A. M. 1971. Classification of the genera and higher categories of the order Tylenchida (Nematoda). *In*: B. M. Zuckerman, W. F. Mai, and R. A. Rohde, eds. Plant-parasitic nematodes, Vol. I. Morphology, anatomy, taxonomy, and ecology. Academic Press, New York and London, pp. 191–232.
Goodey, J. B. 1962. *Tylenchus* (*Cephalenchus*) *megacephalus* n. sbg., n. sp. Nematologica 7:331–333.
Goodey, T. 1932. The genus *Anguillulina* Gerv. and v. Ben. 1859, vel *Tylenchus* Bastian 1865. J. Helminth. 10:75–180.
Goodey, T. 1951. Soil and freshwater nematodes. Methuen, London.

Goodey, T. Revised by J. B. Goodey. 1963. Soil and freshwater nematodes. Methuen, London.

Hagemeyer, Joyce W., and M. W. Allen. 1952. *Psilenchus duplexus* n. sp. and *Psilenchus terextremus* n. sp., two additions to the nematode genus *Psilenchus* de Man 1921. Proc. Helminth. Soc. Wash. 19: 51–54.

Hirschmann, Hedwig. 1960. The Genera *Tylenchus*, *Psilenchus*, *Ditylenchus*, *Anguina*, *Tylenchorhynchus*, *Tetylenchus*, *Trophurus* and *Macrotrophurus*. *In*: J. N. Sasser and W. R. Jenkins, eds. Nematology. Univ. North Carolina Press, Chapel Hill, pp. 171–180.

Husain, S. I., and A. M. Khan. 1967. A new subfamily, a new subgenus, and eight new species of nematodes from India belonging to superfamily Tylenchoidea. Proc. Helminth. Soc. Wash. 34:175–186.

Jairajpuri, M. S. 1965. A redefinition of *Psilenchus* de Man 1921 and *Tylenchus* subgenus *Filenchus* Andrassy 1954 with the erection of *Clavilenchus* n. subgenus under *Tylenchus* Bastian 1865. Nematologica 11:619–622.

Jairajpuri, M. S. 1965. *Basiria kashmirensis* n. sp. (Nematoda: Tylenchida) from India. Labdev. J. Sci. Technol. 3:23–25.

Khan, E., M. L. Chawla, and S. K. Prasad. 1969. *Tylenchus* (*Aglenchus*) *indicus* n. sp. and *Ditylenchus emus* n. sp. (Nematoda: Tylenchidae) from India. Labdev. J. Sci. Technol., Ser. B. 7(4):311–314.

* Kheiri, A. 1970. Two new species in the family Tylenchidae (Nematoda) from Iran, with a key to *Psilenchus* de Man 1921. Nematologica 16:359–368.

Kheiri, A. 1972. *Tylenchus* (*Irantylenchus*) *clavidorus* n. sp. and *Merlinius camelliae* n. sp. (Tylenchida: Nematoda) from Iran. Nematologica 18:339–346.

Knobloch, N. A., and J. A. Knierim. 1969. *Tylenchus vesiculosus* sp. n. (Nematoda: Tylenchidae) from soil in Michigan. Proc. Helminth. Soc. Wash. 36:147–149.

Loof, P. A. A. 1961. *Tylenchus gulosus* Kuhn 1890: proposed suppression under the plenary powers (Nematoda). Bull. Zool. Nom. 18(3):206–207.

Meyl, A. 1961. Die freilebenden Erd- und Susswassernematoden (Fadenwurmer). *In*: Bohmer, Ehrmann, and Ulmer. Die Tierwelt Mitteleuropas. Quelle and Meyer, Leipzig, pp. 1–273.

Savage, H. E., and K. D. Fisher. 1966. Host-parasite relations of *Tylenchus hexalineatus*. Phytopathology 56:898. (Abst.)

Siddiqi, M. R. 1959. *Basiria graminophila* n. g., n. sp. (Nematoda: Tylenchinae) found associated with grass roots in Aligarh, India. Nematologica 4:217–222.

Siddiqi, M. R. 1963. Four new species of the genus *Tylenchus* Bastian 1865 (Nematoda) from North India. Z. Parasitenk. 23(2):170–180.

Singh, S. D. 1971. Studies on the morphology and systematics of plant and soil nematodes mainly from Andhra Pradesh, I. Tylenchoidea. J. Helminth. 45:353–369.

Steiner, G., and Florence E. Albin. 1946. Resuscitation of the nematode *Tylenchus polyhypnus* n. sp., after almost 39 years' dormancy. J. Wash. Acad. Sci. 36(3):97–99.

Szczygiel, A. 1969. A new genus and four new species of the subfamily Tylenchinae de Man 1876 (Nematoda: Tylenchidae) from Poland. Opusc. Zool., Budapest 9(1):159–170.

Szczygiel, A. 1969. Two new species of the genus *Basiria* (Nematoda: Tylenchidae) from Poland. Bull. Acad. Pol. Sci. Cl. II. Ser. Sci. Biol. 17(11/12):679–683.

Szczygiel, A. 1969. Two new species of the genus *Tylenchus* (Nematoda: Tylenchidae) from Poland. Bull. Acad. Pol. Sci. Cl. II. Ser. Sci. Biol. 17(11/12):673–677.

Thorne, G. 1949. On the classification of the Tylenchida, new order (Nematoda: Phasmidia). Proc. Helminth. Soc. Wash. 16:37–73.

Thorne, G. 1961. Principles of nematology. McGraw-Hill, New York.

Thorne, G., and R. B. Malek. 1968. Nematodes of the northern great plains, Part I. Tylenchida (Nemata: Secernentea). S. Dak. Agr. Exp. Stn. Tech. Bull. 31.

Wasilewska, L. 1965. *Tylenchus sandneri* sp. n., a new nematode from Poland (Nematoda: Tylenchidae). Bull. Acad. Pol. Sci. Cl. II. Ser. Sci. Biol. 13:87–89.

Wasilewska, L. 1965. *Tylenchus* (*Tylenchus*) *baloghi* Andrassy 1958 in Poland (Nematoda: Tylenchidae). Bull. Acad. Pol. Sci. Cl. II. Ser. Sci. Biol. 13:163–165.

Wu, Liang-Yu. 1967. Differences of spermatheca and sperm cells in the genera *Ditylenchus* Filipjev 1936 and *Tylenchus* Bastian 1865 (Tylenchidae: Nematoda). Can. J. Zool. 45:27–30.

Wu, Liang-Yu. 1968. *Dactylotylenchus crassacuticulus*, a new genus and new species (Tylenchinae: Nematoda). Can. J. Zool. 46:831–834.

Wu, Liang-Yu. 1969. Dactylotylenchinae, a new subfamily (Tylenchidae: Nematoda). Can. J. Zool. 47: 909–911.

Wu, Liang-Yu. 1969. Five new species of *Tylenchus* Bastian 1865 (Nematoda: Tylenchidae) from the Canadian High Arctic. Can. J. Zool. 47:1005–1010.

Tylodorus

Loof, P. A. A. 1958. Some remarks on the status of the subfamily Dolichodorinae, with description of *Macrotrophurus arbusticola* n. g., n. sp. (Nematoda: Tylenchidae). Nematologica 3:301–307.

Meagher, J. W. 1963. *Tylodorus acuminatus* n. g., n. sp. (Nematoda: Tylenchinae) from Eucalyptus forest in Australia. Nematologica 9:635–640.

Xiphinema

Allen, M. W. 1960. The genera *Xiphinema*, *Longidorus* and *Trichodorus*. *In*: J. N. Sasser and W. R. Jenkins, eds. Nematology. Univ. North Carolina Press, Chapel Hill, pp. 227–228.

Altherr, E. 1958. Nematodes du bassin inferieur de la Weser et des dunes d'Heligoland. Espèces nouvelles ou incompletement decrites. Mem., Soc. Vaudoise Sci. Nat. 12(2):45–63.

Andrassy, I. 1960. Zwei bemerkenswerte Nematoden-Arten aus Belgisch-Kongo. Opusc. Zool., Budapest 3(3/4):101–110. (*Eudorylaimus paracentrocercus* and *Xiphinema obtusum* noted.)

Brown, R. H. 1968. *Xiphinema monohysterum* n. sp. (Nematoda: Dorylaimidae), from southern New South Wales. Nematologica 13:633–637.

Carvalho, J. C. 1962. *Xiphinema itanhaense* n. sp. (Nematoda: Dorylaimidae). Arq. Inst. Biol., Sao Paulo 29:223–225.

Carvalho, J. C. 1965. *Xiphinema paulistanum*—uma nova especie de nematoide. Arq. Inst. Biol., Sao Paulo 32:77–79.

Cobb, N. A. 1913. New nematode genera found inhabiting fresh-water and nonbrackish soils. J. Wash. Acad. Sci. 3:432–444.

Cohn, E., and G. P. Martelli. 1964. Studies on *Xiphinema ingens* Luc and Dalmasso 1964 and the male of *Longidorus brevicaudatus* (Schuurmans Stekhoven 1951) Thorne 1961 (Nematoda: Dorylaimidae). Nematologica 10:192–196.

Cohn, E., and S. A. Sher. 1972. A contribution to the taxonomy of the genus *Xiphinema* Cobb 1913. J. Nematol. 4:36–65.

Coomans, A. 1964. Structure of the female gonads in members of the Dorylaimina. Nematologica 10: 601–622.

Coomans, A. 1964. *Xiphinema basilgoodeyi* n. sp., with observations on its larval stages (Nematoda: Dorylaimina). Nematologica 10:581–593.

Coomans, A., and E. Bracke. 1964. Vergelijkend onderzoek naar het systematisch belang van afmetingen en relatieve groei bij twee nauwverwante *Xiphinema*-soorten (Dorylaimida: Nematoda). Biol. Jaarb. 32:222–236.

Coomans, A., and L. DeConinck. 1963. Observations on spear-formation in *Xiphinema*. Nematologica 9:85–96.

Dalmasso, A. 1969. Etude anatomique et taxonomique des genres *Xiphinema*, *Longidorus*, et *Paralongidorus* (Nematoda: Dorylaimidae). Mem. Mus. Natn. Hist. Nat., Paris, Series A, Zool. 61(2):33–82.

Esser, R. P. 1966. *Xiphinema macrostylum* n. sp. (Nematoda: Longidoridae). Proc. Helminth. Soc. Wash. 33:162–165.

Flegg, J. J. M. 1966. The Z-organ in *Xiphinema diversicaudatum*. Nematologica 12:174.

Goodey, J. B., and D. J. Hooper. 1963. The nerve rings of *Longidorus* and *Xiphinema*. Nematologica 9:303–304.

Goodey, J. B., F. C. Peacock, and R. S. Pitcher. 1960. A redescription of *Xiphinema diversicaudatum* (Micoletzky 1923 and 1927) Thorne 1939 and observations on its larval stages. Nematologica 5:127–135.

Goodey, T. 1936. A new Dorylaimid nematode *Xiphinema radicicola* n. sp. J. Helminth. 14:69–72.

Goodey, T. 1951. Soil and freshwater nematodes. Methuen, London.

Goodey, T. Revised by J. B. Goodey. 1963. Soil and freshwater nematodes. Methuen, London.

Hewitt, W. B., D. J. Raski, and A. C. Goheen. 1958. Nematode vector of soil-borne fanleaf virus of grapevines. Phytopathology 48:586–595.

Heyns, J. 1962. A report on South African nematodes of the families Longidoridae, Belondiridae, and Alaimidae (Nemata: Dorylaimoidea) with descriptions of three new species. Nematologica 8:15–20.

Heyns, J. 1965. Four new species of the genus *Xiphinema* (Nematoda: Dorylaimoidea) from South Africa. Nematologica 11:87–99.

Heyns, J. 1966. Studies on South African *Xiphinema* species, with descriptions of two new species displaying sexual dimorphism of the tail (Nematoda: Dorylaimoidea). Nematologica 12:369–384.

Heyns, J. 1971. Three *Xiphinema* species from the southwestern Cape Province (Nematoda: Longidoridae). Phytophylactica 3(3):107–114.

Khan, E. 1964. *Longidorus afzali* n. sp., and *Xiphinema arcum* n. sp. (Nematoda: Longidoridae) from India. Nematologica 10:313–318.

Loof, P. A. A., and P. W. T. Maas. 1972. The genus *Xiphinema* (Dorylaimida) in Surinam. Nematologica 18:92–119.

Loof, P. A. A., and A. M. Yassin. 1971. Three new plant-parasitic nematodes from the Sudan, with notes on *Xiphinema basiri* Siddiqi 1959. Nematologica 16:537–546.

Loos, C. A. 1949. Notes on free-living and plant-parasitic nematodes of Ceylon, No. 5. J. Zool. Soc. India 1:23–29.

Lopez-Abella, D., F. Jimenez Millan, and F. Garcia-Hidalgo. 1967. Electron microscope studies of some cephalic structures of *Xiphinema americanum*. Nematologica 13:283–286.

Lordello, L. G. E. 1951. *Xiphinema brasiliense*, nova especie de nematoide do Brasil, parasita de *Solanum tuberosum* L. Bragantia 11:87–90.

Lordello, L. G. E. 1951. *Xiphinema campinense*, nova especie (Nematoda: Dorylaimidae). Bragantia 11: 313–316.

* Lordello, L. G. E. 1955. *Xiphinema krugi* n. sp. (Nematoda: Dorylaimidae) from Brazil with a key to the species of *Xiphinema*. Proc. Helminth. Soc. Wash. 22:16–21.

Lordello, L. G. E., and C. P. da Costa. 1961. A new nematode parasite of coffee roots in Brazil. Rev. Bras. Biol. 21:363–366.

Luc, M. 1958. *Xiphinema* de l'Ouest Africain: description de cinq nouvelles espèces (Nematoda: Dorylaimidae). Nematologica 3:57–72.

Luc, M. 1961. Structure de la gonade femelle chez quelques espèces du genre *Xiphinema* Cobb 1913 (Nematoda: Dorylaimoidea). Nematologica 6:144–154.

Luc, M. 1961. *Xiphinema* de l'Ouest Africain (Nematoda: Dorylaimoidea) deuxieme note. Nematologica 6:107–122.
Luc, M., and A. Dalmasso. 1963. Trois nouveaux *Xiphinema* associée à la vigne (Nematoda: Dorylaimidae). Nematologica 9:531–541.
Luc, M., et al. 1964. *Xiphinema vuittenezi* n. sp. (Nematoda: Dorylaimidae). Nematologica 10:151–163.
Luc, M., and A. C. Tarjan. 1963. Note systematique sur le genre *Xiphinema* Cobb 1913 (Nematoda: Dorylaimidae). Nematologica 9:111–115.
Luc, M., and A. C. Tarjan. 1963. Redescription de *Xiphinema rotundatum* Schuurmans Stekhoven and Teunissen 1938 et de *Xiphinema mamillatum* Schuurmans Stekhoven and Teunissen 1938. (Nematoda: Dorylaimidae). Nematologica 9:116–124.
Macara, A. M. 1970. *Xiphinema amarantum* sp. nov. (Nematoda: Dorylaimidae). Revta iber. Parasit. 30(4):649–658.
Martelli, G. P., E. Cohn, and A. Dalmasso. 1966. A redescription of *Xiphinema italiae* Meyl 1953 and its relationship to *Xiphinema arenarium* Luc et Dalmasso 1963 and *Xiphinema conurum* Siddiqi 1964. Nematologica 12:183–194.
* McLeod, R. W., and G. T. Khair. 1971. *Xiphinema australiae* n. sp., its host range, observations on *X. radicicola* Goodey 1936 and *X. monohysterum* Brown 1968 and a key to monodelphic *Xiphinema* spp. (Nematoda: Longidoridae). Nematologica 17:58–68.
Meyl, A. H. 1953. Beitrage zur Kenntnis der Nematoden fauna vulkanisch erhitzter Biotope, I. Mitt., Die Terrikolen Nematoden im Bereich von Fumarolen auf der Insel Ischia. Z. Morph. Okol. Tiere 42:67–116.
Perry, V. G. 1958. Parasitism of two species of dagger nematodes (*Xiphinema americanum* and *X. chambersi*) to strawberry. Phytopathology 48:420–423.
Roggen, D. R. 1966. On the morphology of *Xiphinema index* reared on grape fanleaf virus-infected grapes. Nematologica 12:287–296.
Roggen, D. R., D. J. Raski, and N. O. Jones. 1966. Cilia in nematode sensory organs. Science 152:515–516.
Roggen, D. R., D. J. Raski, and N. O. Jones. 1967. Further electron microscopic observations of *Xiphinema index*. Nematologica 13:1–16.
Schindler, A. F. 1957. Parasitism and pathogenicity of *Xiphinema diversicaudatum*, an ectoparasitic nematode. Nematologica 2:25–31.
Schindler, A. F., and A. J. Braun. 1957. Pathogenicity of an ectoparasitic nematode, *Xiphinema diversicaudatum* on strawberries. Nematologica 2:91–93.
Schuurmans Stekhoven, J. H. 1951. Nematodes saprozoaires et libres du Congo Belge. Mem., Inst. R. Sci. Nat. Belg., Ser. 2, 39:3–79.
Schuurmans Stekhoven, J. H., and R. J. H. Teunissen. 1938. Nematodes libres terrestres. Fasc. 22, Mission (de Witte) (1933–35), Exploration of The Albert National Park.
Siddiqi, M. R. 1959. Studies on *Xiphinema* spp. (Nematoda: Dorylaimoidea) from Aligarh (North India), with comments on the genus *Longidorus* Micoletzky 1922. Proc. Helminth. Soc. Wash. 26:151–163.
Siddiqi, M. R. 1961. On *Xiphinema opisthohysterum* n. sp., and *X. pratense* Loos 1949, two dorylaimid nematodes attacking fruit trees in India. Z. Parasitenk. 20:457–465.
Siddiqi, M. R. 1963. Three new species of Dorylaimoides Thorne and Swanger 1936, with a description of *Xiphinema orbum* n. sp. (Nematoda: Dorylaimoidea). Nematologica 9:626–634.
* Siddiqi, M. R. 1964. *Xiphinema conurum* n. sp. and *Paralongidorus microlaimus* n. sp., with a key to the species of *Paralongidorus* (Nematoda: Longidoridae). Proc. Helminth. Soc. Wash. 31:133–137.
Steiner, G. 1914. Freilebende Nematoden aus der Schweiz. Arch. Hydrobiol. Planktonkunde 9:259–276.
Stoyanov, D. 1964. A contribution to the nematode fauna of the grape vine. Rastit. Zasht. 12:16–24.
Sturhan, D. 1963. Beitrag zur Systematik der Gattung *Xiphinema* Cobb 1913. Nematologica 9:205–214.
Tarjan, A. C. 1956. Known and suspected plant-parasitic nematodes of Rhode Island, II. *Xiphinema americanum* with notes on *Tylencholaimus brevicaudatus* n. comb. Proc. Helminth. Soc. Wash. 23:88–92.
Tarjan, A. C. 1964. Two new American dagger nematodes (*Xiphinema*: Dorylaimidae) associated with citrus, with comments on the variability of *X. bakeri* Williams 1961. Proc. Helminth. Soc. Wash. 31:65–76.
Thorne, G. 1937. Notes on free-living and plant-parasitic nematodes, III. Proc. Helminth. Soc. Wash. 4:16–18.
Thorne, G. 1939. A monograph of the nematodes of the superfamily Dorylaimoidea. Capita Zool. 8(5):261 p.
Thorne, G. 1961. Principles of nematology. McGraw-Hill, New York.
Thorne, G., and M. W. Allen. 1950. *Paratylenchus hamatus* n. sp. and *Xiphinema index* n. sp., two nematodes associated with fig roots, with a note on *Paratylenchus anceps* Cobb. Proc. Helminth. Soc. Wash. 17:27–35.
Thorne, G., and H. H. Swanger. 1936. A monograph of the nematode genera *Dorylaimus* Dujardin, *Aporcelaimus* n. g., *Dorylaimoides* n. g., and *Pugentus* n. g. Capita Zool. 6(4):225 p.
Weischer, B. 1964. Uber Taxonomie und geographische Verbreitung von *Xiphinema* spp. in Deutschland. Nematologica 10:73–74. (Abst.)
Williams, J. R. 1966. The position of the spear guiding ring in *Xiphinema* species. Nematologica 12:467–469.
Williams, T. D. 1961. *Xiphinema bakeri* n. sp. (Nematoda: Longidorinae) from the Fraser River Valley, British Columbia, Canada. Can. J. Zool. 39:407–412.
Wright, K. A. 1965. The histology of the esophageal region of *Xiphinema index* Thorne and Allen 1950, as seen with the electron microscope. Can. J. Zool. 43:689–700.

Zygotylenchus

Braun, A. L., and P. A. A. Loof. 1966. *Pratylenchoides laticauda* n. sp. a new endoparasitic phytonematode. Neth. J. Pl. Path. 72:241–245.

Goodey, T. 1932. The genus *Anguillulina* Gerv. and V. Ben. 1859 vel *Tylenchus* Bastian 1865. J. Helminth. 10:75–180.
Goodey, T. 1940. On *Anguillulina multicincta* (Cobb) and other species of *Anguillulina* associated with the roots of plants. J. Helminth. 18:21–38.
Guiran, G. de. 1964. *Mesotylus*: nouveau genre de Pratylenchinae (Nematoda: Tylenchoidea). Nematologica 9:567–575.
Guiran, G. de, and M. R. Siddiqi. 1967. Characters differentiating the genera *Zygotylenchus* Siddiqi 1963 (Syn.: *Mesotylus* de Guiran 1964) and *Pratylenchoides* Winslow 1958 (Nematoda: Pratylenchinae). Nematologica 13:235–240.
Siddiqi, M. R. 1963. On the classification of the Pratylenchidae (Thorne 1949) nov. grad. (Nematoda: Tylenchida) with a description of *Zygotylenchus browni* nov. gen. et nov. sp. Z. Parasitenk. 23:390–396.
Tarjan, A. C., and B. Weischer. 1965. Observations on some Pratylenchinae (Nemata), with additional data on *Pratylenchoides guevarai* Tobar Jimenez 1963 (Syn.: *Zygotylenchus browni* Siddiqi 1963 and *Mesotylus gallicus* de Guiran 1964). Nematologica 11:432–440.
Tobar Jimenez, A. 1963. *Pratylenchoides guevarai* n. sp., nuevo nematode tylenchido relacionado con el cipres (*Cupressus sempervirens* L.) Revta iber. Parasit. 23:27–36.
Winslow, R. D. 1958. The taxonomic position of *Anguillulina obtusa* Goodey 1932 and 1940. Nematologica 3:136–139.

GENERAL REFERENCES

Baker, A. D. 1962. Check list of the nematode superfamilies Dorylaimoidea, Rhabditoidea, Tylenchoidea, and Aphelenchoidea. Brill, Leiden.

Bird, G. W. 1971. Taxonomy: the science of classification. *In*: B. M. Zuckerman, W. F. Mai, R. A. Rohde, eds. Plant parasitic nematodes, Vol. I. Morphology, anatomy, taxonomy, and ecology. Academic Press, New York, pp. 117–138.

Caveness, F. E. 1964. A glossary of nematological terms. Pacific Printers, Yaba, Nigeria.

Chitwood, B. G., and M. B. Chitwood. 1950. An introduction to nematology, Sect. 1, Anatomy. Monumental Printing, Baltimore.

Clark, W. C. 1962. Measurements as taxonomic criteria in nematology. (Abst.). Nematologica 7:10.

Commonwealth Institute of Helminthology. 1972. Descriptions of plant-parasitic nematodes, Set 1, Nos. 1–15. Commonwealth Institute of Helminthology, St. Albans, United Kingdom.

Commonwealth Institute of Helminthology. 1973. Descriptions of plant-parasitic nematodes, Set 2, Nos. 16–30. Commonwealth Institute of Helminthology, St. Albans, United Kingdom.

Coninck, L. de. 1962. Problems of systematics and taxonomy in nematology today. Nematologica 7:1–7.

Ferris, Virginia R. 1971. Taxonomy of the Dorylaimida. *In*: B. M. Zuckerman, W. F. Mai, R. A. Rohde, eds. Plant parasitic nematodes, Vol. I. Morphology, anatomy, taxonomy, and ecology. Academic Press, New York, pp. 163–189.

Franklin, Mary T. 1971. Taxonomy of Heteroderidae. *In*: B. M. Zuckerman, W. F. Mai, R. A. Rohde, eds. Plant parasitic nematodes, Vol. I. Morphology, anatomy, taxonomy, and ecology. Academic Press, New York, pp. 139–162.

Golden, A. M. 1971. Classification of the genera and higher categories of the order Tylenchida (Nematoda). *In*: B. M. Zuckerman, W. F. Mai, R. A. Rohde, eds. Plant parasitic nematodes, Vol. I. Morphology, anatomy, taxonomy, and ecology. Academic Press, New York, pp. 191–232.

Goodey, J. B. 1962. Taxonomic relatedness in nematology. Ann. Appl. Biol. 50:175–177.

Goodey, T. Revised by J. B. Goodey. 1963. Soil and freshwater nematodes. Methuen, London.

Grasse, P. P., ed. 1965. Traite de zoologie. Anatomie, systematique, biologie, Tome IV, fasc. II. Nemathelminthes (nematodes). Tome IV, fasc. III. Nemathelminthes (nematodes, gordiaces), rotiferes, gastrotriches, kinorhynques. Masson et Cie, Paris. Fasc. II, pp. 1–731; fasc. III, pp. 732–1497.

Hirschmann, Hedwig. 1971. Comparative morphology and anatomy. *In*: B. M. Zuckerman, W. F. Mai, R. A. Rohde, eds. Plant parasitic nematodes, Vol. I. Morphology, anatomy, taxonomy, and ecology. Academic Press, New York, pp. 11–63.

Ichinohe, M. 1961. Bibliography of phytonematology. Japan Plant Protection Association, Tokyo.

Jenkins, W. R., and D. P. Taylor. 1967. Plant nematology. Reinhold, New York.

Southey, J. F. 1970. Laboratory methods for work with plant and soil nematodes. Tech. Bull. Minist. Agr. Fish. Fd., London, No. 2; 5th ed.

Tarjan, A. C. 1960. Check of plant and soil nematodes. Univ. of Florida Press, Gainesville, Florida.

Thorne, G. 1961. Principles of nematology. McGraw-Hill, New York.

Williams, J. R., and J. B. Goodey. 1963. Deposition of type slides at Rothamsted. Nematologica 9:300.

GLOSSARY OF NEMATOLOGICAL TERMS

(Numbers in parentheses refer to plates picturing the characters defined.)

a. Length/greatest width.
Adanal. Pertaining to bursa which does not envelop entire tail. (34A)
Alae. Expansions or projections formed by a longitudinal thickening of the cuticle. Cervical alae are confined to the anterior region of nematodes parasitic in animals. Caudal alae occur in the posterior region of males in a number of genera. Longitudinal alae, usually 4 in number, extend the length of the body sublaterally.
Amphid. Chemo-sensory organ, occurring laterally in pairs, located in the anterior region of the nematode. Among most Secernentea, the amphids appear as small porelike openings on the lips. The openings are postlabial in the Adenophorea, and may be of varied form, i.e., circular, spiral, or cyathiform (pocketlike). Amphids are sometimes called lateral organs.
Annule. Thickened interval between transverse striae. (4k)
Areolated. Divided into small spaces or areolations; usually pertains to cuticle.
Basal region of esophagus. Posterior portion of esophagus which may be variable in shape; i.e., a distinct bulb, a lobe overlapping the intestine, cylindrical, etc. (1, 4m)
Buccal capsule. Structure connecting oral opening with anterior portion of esophagus. The buccal capsule, or stoma as it is sometimes called, is subject to great variation among different nematodes. The shape may be cylindrical or subglobular; it may be toothed or armed with a protrusible spear or stylet. (2)
Bursae. Caudal alae of males used to clasp the female during copulation.
Capitulum. Medial ventral sclerotization of the spicular pouch.
Cardia. Valvular apparatus connecting the esophagus and intestine. Sometimes called the cardiac valve or esophago-intestinal valve. (4f)
Caudal. Pertaining to or located near the posterior region or tail.
Caudal glands. Three to 5 glands located in the tail, and emptying subterminally or terminally through a pore, the spinneret. Caudal glands are only known to occur in some of the Adenophorea. Caudal glands secrete a fluid which hardens to form an attachment thread in water.
Cephalic. Pertaining to or located near the head.
Cervical. Pertaining to the neck region.
Circumoral elevation. Thickened membrane surrounding the oral opening when lips are absent or rudimentary. (19C, 19F).
Cloaca. Common cavity in males, in which the digestive and reproductive systems terminate.
Corpus. Anterior cylindrical part of the esophagus. The basal portion of the corpus at times may be swollen to form a bulb. (1, 4e)
Cuticle. Noncellular, exterior covering of nematodes.
Deirids. Paired, porelike organs located in the lateral fields of many of the Tylenchoidea; believed by some workers to be sensory in nature.
Denticles. Minute teeth or "prickles."
Dioecious. Having the male reproductive organs in one individual, the female in another; sexually distinct.
Diverticulum. Pouch or lobe containing the esophageal glands.
Dorsal gland outlet. That point at which the dorsal gland empties into the lumen of the esophagus. (73b)
Esophagus. Portion of the alimentary canal between the buccal capsule or stoma, and the anterior portion of the intestine. (1, 2)
Excretory pore. Exterior opening of the excretory system. It is generally located on the ventral side of the body near the basal region of the esophagus. It is also known as the orifice of the cervical gland. (4l)
Genital papillae. Tactile or sensory organs located on the tail.
Glottoid apparatus. Sclerotized structure at the base of the stoma. (2b)
Gubernaculum. Spicule guide, sclerotized accessory piece. (75b)

Guiding ring. Sleevelike structure which surrounds and guides the stylet in genera of the Dorylaimoidea. Position varies among the genera from near apex to posterior portion of stylet. (4a, 7a, 8a)
Head. That portion anterior to the base of the stoma or stylet.
Hemizonid. Lenslike structure situated between the cuticle and hypodermal layer on the ventral side of the body just anterior to the excretory pore; generally believed to be associated with the nervous system.
Hypodermis. Thin tissue layer beneath the cuticle which thickens to form the dorsal, lateral, and ventral cords which extend the length of the body.
Incisures. Longitudinal cuticular clefts which divide the lateral fields. Sometimes called involutions. (75f)
Isthmus. Relatively narrow portion of esophagus just anterior to the basal region. (1)
L. Mean length in mm (range).
Labial. Pertaining to the lips.
Lateral field. An interruption of the transverse striae by longitudinal cuticular thickenings situated on top of the lateral cords. The field is divided by longitudinal striae and at times by transverse markings. (65E)
Lumen. Triradiate canal or duct of the esophagus. (4d, 54a)
Metacorpus. Swollen posterior portion of the corpus. Sometimes referred to as the median bulb. (4h)
Micropyle. Minute opening in the membrane of an egg through which the spermatozoa enter.
Monogenic. Producing offspring of only 1 sex.
Mucro. Stiff or sharp point abruptly terminating an organ. (12a)
Neck. That portion of the body occupied by the esophagus.
Nerve ring. Circumesophageal commissure, or, the center of the nervous system of nematodes which encircles the esophagus. It is composed largely of nerve fibers and associated ganglia.
Odontostylet. See stylet. Synonymous with Onchiostylet. (1C)
Oocyte. Female germ cell.
Oral opening. Mouth.
Ovary. Female sexual gland in which the ova or eggs are formed.
Ovoviviparous. Pertaining to production of eggs which hatch within the uterus.
Papillae. Tactile, sensory organs found on various body regions; for example, labial or cephalic papillae.
Parthenogenesis. Process of reproduction by the development of an unfertilized egg.
Perineal pattern. Fingerprintlike pattern formed by cuticular striae surrounding the vulva and anus of the mature *Meloidogyne* female.
Phasmid. Porelike structures located in the lateral fields of the posterior region of nematodes belonging to the class Secernentea. Function is believed to be sensory. Sometimes called precaudal glands. (68a, 75e)
Postuterine sac. Rudimentary or vestigial ovary which may serve to store the spermatozoa. Sometimes known as postvulvar uterine sac.
Probolae. Prominent, sometimes ornate, structures on the head; often modifications of the labial and cephalic regions.
Procorpus. Cylindrical portion of the corpus anterior to the metacorpus. (1)
Prodelphic. Having uteri parallel and anteriorly directed at origin.
Pseudocoel. Body cavity of nematodes containing a fluid in which the various internal organs are suspended. The body fluid apparently functions as a respiratory and circulatory system.
Rays. Genital papillae or tactile organs located within the caudal alae.
Rectal glands. Three to 6 large glands often found in the region of the rectum of some Secernentea. Function unknown.
Retrorse. In a backward or downward direction.
Rhabdions. Plates in the cuticular lining of the stoma which make the walls of the various divisions of the stoma; for example: cheilo-, pro-, proto-, meso-, meta-, and telorhabdions.
s. Stylet length/body diameter measured at base of stylet.
Sclerotized. Pertaining to hardened refractive regions.
Setae. Elongated cuticular structures articulating with the cuticle; in general, tactile sensory organs usually located around the oral openings.
Sexual dimorphism. Species in which the 2 sexes differ morphologically.
Spear. Synonymous with stylet. (1A)
Spicula. Paired sclerotized pieces. Male copulation organs. (75d)
Spicular pouch. Pouch, containing the spicules, which is often called the sheath in parasitic nematodes.
Spinneret. Porelike opening of the caudal glands situated in the tail, located terminally or subterminally on some of the nematodes belonging to the Adenophorea.
Stoma. That portion of the digestive tract between the oral opening and esophagus. (2a)
Striae. Superficial markings of the cuticle, appearing as grooves or clefts. When present, may be seen encircling lips or body. (4j)
Stylet. (Stomatostylet). Hollow protrusible stylet or spear used to puncture plants and animal prey. This type of stylet, thought to have developed gradually through evolution by the coming together of the sclerotizations of the stoma, is commonly found among the nematode species belonging to the order Tylenchida. (4g)
Stylet (Odontostylet; Onchiostylet). Buccal stylet, used to puncture plants and animal prey. This type of stylet represents a large tooth which originates in the esophageal wall. Generally found among some members of the Adenophorea. (4b)

Stylet aperture. Anterior opening of the stylet located dorsally in the Dorylaimoidea and generally ventrally in the Tylenchida.

Stylet extensions. Sclerotized structures extending between the esophagus and the Dorylaimoid spear. (7f)

Tail. That portion of the body between the anus and the posterior terminus. (7e)

Telamon. Rigid sclerotized portion of the cloacal wall which apparently guides the spicules from the spicular pouch into the cloaca.

Titillae. Small projections on either side of the distal end of the gubernaculum.

Triploblastic. Possessing 3 germ layers: ectoderm, mesoderm, and endoderm.

Vestigial. Pertaining to structures or organs that were well developed in an animal's ancestors but have become rudimentary during the course of evolution.

Viviparous. Bearing living young.

Vulva. Exterior portion of the mature female's reproductive system. Generally appears as a transverse slit on the ventral portion of the nematode. (35a, 35b)

INDEX TO GENERA

(First number indicates location of generic description, italic numbers the plates, third set of numbers the Selected References.)

**PICTORIAL KEY TO GENERA
OF PLANT-PARASITIC NEMATODES**

Designed by Richard E. Rosenbaum.
Composed by Syntax International Pte. Ltd.
in 9 point Monophoto Helvetica 765, 1 point leaded,
with display lines in Monophoto Helvetica Medium 766.
Printed offset by Vail-Ballou Press, Inc. on
Warren's Patina Coated Matte, 70 lb. basis.
Bound by Vail-Ballou Press.